UNDERSTANDING OPTIONS FOR AGRICULTURAL PRODUCTION

Systems Approaches for Sustainable Agricultural Development

VOLUME 7

Aims and Scope

The book series *Systems Approaches for Sustainable Agricultural Development* is intended for readers ranging from advanced students and research leaders to research scientists in developed and developing countries. It will contribute to the development of sustainable and productive systems in the tropics, subtropics and temperate regions, consistent with changes in population, environment, technology and economic structure.

The series will bring together and integrate disciplines related to systems approaches for sustainable agricultural development, in particular from the technical and the socio-economic sciences, and presents new developments in these areas.

Furthermore, the series will generalize the integrated views, results and experiences to new geographical areas and will present alternative options for sustained agricultural development for specific situations.

The volumes to be published in the series will be, generally, multi-authored and result from multi-disciplinary projects, symposiums, or workshops, or are invited. All books will meet the highest possible scientific quality standards and will be up-to-date. The series aims to publish approximately three books per year, with a maximum of 500 pages each.

The titles published in this series are listed at the end of this volume.

Understanding Options for Agricultural Production

Edited by

GORDON Y. TSUJI

University of Hawaii-Manoa, Honolulu, Hawaii, USA

GERRIT HOOGENBOOM

Department of Biological and Agricultural Engineering, The University of Georgia, Georgia, USA

PHILIP K. THORNTON

International Livestock Research Institute, Nairobi, Kenya

KLUWER ACADEMIC PUBLISHERS

DORDRECHT / BOSTON / LONDON

in cooperation with

International Consortium for Agricultural Systems Applications

Library of Congress Cataloging-in-Publication Data is available.

ISBN 0-7923-4833-8

Published by Kluwer Academic Publishers,
P.O. Box 17, 3300 AA Dordrecht, The Netherlands.

Sold and distributed in the U.S.A. and Canada
by Kluwer Academic Publishers,
101 Philip Drive, Norwell, MA 02061, U.S.A.

In all other countries, sold and distributed
by Kluwer Academic Publishers,
P.O. Box 322, 3300 AH Dordrecht, The Netherlands

Printed on acid-free paper

Printed in Great Britain.

Contents

Preface

The first premise of this book is that farmers need access to options for improving their situation. In agricultural terms, these options might be management alternatives or different crops to grow, that can stabilize or increase household income, that reduce soil degradation and dependence on off-farm inputs, or that exploit local market opportunities. Farmers need a facilitating environment, in which affordable credit is available if needed, in which policies are conducive to judicious management of natural resources, and in which costs and prices of production are stable. Another key ingredient of this facilitating environment is information: an understanding of which options are viable, how these operate at the farm level, and what their impact may be on the things that farmers perceive as being important.

The second premise is that systems analysis and simulation have an important role to play in fostering this understanding of options, traditional field experimentation being time-consuming and costly. This book summarizes the activities of the International Benchmark Sites Network for Agrotechnology Transfer (IBSNAT) project, an international initiative funded by the United States Agency for International Development (USAID). IBSNAT was an attempt to demonstrate the effectiveness of understanding options through systems analysis and simulation for the ultimate benefit of farm households in the tropics and subtropics.

The idea for the book was first suggested at one of the last IBSNAT group meetings held at the University of Hawaii in 1993. The logic for a book on the IBSNAT project was clear: it had involved many people, eleven years, and millions of dollars. Beyond the final report that was assembled as a contractual obligation for the donor, documentation on IBSNAT was scattered far and wide through the grey literature, in peer-reviewed journal articles, and in unpublished reports.

At the time the idea of the book was mooted, we were sitting in a trailer on the campus at Manoa on the fringes of Honolulu; Krauss Hall, the home of IBSNAT and by far the most interesting and character-laden building on the campus, was undergoing renovation. IBSNAT was itself coming to an end, and the hope then was that there would be a 'project renovation' in terms of a new grant for continuing the work into the twenty-first century. This did not materialize; on the one hand this was (and still is) disappointing; but on the other, the world has changed radically even since the early 1990s, and now we are by no means fully convinced that a second IBSNAT-like project would really have been the most appropriate vehicle for pursuing a truly effective, applied systems approach to international agricultural research. Other mechanisms of collaborative systems research are currently being utilized, and doubtless there are many others that will be tried over the coming years. IBSNAT itself is finished and done with, but many of its ideas and tools are flourishing in a different setting.

The book starts with an overview of the project, written by the Principal Investigator and the Project Manager of IBSNAT. This is followed by seven chapters outlining the tools that were developed largely under the auspices of IBSNAT, although some of the models described have a pedigree that predates IBSNAT by several years. The next nine chapters describe a variety of applications of the DSSAT and crop models. The aim is to show the diversity of technical problems that have been addressed using the crop models both within and without the DSSAT package. The last chapters concern training and education, project documentation, and a synthesis of the lessons learned from the IBSNAT project for future research and development activities.

If the entire IBSNAT project were to be documented thoroughly, the current book is probably a fifth of its appropriate size. We have had to be selective to keep the book to a manageable size and a reasonable cost. This inevitably means that there are many people who made critical contributions to IBSNAT who are not authors of chapters herein; so far as we can tell, however, they are all featured in the reference lists of various chapters. The book had a very long gestation period, for a variety of reasons, but we thank all the authors for producing the goods at the end of the day. We also thank those people who acted as reviewers for particular chapters, and those who helped standardize the figures and tables.

We were geographically separated over 14 time zones as the book neared completion, although via the internet we were able to accomplish many editing tasks by communicating with most of the authors and with each other using faxes and E-mail. A number of people helped us greatly. In particular, we would like to thank Dr Robert Clement Abaidoo for his patience, fortitude, and steadfast dedication in proofreading and checking each manuscript. We would also like to thank Mrs Pamela Keiko Brooks and Mr Daniel Takeshi Imamura for inputting, re-creating and modifying the many graphics received from the authors. Their collective intuitive knowledge of software and graphics layout was invaluable. Finally, we acknowledge USAID, in particular, the former Office of Agriculture, Bureau for Science and Technology, for their shared vision and support of the IBSNAT project.

IBSNAT gave sterling service to the concept of leveraging; USAID's investment of resources was multiplied many times in terms of expertise and input from the core group of institutions and individuals who were involved one way and another. It may also be that the lasting benefits of IBSNAT lie not so much with the software and the models themselves, but with increased awareness among the research and development community at large of the basic rightness of what was proposed, even if it was not successful in all the facets of its implementation. For those of us who were closely involved with the project for a number of years, it was immensely rewarding on a professional level. Equally important, it was also fun: friendships were forged, and many trips on IBSNAT-related business were made all over the world, sometimes involving the running of training courses and field trials in the most bizarre circumstances imaginable. It was a privilege to be a part of it.

Gordon Tsuji
Gerrit Hoogenboom
Philip Thornton

Gainesville, Florida
April 1997

Acronyms

ACSAD	Arab Center for Studies of the Arid Drylands, Damascus, Syria
AEGIS	Agricultural and Environmental Geographical Information System
AID	Agency for International Development
AIDAB	Australian International Development Assistance Bureau
ALES	Automated Land Evaluation System
ARS	Agricultural Research Service
ART	Agricultural Research Trust
ASCII	American Standard Code for Information Interchange
AVRDC	Asian Vegetable Research and Development Center
BARC	Bangladesh Agricultural Research Council
CATIE	Centro Agronmico Tropical de Investigacióñ y Enseanza
CENIAP	Centro Nacional de Investigaciones Agropecuarias
CERES	Crop-Environment Resource Synthesis
CFC	Chlorofluorocarbons
CIAT	Centro Internacional de Agricultura Tropical
CIMMYT	Centro Internacional de Mejoramiento de Maíz y Trigo
CIP	Centro Internacional de la Papa
CODATA	Committee on Data for Science and Technology
COMCIAM	Climate Impact Assessment and Management Program for Commonwealth Countries
CPF	Cumulative Probability Function
CROPGRO	Generic crop model based on the SOYGRO, PNUTGRO, and BEANGRO models
DAS	Days After Sowing
DSSAT	Decision Support System for Agrotechnology Transfer
EPA	Environmental Protection Agency, United States of America
EPIC	Environmental Productivity Impact Calculator
FAO	Food and Agriculture Organization of the United Nations
FFTC/ASPAC	Food and Fertilizer Technology Center for the Asia and Pacific Region, Republic of China
FOM	Fresh Organic Matter
FONAIAP	Fondo Nacional de Investigaciones Agropecuarias, Maracay, Venezuela
FSR	Farming Systems Research
GCM	General Circulation Model
GFDL	Geophysical Fluid Dynamics Laboratory
GIS	Geographic Information System
GISS	Goddard Institute for Space Studies

GPS	Global Positioning System
HERB	Decision model for post-emergence weed control in soybean
IBSNAT	International Benchmark Sites Network for Agrotechnology Transfer
ICASA	International Consortium for Agricultural Systems Applications
ICIS	International Crop Information System
ICRISAT	International Crops Research Institute for the Semi-Arid Tropics
IFDC	International Fertilizer Development Center
IIEA	International Institute for Advanced Studies, Caracas, Venezuela
ILRI	International Livestock Research Institute
INIA	Instituto Nacional de Investigaciones Agrarias
IPAR	Intercepted Photosynthetically Active Radiation
IPCC	Intergovernmental Panel on Climate Change
IPM	Integrated Pest Management
IRRI	International Rice Research Institute
IWIS	International Wheat Information System
LAI	Leaf Area Index
MARDI	Malaysian Agricultural Research and Development Institute
MDS	Minimum Data Set
MG	Generic Maturity Group
MRG	Management Review Group
NDU	National Defense University
NifTAL	Nitrogen Fixation by Tropical Agricultural Legumes
PAN-EARTH	Predictive Assessment Network for Ecological and Agricultural Responses to Human Activities
PAR	Photosynthetically Active Radiation
PCARRD	Philippine Council on Agriculture Forestry and Resources Research and Development
PFD	Photon Flux Density
PID	Pest Identifiers
PLA	Plant Leaf Area
RISSAC	Research Institute of Soil Science and Agricultural Chemistry, Hungary
RLWR	Root Length Weight Ratio
RUE	Radiation Use Efficiency
RWU	Root Water Uptake
SAFGRAD	Semi-Arid Food Grain Research & Development,
OAU	Organization of African Unity
SCS	Soil Conservation Service (now Natural Resource

	Conservation Service) U.S. Department of Agriculture
SD	Stochastic Dominance
SLA	Specific Leaf Area
SLW	Specific Leaf Weight
SMSS	Soil Management Support Service
SOM	Soil Organic Matter
SOYWEED	Simulation model of soybean and common cocklebur growth and competition
SSDS	Soil Survey Division Staff
SUBSTOR	Simulation of Underground Bulking Storage Organs
TAC	Technical Advisory Committee
UF	University of Florida
UKMO	United Kingdom Meteorological Office
UNED	Universidad Estatal a Distancia
USA	United States of America
USAID	United States Agency for International Development
USDA	United States Department of Agriculture
USLE	Universal Soil Loss Equation
VPD	Vapor Pressure Deficit
WEEDING	Multi-species weed competition program
WFM	Whole Farm Model
WUE	Water Use Efficiency
WUF	Water Uptake Fraction

Overview of IBSNAT

G. UEHARA and G.Y. TSUJI
Department of Agronomy and Soil Science, 1910 East West Road, University of Hawaii, Honolulu, Hawaii 96822, USA

Key words: systems analysis, simulation, crop models, technology transfer, decision support system, collaborative network

Abstract
The objective of the IBSNAT or International Benchmark Sites Network for Agrotechnology Transfer project was to apply systems analysis and simulation to problems faced by resource-poor farmers in the tropics and sub-tropics, specifically in the area of evaluating new and untried agricultural technologies. Outputs from the project include the portable decision support system, DSSAT, the minimum data set necessary to drive the system and a collaborative network composed of an international, interdisciplinary team of scientists from more than 25 countries. Network members joined together to develop, assemble, test and use one of the most widely used decision support systems in the world today. The decision support system products enable users to match the biological requirements of crops to the physical characteristics of land to provide them with management options for improved land use decisions.

Introduction

The purpose of the IBSNAT project was to enable agricultural decision makers to explore the future with simulated outcomes of alternative practices and policies. The aim was to empower farmers and policy makers with a capacity to choose a better future for themselves and society.

To achieve its purpose, the project mobilized researchers and resources to focus on the following objectives:

(1) Production of a decision support system capable of simulating the risks and consequences of alternative choices.
(2) Definition of the minimum amount of data required to make simulations.
(3) Testing and application of the product on global agricultural problems requiring site-specific yield simulations.

Assembling the decision support system

In 1982 an international meeting of agricultural and systems scientists was convened at the International Crops Research Institute for the Semi-Arid Tropics (ICRISAT) in Hyderabad, India, to design a decision support system for agrotechnology transfer (Kumble, 1984). The participants were asked to

G.Y. Tsuji et al. *(eds): Understanding Options for Agricultural Production, 1–7*

focus on systems analysis and crop simulation models as the primary means to match crop requirements with land characteristics. The aim was to develop a solid foundation for dealing with the genotype by environment interactions in the soil-plant-atmosphere continuum so that strong links between the biophysical outcomes and socioeconomic needs could later be forged. The scope of work was limited to ten food crops considered critical to food security, including four cereals (maize, rice, sorghum, and wheat), three grain legumes (dry beans, groundnut, and soybean), and three root crops (aroid, cassava, and potato).

When the conference was convened in 1982, dynamic, process-based, whole crop models for only four of the ten crops (wheat, maize, soybean, and groundnut or peanut) were available in partially or fully developed forms, but none had been tested in the tropics or subtropics. Participants from the developing countries agreed to install field experiments to test the reliability of the models. The modelers, in turn, agreed to distribute existing models to all collaborators and to assist them in generating the simulated results so that observed and simulated results could be compared.

The project held its first crop modeling workshop in Venezuela in 1983. This workshop revealed the need to standardize the input and output structure of existing models to enable all of them to access a common database. Following this workshop a small group of modelers and system scientists formulated a plan to combine the models (programmed in FORTRAN), databases (.dbf or dBASE format) and an application program (in BASIC) into a computer software called the Decision Support System for Agrotechnology Transfer (DSSAT).

Five years after the meeting in Hyderabad the first prototype Decision Support System for Agrotechnology Transfer was assembled (IBSNAT, 1989). This DSSAT provided users with easy access to soil, crop, and weather information as well as to crop models and application programs to simulate, analyze, and display outcomes of alternative management strategies specified by the user.

A typical scenario simulated by DSSAT may involve a comparison of a new variety with the one grown by the farmer, planted on the same day, on a sandy and clayey section of the farm, and supplied with 25 or 50 kilograms of nitrogen per hectare, once at planting and again at flowering. The outcome of this treatment combination would then be simulated for 50 different weather years to assess stability and productivity of the cropping system. An experiment that would nearly consume an agronomist's entire professional career could now be completed in a few minutes on a desktop computer, but the advantage of DSSAT did not end there. In real experiments there was no going back to explore 'what if' scenarios.

What if the planting data had been delayed or advanced a fortnight, or the plant population and nitrogen rates increased, or a green manure substituted for chemical fertilizer, or even a totally new crop planted? Farmers with the help of extension agents could explore many options for achieving their farming objectives.

Defining the minimum data set

The inaugural meeting in Hyderabad, India, also achieved the second of its two principal objectives. It reached agreement on the minimum set of crop, soil, weather and management data with which to run the models. A technical report (IBSNAT, 1984) on the Minimum Data Set (MDS) was published and revised two years later (IBSNAT, 1986) and again in 1988 (IBSNAT, 1988).

Reaching agreement on the MDS was a major achievement of the project. Fortunately, the state of crop models existing at the time enabled the participants to focus on data needed by the models. By 1983, most crop models were operating on a daily time step and considerable convergence in their data requirements had already occurred. Thus the MDS was largely predetermined by the models themselves. Even so, it took great restraint on the part of even the modelers to keep the data set to the bare minimum. There was great temptation, for example, to add relative humidity, wind velocity and direction, and pan evaporation to the weather MDS of maximum and minimum air temperature, solar radiation and rainfall. In the beginning, what was important was not so much the number and kinds of variables that were included in the MDS, but the simple fact that the MDS, imperfect as it was, existed. The initial set was flawed; this is clear from the number of revisions it has undergone. The MDS is imperfect now and will always be imperfect because knowledge of the processes that scientists are trying to simulate is imperfect. The simple fact that a MDS existed, however, gave a sense of unity and coherence to a dispersed and decentralized project operating in more than fifteen countries.

In the final analysis it was the utilitarian goals of the IBSNAT Project that determined the size and the nature of the MDS. It was reasoned that a large and complex MDS would merely add to the burden of data acquisition in developing countries. There was full agreement among all involved that the project would not assemble decision aids, however powerful they might be, which would be rejected by the very clients they were meant for, simply because the input data to use them were unavailable or too difficult and costly to obtain.

Validation and application on a global scale

Between 1974 and 1985 standardized field experiments were conducted under a project known as the Benchmark Soils Project (BSP) on soils of three taxonomically defined soil families in Brazil, Cameroon, Indonesia, Hawaii, Philippines and Puerto Rico. The purpose of this research was to test the hypothesis that the behavior and performance of soils belonging to a common family would be sufficiently alike to enable crop production technology to be transferred among widely separated locations on the basis of soil nomenclature (Silva, 1975). The experimental results generated by the Benchmark Soils Project gave IBSNAT the opportunity to test its models with a large independent data set. A comparison of observed and predicted grain yields, and number of days to silking and maturity, is shown in Figure 1.

The agreement between observed and simulated outcomes confirmed what modelers have been saying for a long time, namely, that generic models can mimic growth and development of crops on a site-specific basis anywhere in the world. This result showed that processed-based models could operate globally and at the same time operate locally through the use of location-specific data.

Testing of a crop model is one thing, applying it is another. If the yield of an experimental crop can be simulated, can a model predict yields for the coming year? The answer, unfortunately, is no. We are unable to simulate future yields because of our inability to predict future weather. We were able to simulate the yields obtained by the Benchmark Soils Project because temperature, rainfall and solar radiation data along with information on planting date, location, row spacing, plant populations, crop variety, soil condition and amount and type of fertilizer applied were kept. Temperatures, rainfall and solar radiation are the random variables that keep us from predicting next season's crop yield.

If the models are unable to predict next year's yield, they can still generate probabilities concerning what is likely to occur for the user. This is obtained by supplying the model with long-term historical weather data and instructing it to simulate yield not for a single year but for as many years as weather data are available. The result is a whole probability distribution of yields from which the mean yield, variance, and other descriptive statistics can be calculated. From this the model user is able to visualize the weather-related risks associated with growing a crop in a particular way, and thus some of the risks associated with the adoption of a new crop or practice (Anderson, 1974).

It is this capacity to generate whole probability distributions of outcomes that give simulated results a distinct advantage over experimental observations. The element of risk which so affects decision making resides in the tail of probability distributions. While models cannot predict the future, they enable users to explore the future with viable options. A good illustration of this is given by Rosenzweig et al. (1995), in which the crop models in DSSAT are employed to examine the impact of climate change on world food production and international trade.

The collaborators

A unique feature of the IBSNAT project has been the role played by the collaborating scientist to design, manage and implement project activities. This project quickly evolved into a participatory effort out of practical necessity. Unlike projects that deal with a single crop or a single component such as soil, water or climate, IBSNAT, designed as a systems project, was intended to deal

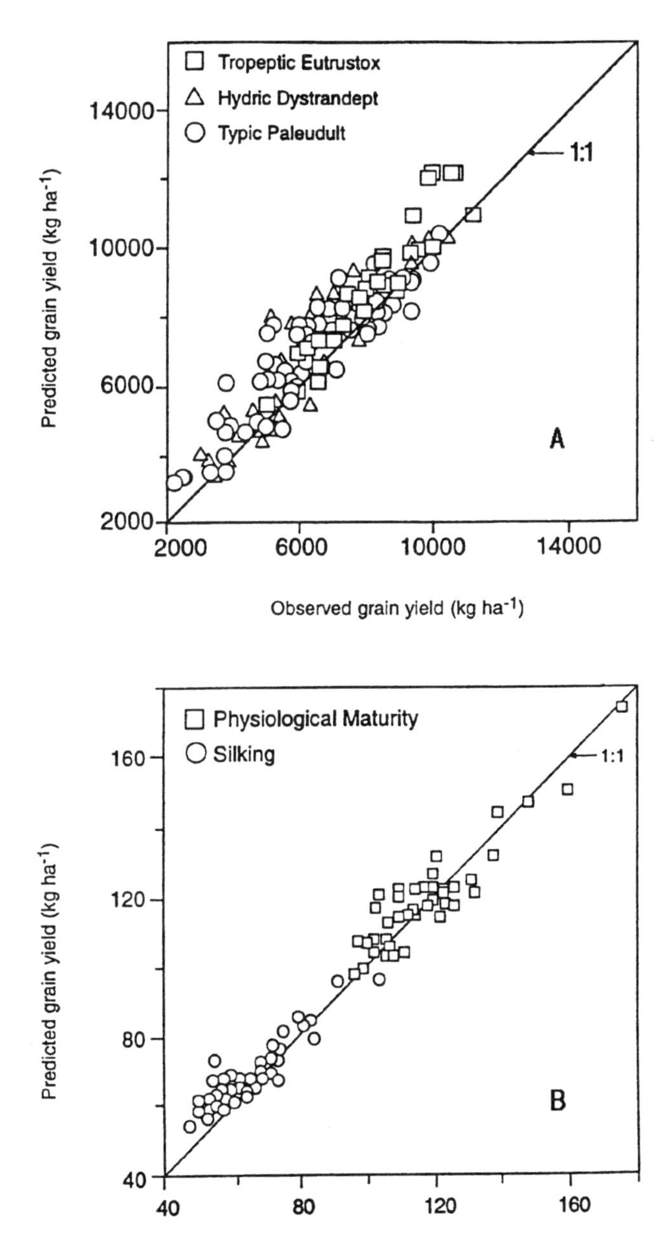

Figure 1. Comparison of model outputs from CERES-Maize and field data for experiments conducted in seven locations in three soil types in Hawaii, Indonesia, and the Philippines over a period of eight years. (A) Simulated and observed maize grain yield. (B) Simulated and observed number of days to silking and to physiological maturity.

with many crops and many key biophysical and socioeconomic aspects of agricultural systems. The task was to find suitable individuals to lead sub-programs to make up the whole. The first two participants were selected by the principal investigator, but thereafter, all others were jointly identified and selected by the existing group. This method of inviting new participants into the group had two advantages. First, the group as a whole provided a larger base from which to choose competent candidates, and second, the close and intimate working relations required for interdisciplinary work made it imperative that a new member was the choice of the group rather than an individual. One danger of forming research teams in this way is that it tends to bring together individuals who think alike, and runs the risk of reaching consensus on a flawed concept. True inter-disciplinary teams, however, are composed of individuals with such diverse backgrounds and training that some convergence in thinking, particularly with respect to agreement on project goals, research methodology, and client orientation, is not only desirable but necessary. A strong commitment to a client-oriented effort, driven and guided by client-needs, was a key element shared by team members.

A commitment to client-oriented, problem-solving project has probably been the single characteristic that has united and sustained the team members. They had long realized that the problems faced by clients ranging from farmers to policy makers were not disciplinary problems but systems problems that could only be diagnosed and solved through interdisciplinary effort. Another characteristic of the collaborators was their willingness to adjust their on-going programs to accommodate project goals and objectives, thereby greatly leveraging the overall effort. The incentive to operate in this way came from knowledge that the value and quality of the product assembled by the team would far exceed anything a single member working alone could produce. The participants of the IBSNAT project were, therefore, not individuals who needed additional resources to do research, but leaders of on-going research programs who were themselves looking for opportunities to work in a more integrated manner with workers in other disciplines. The IBSNAT project gave them this opportunity.

References

Anderson J R (1974) Risk efficiency in the interpretation of agricultural production research. Review of Marketing and Agricultural Economics 42:131–184.

International Benchmark Sites Network for Agrotechnology Transfer (IBSNAT) (1984) Experimental design and data collection procedures for IBSNAT: the minimum data set for systems analysis and crop simulation. Technical Report 1, Department of Agronomy and Soil Science, University of Hawaii, Honolulu, Hawaii, USA.

International Benchmark Sites Network for Agrotechnology Transfer (IBSNAT) (1986) Experimental design and data collection procedures for IBSNAT: the minimum data set for system analysis and crop simulation. Technical Report 1, second edition. Department of Agronomy and Soil Science, University of Hawaii, Honolulu, Hawaii, USA.

International Benchmark Sites Network for Agrotechnology Transfer (IBSNAT) (1988) Experimental design and data collection procedures for IBSNAT: the minimum data set for

system analysis and crop simulation. Technical Report 1, third edition. Department of Agronomy and Soil Science, University of Hawaii, Honolulu, Hawaii, USA.
International Benchmark Sites Network for Agrotechnology Transfer (IBSNAT) (1989) Decision support system for agrotechnology transfer v2.1 (DSSAT v2.1). Department of Agronomy and Soil Science, University of Hawaii, Honolulu, Hawaii, USA.
International Benchmark Sites Network for Agrotechnology Transfer (IBSNAT) (1990) Documentation for IBSNAT crop model input and output files for the Decision Support System for Agrotechnology Transfer (DSSAT V2.1). Technical Report 5, version 1.1, Department of Agronomy and Soil Science, University of Hawaii, Honolulu, Hawaii, USA.
Kumble V (Ed.) (1984) Proceedings of the international symposium on minimum data sets for agrotechnology transfer, 21–26 March 1983, ICRISAT, Patancheru, AP, India.
Rosenzweig C, Allen L H Jr, Harper L A, Hollinger E, Jones J W (1995) Climate change and agriculture: analysis of potential international impacts. ASA Special Publication Number 59, American Society of Agronomy, Madison, Wisconsin, USA.
Silva J A (Ed.) (1985) Soil based agrotechnology transfer. Benchmark Soils Project, Department of Agronomy and Soil Science, Hawaii Institute of Tropical Agriculture and Human Resources, University of Hawaii, Honolulu, Hawaii, USA.

Data for model operation, calibration, and evaluation

L.A. HUNT[1], K.J. BOOTE[2]

[1]*Department of Crop Science, University of Guelph, Guelph, Ontario, Canada, N1G 2W1*

[2]*Department of Agronomy, University of Florida, Gainesville, Florida 32611, USA*

Key words: minimum data sets, crop models, data checking, data storage, data exchange

Abstract

The International Benchmark Sites Network for Agrotechnology Transfer (IBSNAT) Project has recognized that the application of models depends not only on the availability of models and application software, but also on the availability of data that facilitate model calibration, evaluation and application. IBSNAT, therefore, has devoted considerable attention to developing and clarifying procedures that relate to data acquisition, storage, exchange, and use. For model operation, IBSNAT has emphasized the need for a balanced set of information that includes information on the site where the experiment was conducted, on the weather during the growing cycle, on the characteristics of the soil at the start of the growing cycle, on the management of the crop, and on new cultivar traits. For each of these, IBSNAT has defined a minimum amount of data that is necessary for model operation. This minimum amount of information has been termed a 'Minimum Data Set', a phrase that is applicable to data sets for model operation as well as calibration and evaluation. For the latter, data on the date of occurrence of the main phenological events, on yield and its components, and on biomass at final harvest are necessary as a minimum addition to the data for model operation. Within-season measurements of some growth characteristics may also be necessary for calibrating models for new situations. Information from such studies is easily lost unless specific steps are taken to ensure that it is conserved. To facilitate both conservation and use, IBSNAT has developed some simple, standard experiment documentation files that can be established and edited easily, and that can also be transferred directly among workers without the need for 'retrieval' from a central database. The widespread use of the data structures developed by IBSNAT would make possible a 'dispersed' but nonetheless standard database for model calibration and evaluation, as well as documentation of experiments.

Introduction

At an early date, IBSNAT recognized that the application of crop simulation models to problems in the real world would depend not only on the availability of models and application software, but also on the availability of information that would make it possible both to run models for particular scenarios and to specify the accuracy of the models for particular target regions. Further, IBSNAT acknowledged that this latter requirement would only be satisfied when users were able to calibrate models for use with cultivars and soils of their own target regions, and to evaluate the accuracy of the model calculations for their own particular concerns. With this in mind, IBSNAT devoted attention from an early stage of its existence to clarifying appropriate data needs, to specifying data collection and experimental procedures, to defining simple data

G.Y. Tsuji et al. *(eds): Understanding Options for Agricultural Production, 9–39*

handling structures, and to developing analytical approaches and software that can be applied by users to calibrate and evaluate models for their own applications. The objective of this chapter is to present an overview of experimental data required for crop model calibration and evaluation.

Data requirements

Model operation

All crop simulation models require information on one or more aspects of the aerial and soil environments (see Hunt, 1994). For some models, this basic information on environment must be supplemented with information on starting conditions and on other aspects of the system – for instance, on the characteristics of the genotype(s) and on crop management (e.g., planting and/or emergence date, row spacing, plant population, amount and timing of fertilizer application, and irrigation applications). The number of ways in which the basic information could be supplemented is almost infinite, as also is the detail required for definition of the environment, with each model having specific requirements. IBSNAT recognized this and endeavored to avoid an ever-extending array of data requirements by directing attention to a hierarchy of data sets that could be considered appropriate for the operation of models of different degrees of complexity. Three levels of complexity were outlined by Nix (1984), who defined them in terms of the frequency with which weather data would be required. This included weekly, daily and hourly data.

With time and experience it was recognized, however, that models operating at a daily time step were perhaps the most appropriate for application to crop production and environmental sustainability problems. Efforts were thus concentrated on defining a set of data that could be widely used for models at this level of complexity, a set that could be regarded as a Minimum Data Set for the operation of models destined for use in agrotechnology transfer. In these efforts, one thought was uppermost in the minds of IBSNAT scientists – namely, that models should be balanced in the degree to which they incorporate details of the different components of the soil-crop-atmosphere system (Jones, 1984; IBSNAT, 1986a, 1988; Hunt et al., 1994b), and that data for the operation of models should be equally balanced over the different components of the system. The IBSNAT Minimum Data Sets for model operation (Table 1a), therefore, came to include data on the site where the model is to be operated (e.g., slope and aspect – characteristics that impact on the amount of radiation actually received by the crop, and on rainfall runoff), on the daily weather during the growing cycle, on the characteristics of the soil at the start of the growing cycle, and on the management of the crop (e.g., seeding rate, fertilizer applications, irrigations).

A documentation of the characteristics of the site, which could be an individual plot, a field or even a region, is the first essential aspect of the IBSNAT

Table 1. Contents of minimum data sets for model operation, calibration, and evaluation

A. Operation

1. Site
   - Latitude and longitude, elevation; average annual temperature; average annual amplitude in temperature
   - Slope and aspect; major obstruction to the sun (e.g. nearby mountain); drainage (type, spacing and depth); surface stones (coverage and size)
2. Weather
   - Daily global solar radiation, maximum and minimum temperatures, precipitation
3. Soils
   - Classification using the local system and (to family level) the USDA-SCS taxonomic system
   - Basic profile characteristics by soil layer: *in-situ* water release curve characteristics (saturated, drained upper limit, lower limit); bulk density, organic carbon; pH; root growth factor; drainage coefficient
4. Soil analysis
   - Surface layer(s) measurements of bulk density, organic carbon, organic nitrogen, pH, P, and K
5. Initial conditions
   - Previous crop, root, and nodule amounts; numbers and effectiveness of rhizobia (if for nodulating crop)
   - Water, ammonium and nitrate by soil layer
6. Management
   - Cultivar name and type
   - Planting date, depth and method; row spacing and direction; plant population
   - Irrigation and water management, dates, methods and amounts or depths
   - Fertilizer (inorganic) and inoculant applications
   - Residue (organic fertilizer) applications (material, depth of incorporation, amount and nutrient concentrations)
   - Chemical (e.g. pesticide) applications (material, amount)
   - Tillage
   - Environment (aerial) adjustments
   - Harvest schedule

Continued

minimum data set for model operation. Information on slope and orientation of slope with respect to North is required to adjust basic radiation data, which relate to a flat surface, and for consideration of run-off. Some information on the physiographic features of the site is also required to identify those cases where experiments are conducted in valleys or near forests, where the sun's direct rays are obstructed in the morning or evening. Further information on the presence or absence of soil drains is required to allow for modifications of the soils inherent drainage characteristics. A few other defining aspects are also required (e.g., latitude and longitude) as listed in Table 1.

Required weather data (Table 1A) encompass daily records of total solar radiation incident on the top of the crop canopy, maximum and minimum air temperature above the crop, and rainfall. Some models (e.g., the IBSNAT cassava and legume models) can make use of the water vapor content of the air, and it has been argued that some measure of humidity should be included

Table 1. Continued.

B. Calibration
All of the above plus:
7. Crop performance
- Date of emergence
- Date of flowering or pollination (where appropriate)
- Date of onset of bulking in vegetative storage organ (where appropriate)
- Date of physiological maturity
- Leaf area index and canopy dry weight at 3 stages during the life cycle
- Canopy height and breadth at maturity
- Yield of appropriate economic unit (e.g. kernels) in dry weight terms
- Canopy (above-ground) dry weight or harvest index (plus shelling percentage for legumes)
- Harvest product individual dry weight (e.g. weight per grain, weight per tuber)
- Harvest product number per unit at maturity (e.g. seeds per spike, seeds per pod)
- Damage level of pest (disease, weeds, etc.) infestation (recorded when infestation first noted, and at maximum)
- Number of leaves produced on the main stem
- N percentage of economic unit
- N percentage of non-economic parts

Data required from a number of experiments in which the same cultivar(s) were included, and which encompass a range of environmental conditions.

C. Evaluation
All aspects required for model operation plus field information on the aspect(s) for which the model is being validated (e.g. anthesis date, maturity date, grain yield, grain protein concentration, canopy dry weight). Data are required from a number of experiments covering the complete range of environments for which information on the precision/accuracy of the model is required.

in any minimum set. Further, data on rainfall intensity and duration would help improve algorithms dealing with runoff and erosion, and arguments for their inclusion have also been made. IBSNAT, bearing in mind the problems of measurement, and the possibilities for computation of surrogate values, recognized such arguments but has preferred to keep the defined minimum data set as simple as possible.

Even though the minimum set has been kept as simple as possible, IBSNAT has recognized that all required weather data for a particular site and a particular time period are often not available. In such cases, it has been suggested that the integrity of the minimum data set be maintained by calculating surrogate values or using data from nearby sites. Required radiation values, for example, could be estimated from computations of solar radiation above the atmosphere, a function of day of year and latitude, and measurements of the hours of bright sunshine (Selirio et al., 1971) or of cloud cover and visibility type (Davies and McKay, 1988, 1989). If the data necessary for a calculation are not available, then data from a nearby site could possibly be used for most variables except precipitation, provided that climatic gradients are not too pronounced. Where climatic gradients are pronounced, however, errors can be considerable (e.g., Schroedter, 1983), and caution would be necessary in interpreting outputs from a model run.

In the same vein, the long-term runs of daily weather data necessary for some model applications (Godwin et al., 1990) are often not available. Such runs, however, can be generated from statistics of the weather at a particular site. A program to generate daily data from statistics of monthly data of the primary weather variables (solar radiation, minimum and maximum temperature, rainfall) has been developed by IBSNAT co-operators using the framework of Richardson (1985) and Geng et al., (1988) and incorporated into the IBSNAT models so that these can use climate statistics to compute the daily weather data required for operation.

Obtaining all the soil data needed for model operation is often quite difficult. IBSNAT has endeavored, therefore, to specify a Minimum Soil Data Set, and to make available procedures by means of which surrogate values can be obtained by calculation from more readily available data (Ritchie et al., 1990). This minimum soil data set does not include information on specific adverse aspects of the soil (e.g., sodium content, salinity, and heavy-metal content), the need for which could increase as attempts are made to apply models in more marginal agricultural regions. The data does include, however, a root weighting factor that can accommodate the impact of several adverse soil factors on root growth in different soil layers.

Information for the minimum soil data set can often be obtained from soil survey publications. However, because some aspects measured during soil profile characterization (e.g., organic carbon, bulk density) may change slowly over time, particularly in the surface (plough) layer, the soil profile data may need to be supplemented with soil analysis information for the surface layer. A recent data set, not necessarily one taken immediately prior to the experiment being reported, is all that is required in most cases. For experiments specifically dealing with soil aspects (e.g., phosphorus), an analysis immediately prior to the experiment may be required, however.

In contrast to the aspects generally measured in a soil analysis (e.g., pH, P, K), the water and inorganic nitrogen contents of the soil change rapidly. A measurement or estimate of soil water, and of the two inorganic nitrogen components, at the start of any experiment, together with an estimate of the below ground residues from the previous crop, is thus regarded as necessary. Sampling for this purpose should be conducted within the month prior to planting. Each soil layer sampled should not exceed 20 cm in depth, and if possible, the first layer designated should not exceed 10 cm. Preferably, layers should match those reported for the soil profile. The total measured depth of the soil should be at least 1 m unless bedrock or other impermeable layers occur at a more shallow depth, and in all cases should be greater than the estimated depth of root activity. In cases where measurement of the initial conditions is not possible, estimates (using a 0 to 5 or other scale) should be provided for both the plough layer and the sub-plough layers as a group.

Following a definition of the initial conditions, all aspects of crop management including modifications to the environment (e.g., photoperiod extension), as imposed in some crop physiology studies, should be reported. Typical crop management factors include planting date, planting depth, row spacing, plant population, fertilization, irrigation and inoculation. Some simple models may require no management data other than the planting date, whereas more complex models will require information on all the aspects mentioned and possibly also on plant bed configuration and bund height, if appropriate. Requirements for the IBSNAT minimum data set cover most significant aspects, as documented in Table 1.

Crop cultivar information required for model operation also varies greatly among models. Some models do not require any crop cultivar (genotype) details. This may be quite acceptable with models designed for conditions in which productivity is determined by one overriding environmental factor, or for rather general applications. It would hardly be acceptable, however, for models designed for wider application. Such models would have to be recalibrated for genotypes differing from the one used during model construction. To overcome this problem, the IBSNAT models have been developed to use input data that document the characteristics of the genotypes involved (see Hunt et al., 1990; Ritchie, 1993). Such data are sometime termed 'Genetic coefficients'. These coefficients, the number of which varies among the different IBSNAT models, are not generally regarded as forming part of the minimum data set for model operation even though they are essential if a model is to be run. What is considered essential, however, is a specification of the cultivar name and type. With information on the latter, it should be possible to operate a model by selecting a similar cultivar and/or cultivar from the cultivar database.

Model calibration

Calibration requires data sets with all information needed as inputs to run the model (e.g., crop management, aerial and soil environments, and genotype characteristics), together with some data on plant performance or soil conditions. For many models, the following have been found most useful: (1) the time of occurrence of the major stages of plant development; (2) the dry weights of the major organs at various times throughout the growing season; (3) the final yield and its components; (4) the number of branches, leaves, fruits, and other organs; and (5) the main stem and branch heights. For water and nutrient stress, a record of the distribution of roots, water, and inorganic nitrogen in the soil profile during the course of the growing season is useful. Measurement data requirements that are regarded as a minimum by IBSNAT are documented in Table 1.

Model evaluation

Evaluation involves comparison of the outputs of a fully calibrated model to real data and a determination of suitability for an intended purpose (Lemon,

1977). In this context, the end-point of a evaluation exercise is a stamp of approval (or disapproval!) for a particular use. However, a number of IBSNAT co-operators have found it more useful to think of the end-point as a documentation of the precision and accuracy of the model for specified predictions in specified environments. Thus, if it is desired to predict grain yield, the evaluation end-point should encompass information on the relationship between predicted and actual grain yields, on the environments involved, and on specific aspects that could affect interpretation (e.g., possible errors in input variables or evaluation data). In this connection, environments could be specified in a general way by using an environmental index (e.g., grain yield, as used in some plant breeding analysis), in terms of some agronomic factor (e.g., planting dates, plant populations), or in terms of the major physical aspects of the environment (e.g., soil textures, soil depths, mean temperatures, daylengths).

Essential parts of any minimum data set for evaluation are: (1) a complete record of the information required to run the model – on the aerial and soil environments, on the starting conditions, on the cultivars used, and on crop management; and (2) field information on the aspect(s) that the model is desired to predict, and for which the model is being validated. The data sets should not have been used previously for calibration and should represent the complete array of environments in which the model will be applied. In the past, it has often been difficult to obtain enough data sets for effective evaluation, and techniques that make it possible to extract the maximum of information from a limited number of data sets have been used. One such technique is 'jack-knifing', a technique in which available data sets are used in different combinations for calibration and evaluation (Tichelaar and Ruff, 1989). For example, of six data sets, five sets could be used in different combinations for calibration, and one for evaluation. Within the IBSNAT group, Jintrawet (1991) has used this technique to examine the performance of a rice model. Batchelor et al. (1994) used it to test the ability to predict optimum harvest date for peanut relative to pod loss predictions.

Data acquisition

Weather

The weather data required for model operation can be obtained using standard meteorological procedures (World Meteorological Organization, 1983). However, IBSNAT recognized that some of the standard instrumentation may not be the most appropriate for use at isolated experimental sites, and fostered the development of compact instrumentation packages designed for measurement of the minimum weather data set. A number of such packages are now available commercially.

Soil

Much of the required soil data can be obtained using standard soil analysis procedures. This is not the case, however, for the lower moisture limits. Ritchie (1981) has described procedures that involve soil moisture measurements after a crop has ceased extracting water at a site where all rainfall is intercepted, but these are generally difficult to apply. Alternative procedures that either involve the use of algorithms to compute the limit from textural data, or the use of look-up tables, have thus been developed. A set of look-up tables has been included in the latest software package (DSSAT v3) released by IBSNAT.

Crop performance

Standard field experiments have proven to be the most useful source of crop and soil data for model calibration and evaluation. When these are used, however, researchers face the general problem of endeavoring to obtain good and representative data in the face of differences in soil fertility, water availability, insect, disease and bird damage, tillage operations, and many other factors that influence plant growth. To minimize these problems, plant breeders and experimental agronomists have, over the years, introduced various techniques and approaches; IBSNAT has highlighted many of these. Attention has been drawn to the problems that could arise, at least partly, because of mistakes both in site selection, and in the general management of experiments, and information that would help eliminate mistakes in these categories has been presented and discussed in a number of workshops/publications.

Sites

In conventional agronomic research, the objective of experimentation is generally to collect data that will make it possible to predict the performance in agricultural practice of a new cultivar or agronomic practice in comparison with some established standard cultivar or practice. It is generally desired that performance be relevant over a reasonably wide region that would represent a number of different soil types, fertility levels, etc., and over a number of years.

It is generally impossible to select one site that is representative of the agriculture of a wide region because differences in fertility level, etc. generally produce differences in cultivar or treatment rankings. Because of this, it has become axiomatic that agronomic experiments be carried out at a number of sites chosen to represent the array of conditions encountered in the target region, with particular emphasis on the sites at the upper and lower productivity levels. IBSNAT has emphasized that this principle applies equally to experimentation for model calibration or evaluation.

The question of performance over a number of years is also of significance relative to site selection. As when considering sites in one year, cultivars and treatments often rank differently when tested in different years under varying weather conditions, disease incidences, etc. Traditionally, attempts have been made to continue experimentation over a number of years to permit evaluation of this year effect, but in practice such a course has had limitations because of the delay in release of improved cultivars or information to the farming community. An alternative approach in conventional work has been to examine the causes of a differential performance over years and, if possible, to try to select sites where the causes of differential performance are present each year. Such examination has shown that differential performance arises most often when cultivars or treatments produce differences in heading and maturity dates (late types being favored by good late season conditions), in lodging resistance, in resistance to disease and insect pests, and in sensitivity to moisture and temperature stress particularly at key developmental stages. The selection of sites where lodging or diseases are prevalent each year, or where moisture or temperature stress occurs regularly, is thus of considerable value in allowing a complete evaluation of a new cultivar or treatment, and in facilitating prediction of how this cultivar or treatment would perform in practice over a number of years. A second principle of site selection in conventional agronomic research, therefore, has been that some sites should be chosen to represent those extreme environmental conditions that generally are not encountered each year. This second principle has been emphasized by IBSNAT as applying equally to experimentation for model calibration and/or evaluation.

For all trials, whether they be under conditions of high productivity or low, disease or environmental stress, a third principle in conventional research is that chosen sites should be as uniform as possible, both inherently and in recent cropping history. IBSNAT recognized this aspect and emphasized that even though it is impossible to obtain perfect uniformity, researchers endeavoring to collect data for model calibration/evaluation should pay attention to aspects that affect uniformity over and above soil type and variability in inherent depth. Some aspects that warrant attention, and that have been raised at various times in IBSNAT workshops and publications, are: slope, drainage, grading, fertilizing, weed infestation, previous experimentation, trees, poles, building, neighboring vegetation, exposure and roadways.

Management

In all experimentation, attention should be paid to the details of how experiments are conducted so as to increase precision and, in turn, the return in useful information. There are a number of considerations that should be taken into account, and some of these have been highlighted by IBSNAT as being particularly relevant to experimentation for model calibration/evaluation, for which an absolute rather than relative (to 'check' cultivars or treatments) measurement is required.

Plot size measurement. The most common error in plot work is inaccurate measurement of plot dimensions. Errors of measurement may average out from plot to plot, but variability is increased and the precision of the test is decreased. A basic operating principle of test management, therefore, is that plot size should be large enough to facilitate easy identification of errors in measurement that could reduce precision below that desired for overall yield measurement.

Border effects. Adjacent plots and alleyways affect the performance of neighboring cultivars and treatments. Competition for light, water and nutrients occurs between plants, so that tall cultivars growing next to a short cultivar or to a weak plot are benefitted, as are all plants growing next to the alleyways. The effects of inter-plot competition are not overcome by a random arrangements of plots, because (a) the shortest cultivar in a trial is always surrounded by taller material, and (b) in a trial with few replicates the variability of some cultivars or treatments is increased, thereby reducing the precision of the experiment. Such effects can be minimized, however, by grouping varieties of different heights, if an appropriate statistical approach is taken.

Row spacing. Variation in row width cannot be corrected for, and merely adds to overall variability. Care should thus be taken at planting to maintain row spacing as constant as possible. Because of the inevitable variability in row width, however, yield data obtained by harvesting a run of plants in one row is likely to be more variable, and perhaps also biased, compared to yield data obtained from plants in several adjacent rows.

Plant counts. With crops that branch profusely it is difficult to separate plants once the early growth period has been completed, and plant counts made late in the season are often highly variable. Data that are obtained by multiplying plant measurements by plant population are thus likely to be equally variable, and possibly also biased, with such crop types. Yield and biomass determination made in this way are particularly suspect.

Coding. IBSNAT has emphasized that standard coding schemes be used for defining growth stages. At first, attention was drawn to various scales published for staging the growth and development of different crops (e.g., peanut (*Arachis hypogea* L.) by Boote, 1982; soybean (*Glycine max* L. Merr.) by Fehr et al., 1971; maize (*Zea mays* L.)by Ritchie and Hanway, 1982; sorghum (*Sorghum bicolor* L.) by Vanderlip and Reeves, 1972; and cereals by Zadoks et al., 1974). Throughout, however, it was recognized that some of the widely used scales included both letters and numbers and thus presented difficulties when attempts were made to graph simulated data against field recorded data. Because of this, co-operators have been encouraged to use a decimal scale, perhaps by modifying the Zadoks' scale for crops other than the cereals for which it was originally conceived. The BBCH scale (Lancashire et al., 1994) has been introduced recently in an attempt to standardize and decimalize growth stage recording over a wide array of crops and weed species, and it's widespread adaptation would facilitate the interchange and use of data in model calibration/evaluation efforts.

In the same vein, IBSNAT has recognized that plant and soil system components often need to be referred to by code rather than by name. Suggested codes for both measured and simulated plant and soil data assigned in accord with a defined convention, have been included in files in the DSSAT v3 software package, and it has been suggested that researchers use these as far as possible. However, it is an accepted fact that researchers often prefer to use their own set of codes. To allow for this and yet maintain simplicity in data interchange, IBSNAT has emphasized that a definition of all unique codes should be included in data files.

Replication. Replication makes it possible both to reduce the effects of random variation and to measure the degree of this random variation. Theoretically, an increase in replication increases the precision of an experiment, to an extent largely proportional to $\sqrt{r}$ where r is the number of replicates. The degree of improvement with replication thus falls off rapidly – to double the precision of a 4 – rep test would require 16 reps, other things being equal. In practice, other things are not equal because an increase in replication requires more land, with a consequent (generally) decrease in uniformity of the experimental area. The real effect of an increase in replication is thus usually less than the theoretical effect. Because of these considerations it has generally been considered that for plots of a size suitable for mechanization, and where yield is the aspect of concern, 3 or 4 replications is best; for smaller plots, as for hand harvesting, 5 or 6 replications may be best.

In considering replication, however, it has to be recognized that where cultivars or treatments are characterized by a differential response to sites and/or seasons, replications of plots in one individual trial may have little effect on the quality of information collected from a range of trials conducted to characterize performance over a number of years for a region. Efforts in such a case should be directed to obtaining information relative to performance over the region and over a span of years. To this end, plant breeders often adopt a trial strategy in which the degree of replication of individual trials is reduced and the number of sites at which trials are conducted is increased. The principle underlying such a strategy is of considerable relevance to obtaining data for model calibration/evaluation, and is one that led Nix (1984) to argue for the use of multifactorial, nonrandomized, nonreplicated experiments that he, in an earlier paper (Nix, 1980), had termed 'omnibus' experiments. The objective in undertaking such experiments would be to obtain data from a set of treatments that span the widest possible range of cultivar by environment by management interactions in the shortest possible time and with the most economical use of available resources. Every effort would be made to identify the major sources of potential variation in crop performance and to test the whole gradient of values of one or more factors.

Such a strategy would very quickly give rise to hundreds of individual treatments. Obviously, if these were to be randomized and replicated in a conventional design, the strategy would become unworkable. Nix (1984) argued that this should not result in abandonment of the strategy, but rather, in a reduction in randomization and replication in the interest of obtaining a wider range of treatment combinations. He pointed out, however, that each treatment would need to be monitored to provide minimum data set information adequate for explanation of variations in performance.

Environmental manipulation. The adoption of experimental strategies that allow for the examination of a wide array of treatments in one experiment is desirable, particularly to evaluate response to different climatic conditions. One such strategy is to use planting date experiments in which a whole array of cultivar or other treatment plots are seeded at weekly or longer intervals to expose cultivars and other experimental treatments to a wide range of climatic conditions at one location. Such an approach does not allow, however, for exposure to a complete array of environmental conditions at all locations. Experiments in which planting date has been varied may thus need complementation with other experiments in which the environment is artificially modified in some way. Extending the daylength with artificial illumination is an approach that has been used by IBSNAT (Hunt et al., 1990; Ogoshi, 1994); line-source sprinklers or rain-out shelters, and CO_2 supplementation through feeder pipes, can also be used and are appropriate for model calibration/evaluation under field conditions. Planting on the same date at sites differing in elevation is another approach that can be used to obtain a temperature range at similar daylengths (e.g., Sexton et al., 1994).

Data handling

Storage and transfer

Information from experiments in which minimum data set requirements have been observed is a valuable resource that not only should be conserved for future use, but also should be used widely. To facilitate these activities, IBSNAT has developed some simple, standard files that can be established and edited easily, and that can also be transferred directly among workers without the need for 'retrieval' from a central database (Hunt et al., 1994a; Jones et al., 1994). Further, because the files are based on simple ASCII characters, software that reads them directly can easily be written.

The initial set of IBSNAT files, which were used for models and application programs integrated into v2.1 of the software package known as DSSAT (Decision Support System for Agrotechnology Transfer, IBSNAT, 1989b) were documented earlier (IBSNAT, 1986b, 1989a). They were found to be useful for running and evaluating the performance of models, for conducting sensitivity analysis, and for evaluating the variabilities and risks of different management strategies for a range of locations specified by soil and weather data. Experience gained in using these general files and formats demonstrated the utility of the endeavor, but also revealed several deficiencies in the extent to which they were able to accommodate (a) some aspects of the Minimum Data Set and (b) additional aspects documented in a particular study. Further, the large number of small files originally defined presented difficulties to many workers. Effort was directed, therefore, to the development of a more universal set of files.

The new files, in DSSAT v3, are sufficiently flexible to allow addition of variables that may be required by specific modelers, or that may be essential to document a specific experiment. They have been designed to accommodate a diversity of crop models and applications, to facilitate the exchange of data among users, and to be used as direct input to crop models.

The files are organized into input (environment, experimental details, and cultivar characteristics), output, and crop performance categories (Table 2). The crop performance files, which are needed for storage of results, are often used during model runs to ensure that one or more of the model output files contain simulated results along with data from the field. In some cases, however, they can be used as input files to 'reset' some variables during the course of a simulation run. For example, they could be used to record time series of pests or pest damage to the crop which, as with other data, could be used as input to crop models. The model output files are organized into categories that allow users to select information needed for a particular application. Similarly, model inputs are organized into sections to allow flexibility in their use with models that may not require the complete Minimum Data Set. For example, the soil nutrient management section in the experiment details file could be eliminated when a crop model does not include a soil fertility component.

All files contain headings for different sections, and header lines that indicate the nature of the following data items, and can also contain notes relating to some aspects of the quality or source of the data. These inserts are identified by symbols placed as the first item on a line. The symbols used in include '*' for Section heading, '@' for header line specifying variables occurring below and '!' for a note. Of these inserts, the header line is highly significant. Information on this line identifies the data items (variables) that occur below, and could be used to avoid the need for specific format statements when reading data. Information on the header line, as for the data below, should be separated by one or more spaces. Such space delimiters, together with the '*', '@', and '!' symbols, constitute the basic structural elements of the files.

To facilitate recognition of the category of data in a file, the adoption of a standard file naming convention has been recommended by IBSNAT. The suggested standard convention has two parts – the file extension, used to specify the type of file, and the prefix, used to identify the source of data. The file extensions currently used in DSSAT v3 have been documented by Hunt et al., (1994a).

Table 2. Files used for the storage and transfer of data relevant to model operation, calibration, and evaluation.

Reference Name	File Name(s) (Example)	Description
MODEL INPUT FILES		
Experimental details		
FILE_X	UFGA8201.MZX[1]	Experimental details for a specific (e.g. UFGA8201MZ) study/model run: field conditions, crop management.
Weather and soil		
FILE_W	UFGA8201.WTH	Weather data, daily, for a specific (e.g., UFGA) station, or for a specific station and time period (e.g., for one year)
FILE_S	SOIL.SOL	Soil profile data for a group of experimental sites in general UF.SOL (SOIL.SOL) or for a specific institute (e.g., UF.SOL).
Crop and cultivar		
FILE_C	MZCER960.CUL	Cultivar specific coefficients for a particular model and crop species, e.g., maize for the 'CER' model, version 96 (i.e., released in 1996).
MODEL OUTPUT FILES		
OUT_O	OVERVIEW.OUT[2]	Overview of inputs and major crop and soil output variables.
OUT_S	SUMMARY.OUT	Summary information: crop and soil input and output variables, one line for each crop cycle or model run
		Detailed time-sequence information on:
OUT_G	GROWTH.OUT	Growth
OUT_W	WATER.OUT	Water balance
OUT_N	NITROGEN.OUT	Nitrogen balance
SYSTEM PERFORMANCE FILES		
FILE_A	UFGA8201.MZA	Average values of performance data for a specific (e.g. UFGA8201MZ) experiment. (Used for comparison with summary model results.)
FILE_T	UFGA8201.MZT	Time course data (averages) for a specific (e.g. UFGA8201MZ) experiment. (Used for graphical comparison of measured and simulated time course results.)

[1] These names reflect a standard naming convention in which the first two spaces are for the crop code, the next five characters are for the model name, beginning at position 3, and the final one is a file identifier that in general is set to zero.
[2] The example names for the output files (e.g., GROWTH.OUT) are for temporary files that are re-written each simulation run. Outputs can be saved however, and in this case the file names could be made up of the usual institute, site, experiment and crop identifiers, with a final letter, G, W, etc., to designate growth, water or other data types.

File prefixes, for most model input and system performance files, are constructed from an institute or group code (2 characters), a site code (2 characters), the year in which the experiment was planted (2 characters) and an experiment number (2 characters). Prefix codes can be assigned by a user; however, to avoid duplication, IBSNAT has designated a number of institute and site codes over the past 5–10 years. Using the designated codes, an experiment conducted by the University of Florida (UF) at Gainesville (GA) in 1988 (88) would yield a file prefix of UFGA8801. For other files (e.g., output files and genotype coefficient files), suggested file prefix conventions are shown in Table 2.

Input files are divided into those dealing with the experiment or specific model run, the weather, the soil, and the characteristics of different cultivars. These will be dealt with separately below.

Experimental details file. The experimental details file documents model inputs for each 'experiment' to be simulated. Each experiment could be a real one for which there is corresponding field data, or a hypothetical one defined solely for simulation. Thus, data for many real and hypothetical experiments can be stored for use at different times, and each set of data is available for interchange with co-operators without the need for any re-organization. The file heading contains the experiment code and name, and file sections contain information on the treatment combinations, and details of the experiment (cultivars, field characteristics, soil analysis data, initial soil water and inorganic nitrogen conditions, seedbed preparation and planting geometries, irrigation and water management, fertilizer management, organic residue applications, chemical applications, environmental modifications, tillage, harvest management, and if required, disease, pest and weed incidences at the start or during the crop cycle). The experimental details file uses the same file naming convention to provide information on institute, site, planting year, experiment number, and crop. For example, UFGA8201MZ is the code for maize experiment 01, planted in 1982 by the institute designated by UF at site GA. For real experiments, the file should also contain a section of general information with the names of the people supplying the data set, addresses, site name(s), information on the plot sizes, and so on, used in the experiment. It may also contain notes on any incidents that occurred during the course of the experiment that may affect the interpretation of the data. These latter items are not normally used by simulation models, but are provided for reference and assistance in interpreting simulation results.

The structure of the file has been designed with the goal of maximizing the flexibility of input configurations while preserving the concept of entering only a minimum of inputs to run a simulation. This has been accomplished by defining an experimental treatment in terms of constituent factors that deal separately with information on planting, fertilization, irrigation, cultivars, fields, etc. For an experiment dealing with cultivars across locations without water, nutrient and pest limitations, the factors would be cultivars and fields. The inclusion of fields as a factor enables an experiment to utilize multiple weather data sets, a facility not possible when using the earlier model inputs and outputs (IBSNAT, 1989a).

The information on specific factors is contained in dedicated file sections. Only those sections required for the particular experiment or simulation run

Table 3. Example of an experimental details file for a maize irrigation and nitrogen experiment. A standard file name for this experiment would be UFGA8102.MZX.

```
*EXP.DETAILS: UFGA8201MZ NIT X IRR, GAINESVILLE 2N*3I
*GENERAL
@ PEOPLE
 BENNET, J.M. ZUR, B. HAMMOND, L.C. JONES, J.W.
@ ADDRESS
 UNIVERSITY OF FLORIDA, GAINESVILLE, FL, USA
@ SITE
 IRR.PARK,UF.CAMPUS 29.63;-82.37;40.;FLA

*TREATMENTS                                   FACTOR LEVELS
                                   ------------------------------------------------
@N R O C TNAME                     CU FL SA IC MP MI MF MR MC MT ME MH SM
 1 1 0 0 RAINFED LOW NITROGEN       1  1  0  1  1  1  1  0  0  0  0  0  1
 2 1 0 0 RAINFED HIGH NITROGEN      1  1  0  1  1  1  2  0  0  0  0  0  1
 3 1 0 0 IRRIGATED LOW NITROGEN     1  1  0  1  1  2  1  0  0  0  0  0  1
 4 1 0 0 IRRIGATED HIGH NITROGEN    1  1  0  1  1  2  2  0  0  0  0  0  1
 5 1 0 0 VEG STRESS LOW NITROGEN    1  1  0  1  1  3  1  0  0  0  0  0  1
 6 1 0 0 VEG STRESS HIGH NITROGEN   1  1  0  1  1  3  2  0  0  0  0  0  1

*CULTIVARS
@C CR INGENO CNAME
 1 MZ IB0035 McCurdy 84aa

*FIELDS
@L ID_FIELD WSTA....  FLSA  FLOB  FLDT  FLDD  FLDS  FLST SLTX  SLDP  ID_SOIL
 1 UFGA0002 UFGA     -99.0     0 DR000     0     0 00000  -99   180  IBMZ910014
@L ...........XCRD ...........YCRD .....ELEV .............AREA .SLEN .FLWR .SLAS
 1         0.00000         0.00000      0.00               0.0     0   0.0   0.0
```

Continued

need be present. Furthermore, for any particular section, data may only be required for the first treatment of the experiment. This would be the case when data are common to all subsequent treatments. Additional data would be needed, however, when the conditions for a treatment differ from those coded for the first one. As an example, if an experiment was examining the effect of two nitrogen rates, the experimental details file would contain sections for planting details, initial conditions and fertilizers. These sections would contain one set of information required for the first treatment. For the second treatment, the planting details and initial conditions would not be repeated but a second set of information would appear in the fertilizer section. The treatment itself would be linked to this information by means of a 'pointer' or 'level' indicator that would differ from that used for the first treatment. The use of these 'pointers' is illustrated in Table 3, which documents a maize irrigation and nitrogen experiment. The meaning of the individual headings for the level indicators have been documented by Hunt et al. (1994a). The experimental details file could contain as many as eighteen sections, although generally there would be fewer.

Table 3. Continued.

```
*INITIAL CONDITIONS
@C   PCR  ICDAT   ICRT   ICND   ICRN   ICRE   ICWD  ICRES  ICREN  ICREP
 1    MZ  82056    100    -99   1.00   1.00  -99.0   1000   0.80   0.00
   ICRIP  ICRID
     100     15
@C  ICBL   SH2O   SNH4   SNO3
 1     5  0.086    1.5    0.6
 1    15  0.086    1.5    0.6
 1    30  0.086    1.5    0.6
 1    60  0.086    1.5    0.6
 1    90  0.076    0.6    0.6
 1   120  0.076    0.5    0.6
 1   150  0.130    0.5    0.6
 1   180  0.258    0.5    0.6

*PLANTING DETAILS
@P PDATE  EDATE   PPOP   PPOE   PLME   PLDS   PLRS   PLRD   PLDP   PLWT
 1 82057    -99    7.2    7.2      S      R     61      0    7.0    -99
    PAGE   PENV   PLPH   SPRL
     -99  -99.0  -99.0    0.0

*IRRIGATION AND WATER MANAGEMENT
@I  EFIR   IDEP   ITHR   IEPT   IOFF   IAME   IAMT
 1  1.00    -99    -99    -99    -99    -99    -99
@I IDATE   IROP  IRVAL
 1 82063  IR001     13
@I  EFIR   IDEP   ITHR   IEPT   IOFF   IAME   IAMT
 2  1.00    -99    -99    -99    -99    -99    -99
@I IDATE   IROP  IRVAL
 2 82063  IR001     13
 2 82077  IR001     10
 2 82094  IR001     10
 2 82107  IR001     13
 2 82111  IR001     18
 2 82122  IR001     25
 2 82126  IR001     25
 2 82129  IR001     13
 2 82132  IR001     15
 2 82134  IR001     19
 2 82137  IR001     20
 2 82141  IR001     20
 2 82148  IR001     15
 2 82158  IR001     19
 2 82161  IR001      4
 2 82162  IR001     25
@I  EFIR   IDEP   ITHR   IEPT   IOFF   IAME   IAMT
 3  1.00    -99    -99    -99    -99    -99    -99
@I IDATE   IROP  IRVAL
 3 82063  IR001     13
 3 82077  IR001     10
 3 82094  IR001     10
 3 82107  IR001     13
 3 82111  IR001     18
 3 82133  IR001     30
 3 82134  IR001      4
 3 82137  IR001     20
 3 82141  IR001     20
 3 82148  IR001     15
 3 82158  IR001     19
 3 82161  IR001      4
 3 82162  IR001     25
```

Continued

Table 3. *Continued.*

```
*FERTILIZERS (INORGANIC)
@F FDATE  FMCD  FACD  FDEP  FAMN  FAMP  FAMK  FAMC  FAMO  FOCD
1  82097  FE001 -99   10    27    -99   -99   -99   -99   -99
1  82102  FE001 -99   10    35    -99   -99   -99   -99   -99
1  82137  FE001 -99   10    54    -99   -99   -99   -99   -99
2  82074  FE001 -99   10    56    -99   -99   -99   -99   -99
2  82089  FE001 -99   10    52    -99   -99   -99   -99   -99
2  82102  FE001 -99   10    75    -99   -99   -99   -99   -99
2  82118  FE001 -99   10    37    -99   -99   -99   -99   -99
2  82127  FE001 -99   10    55    -99   -99   -99   -99   -99
2  82137  FE001 -99   10    126   -99   -99   -99   -99   -99
*SIMULATION CONTROLS
@N GENERAL     NYERS NREPS START SDATE RSEED SNAME......
 1 GE              1     1     S 82056  2150 N X IRRIGATION, GAINESVIL
@N OPTIONS     WATER NITRO SYMBI PHOSP POTAS DISES  CHEM  TILL
 1 OP              Y     Y     Y     N     N     N     N     N
@N METHODS     WTHER INCON LIGHT EVAPO INFIL PHOTO  HYDRO
 1 ME              M     M     E     R     S     C     R
@N MANAGEMENT  PLANT IRRIG FERTI RESID HARVS
 1 MA              R     R     R     N     M
@N OUTPUTS     FNAME OVVEW SUMRY FROPT GROUT CAOUT  WAOUT  NIOUT MIOUT DIOUT LONG
 1 OU              N     Y     Y     3     Y     N      Y      Y     N     N    Y

@  AUTOMATIC MANAGEMENT
@N PLANTING    PFRST PLAST PH2OL PH2OU PH2OD PSTMX  PSTMN
 1 PL          82050 82064    40   100    30    40     10
@N IRRIGATION  IMDEP ITHRL ITHRU IROFF IMETH IRAMT  IREFF
 1 IR             30    50   100 GS000 IR001    10   1.00
@N NITROGEN    NMDEP NMTHR NAMNT NCODE NAOFF
 1 NI             30    50    25 FE001 GS000
@N RESIDUES    RIPCN RTIME RIDEP
 1 RE            100     1    20
@N HARVEST     HFRST HLAST HPCNP HPCNR
 1 HA              0 83057   100     0
```

It should be emphasized, however, that for any particular simulation, only a few of these sections may be needed. The minimum number for an experiment in which there were no soil water, nutrient or pest limitations would be the treatment, cultivar, field, planting details and possibly also, simulation controls sections.

Weather data. Weather data, preferably daily, must be available for the duration of the growing season, beginning with the day of planting and ending at crop maturity. Ideally, for a particular simulation, the weather file should contain data collected from before planting to after maturity. This would allow a simulation to be started before planting, thus providing an estimate of soil conditions at planting time. Additional weather data would also allow for the selection of alternate planting dates, and for simulation based on weather and soil conditions at planting, and for the simulation of longer duration crop cultivars for model sensitivity analysis.

The first lines in each weather data file, regardless of its length, contain some details of the site (name, country, annual average temperature and amplitude

Table 4. Example of a Weather Data File for Gainesville, Florida.

```
*WEATHER DATA: Gainesville, Florida, USA
@INSI    LAT     LONG   ELEV   TAV   AMP  REFHT  WNDHT
 UFGA  29.630  -82.370    10  20.9   7.5   2.00   3.00
@DATE  SRAD  TMAX  TMIN  RAIN
```

@DATE	SRAD	TMAX	TMIN	RAIN
82001	5.9	24.4	15.6	19.0
82002	7.0	22.2	15.0	0.0
82003	9.0	27.8	17.2	0.0
82004	3.1	26.1	15.0	9.4
82005	12.9	17.2	2.2	0.0
82006	11.4	25.0	4.4	0.0
82007	8.5	26.7	11.7	0.0
82008	3.0	22.8	14.4	5.3
82009	14.4	16.7	8.9	0.0
82010	12.0	14.4	1.1	0.0
82011	15.0	14.4	−6.7	0.0
82012	10.8	10.0	−7.8	0.0
82013	4.8	20.0	3.9	19.8
82014	6.2	18.9	6.1	81.0
82015	14.4	12.2	−3.3	0.0
82016	12.6	19.4	−1.7	0.0
82017	14.6	19.4	5.6	2.8
82018	11.2	20.6	1.1	0.0
82019	13.3	25.6	8.9	0.0
82020	13.4	27.8	8.3	0.0
82021	11.6	27.8	10.0	0.0
82022	10.7	25.6	12.2	0.0
82023	10.1	26.1	14.4	0.0
82024	12.9	23.9	10.0	0.0
82025	14.1	21.1	0.0	0.0
82026	14.2	20.0	4.4	0.0
82027	13.9	15.6	−1.1	0.0
82028	13.8	21.7	2.2	0.0
82029	13.3	22.2	3.3	0.0
82030	12.5	25.6	8.9	0.0

of its monthly averages, latitude and longitude, elevation). On all subsequent lines, there are data for different weather aspects at daily intervals. Generally, the file should contain data on solar radiation, maximum and minimum temperature, and precipitation, but more variables could be included as long as abbreviations for all variables are included in the header line. An example of an abbreviated weather file currently used in DSSAT v3 is presented in Table 4.

Soil data. The soil file contains data on the surface and profile characteristics. These data are used in the soil water, nitrogen, and root growth sections of the crop models. The file generally contains information that is available for the soil at a particular experimental site, and supplementary information extracted from a soil survey database for a soil of the same taxonomic classification as that at the experimental site. Occasionally, when a detailed soil analysis has been performed at the experimental site, the file will contain no information from a survey database. In the file (Table 5), the first line of data contains the

Table 5. Example of a file for soils data for Gainesville, Florida.[1]

*IBMZ910014		Gainesville FSA			180 Millhopper Fine Sand								
@SITE		COUNTRY		LAT	LONG SCS FAMILY								
Gainesville		USA		29.630	−82.370 Loamy, silic, hyperth Arenic Paleudult								
@SCOM	SALB	SLU1	SLDR	SLRO	SLNF	SLPF	SMHB	SMPX	SMKE				
−99	0.18	2.0	0.65	60.0	1.00	0.80	IB001	IB001	IB001				
@SLB	SLMH	SLLL	SDUL	SSAT	SRGF	SSKS	SBDM	SLOC	SLCL	SLSI	SLCF	SLNI
5	−99	0.026	0.096	0.230	1.000	−99	1.30	2.00	−99	−99	−99	−99
15	−99	0.025	0.086	0.230	1.000	−99	1.30	1.00	−99	−99	−99	−99
30	−99	0.025	0.086	0.230	0.800	−99	1.40	1.00	−99	−99	−99	−99
60	−99	0.025	0.086	0.230	0.200	−99	1.40	0.50	−99	−99	−99	−99
90	−99	0.028	0.090	0.230	0.100	−99	1.45	0.10	−99	−99	−99	−99
120	−99	0.028	0.090	0.230	0.050	−99	1.45	0.10	−99	−99	−99	−99
150	−99	0.029	0.130	0.230	0.002	−99	1.45	0.04	−99	−99	−99	−99
180	−99	0.070	0.258	0.360	0.000	−99	1.20	0.24	−99	−99	−99	−99

[1] Truncated for presentation purposes.

soil identifiers, information on soil texture and depth, a description that could equate to the soil classification according to a specified, locally used system (such as the Canadian classification system), while the second line contains geographic data together with taxonomic information presented according to the USDA-SCS Soil Taxonomy (1975) system. The third line contains information on soil surface properties such as albedo and on measurement techniques, while the fourth and subsequent lines contain data for each layer in the profile. The percentage of sand is assumed to be 100 minus the percentages of clay and silt, and thus is not included as an input. The number of layers, and the thickness of each layer should be the same as those in the soil analysis and initial conditions sections of the experiment file whenever possible. Properties for several soils may be included by appending data from several soils, each with its own 'soils' code number. The file may thus contain properties for several soils of the same classification. An example file is shown in Table 5.

The data in the soils file are arranged so that entries need be made only for the aspects required by a specific model, or for a specific model run. For example, if only water aspects are to be simulated, only those variables described as physical characteristics need be supplied. If only water and nitrogen aspects are to be simulated, then the physical, N and pH variables need to be entered. If phosphorus is to be considered, then all these latter variables plus all P variables must be entered. Additional variables can be added as long as they are accompanied by an appropriate entry in the header line.

Cultivar data. The content and organization of files that contain the genotype specific inputs required for simulation currently vary greatly among crop models and crops. No attempt has been made, therefore, to document contents. The use of at least one genotype file (a cultivar file) is highly recommended. For such a file, IBSNAT has defined a standard format so that models can be

Table 6. Example of a Cultivar File used by the CROPGRO model.[1]

*SOYBEAN GENOTYPE COEFFICIENTS – CRGRO960 MODEL
!

! COEFF	DEFINITIONS
! ECO#	Code for the ecotype to which this cultivar belongs (see *.eco file)
! CSDL	Critical short day length below which reproductive development progresses with no daylength effect (for shortday plants) (hour)
! PPSEN	Slope of the relative response of development to photoperiod with time (positive for shortday plants) (1/hour)
! EM-FL	Time between plant emergence and flower appearance (R1) (photothermal days)
! FI-SH	Time between first flower and first pod (R3) (photothermal days)
! FI-SD	Time between first flower and first seed (R5) (photothermal days)
! SD-PM	Time between first seed (R5) and physiological maturity (R7) (photothermal days)
! FI-LF	Time between first flower (R1) and end of leaf expansion (photothermal days)
! LFMAX	Maximum leaf photosynthesis rate at 30°C, 350 vpm CO_2, and high light (mg CO_2/m^2 s)
! SLAVR	Specific leaf area of cultivar under standard growth conditions (cm^2/g)
! SIZLF	Maximum size of full leaf (three leaflets) (cm^2)
! XFRT	Maximum fraction of daily growth that is partitioned to seed + shell
! WTPSD	Maximum weight per seed (g)
! SFDUR	Seed filling duration for pod cohort at standard growth conditions (photothermal days)
! SDPDV	Average seed per pod under standard growing conditions (#/pod)
! PODUR	Time required for cultivar to reach final pod load under optimal conditions (photothermal days)

Continued

used with software that facilitates the calculation of cultivar coefficients (Hunt et al., 1993; Hunt and Pararajasingham, 1994). The standard suggested begins each line with 6 spaces for a cultivar identification code (the first two items should be a code for the Institute or Person that assigned the number), a blank, 16 spaces for the cultivar name, a blank, 6 spaces for a type identifier (e.g., an identifier for highland or lowland bean ecotypes), and then data in a (1X, F5.?) format (i.e., 1 blank, followed by 5 spaces for a real variable with the required number of decimals). An example file used by the CROPGRO model for soybean is shown in Table 6.

Output data. A number of output files for each simulation run, which may encompass several experiments, are described in Table 2. The first file provides an overview of input conditions and crop performance, and a comparison with actual data if available. This file presents information that uniquely describes the simulated data set, a summary of soil characteristics and cultivar coefficients, the crop and soil status at the main developmental stages, a comparison of simulated and measured data for major variables, and information on simulated stress factors and weather data during the different developmental phases (as appropriate to the crop). Actual contents and format of most of the different

Table 6. Continued.

@VAR#	VRNAME......		ECO#	CSDL	PPSEN	EM-FL	FL-SH	FL-SD	SD-PM	FL-LF	LFMAX
!				1	2	3	4	5	6	7	8
990101	M GROUP 1	Early	SB0101	13.97	0.187	17.0	6.0	13.0	32.00	26.00	1.030
990001	M GROUP 1		SB0101	13.84	0.203	17.0	6.0	13.0	32.00	26.00	1.030
990201	M GROUP 1	Late	SB0101	13.71	0.226	17.0	6.0	13.0	32.00	26.00	1.030
990102	M GROUP 2	Early	SB0201	13.71	0.226	17.4	6.0	13.5	33.00	26.00	1.030
990002	M GROUP 2		SB0201	13.59	0.249	17.4	6.0	13.5	33.00	26.00	1.030
990202	M GROUP 2	Late	SB0201	13.50	0.267	17.4	6.0	13.5	33.00	26.00	1.030
990103	M GROUP 3	Early	SB0301	13.49	0.267	19.0	6.0	14.0	34.00	26.00	1.030
990003	M GROUP 3		SB0301	13.40	0.285	19.0	6.0	14.0	34.00	26.00	1.030
990203	M GROUP 3	Late	SB0301	13.26	0.289	19.0	6.0	14.0	34.00	26.00	1.030
990104	M GROUP 4	Early	SB0401	13.24	0.290	19.4	7.0	15.0	34.50	26.00	1.030
990004	M GROUP 4		SB0401	13.09	0.294	19.4	7.0	15.0	34.50	26.00	1.030
990204	M GROUP 4	Late	SB0401	12.96	0.298	19.4	7.0	15.0	34.50	26.00	1.030
990105	M GROUP 5	Early	SB0501	12.96	0.299	19.8	8.0	15.5	35.00	18.00	1.030
990005	M GROUP 5		SB0501	12.83	0.303	19.8	8.0	15.5	35.00	18.00	1.030
990205	M GROUP 5	Late	SB0501	12.71	0.307	19.8	8.0	15.5	35.00	18.00	1.030
990106	M GROUP 6	Early	SB0601	12.70	0.307	20.2	9.0	16.0	35.50	18.00	1.030
990006	M GROUP 6		SB0601	12.58	0.311	20.2	9.0	16.0	35.50	18.00	1.030
990206	M GROUP 6	Late	SB0601	12.46	0.315	20.2	9.0	16.0	35.50	18.00	1.030
990107	M GROUP 7	Early	SB0701	12.45	0.316	20.8	10.0	16.0	36.00	18.00	1.030
990007	M GROUP 7		SB0701	12.33	0.320	20.8	10.0	16.0	36.00	18.00	1.030
990207	M GROUP 7	Late	SB0701	12.20	0.325	20.8	10.0	16.0	36.00	18.00	1.030
990108	M GROUP 8	Early	SB0801	12.20	0.325	21.5	10.0	16.0	36.00	18.00	1.030
990008	M GROUP 8		SB0801	12.07	0.330	21.5	10.0	16.0	36.00	18.00	1.030

[1] Truncated for presentation purposes

sections of this file vary somewhat at a modelers discretion, excepting those cases where software that makes use of the file (e.g., GENCALC, Hunt et al., 1993) is to be used.

A second output file provides a summary of outputs for use in applications programs with one line of data for each simulation run, while subsequent files contain detailed simulation results, including simulated seasonal (at daily or less frequent intervals) growth and development, carbon balance, soil water balance, nitrogen balance, and mineral nutrient (e.g., P, K) aspects. These files are required for detailed graphic comparison of simulated results with data that may have been collected periodically during the growing season. They can be saved in files named according to the code of the first experiment in the simulation session, but with a final letter to indicate the aspect dealt with in the file, or with a sub-prefix of choice entered in the simulation controls. An example of a file dealing with growth aspects is given in Table 7. The contents of such a file will inevitably vary depending on the particular interests and requirements of the modeler/modeling group. All files, however, should contain the required structural indicators – a 'section' or 'run' heading, designated by a '*' in column 1, and a header line with abbreviations for the variables presented in the columns below the header. As for input files, these header lines must be designated with a '@' in column 1.

Table 7. Example of an output file presenting time-sequence data for various growth related aspects.

```
GROWTH ASPECTS OUTPUT FILE
*RUN 1              :  IRRIGATED HIGH NITROGEN
 MODEL              :  GECER960 – MAIZE
 EXPERIMENT         :  UFGA8201 MZ  NIT X IRR, GAINESVILLE 2N*3I
 TREATMENT 4        :  IRRIGATED HIGH NITROGEN

 CROP               :  MAIZE              CULTIVAR: McCurdy 84aa
 STARTING DATE      :  FEB 25 1982
 PLANTING DATE      :  FEB 26 1982        PLANTS/m2  7.2 ROW SPACING: 61.0 cm
 WEATHER            :  UFGA 1982
 SOIL               :  IBMZ910014         TEXTURE: FSA – Millhopper Fine Sand
 SOIL INITIAL C     :  DEPTH:180 cm EXTR. H2O: 160.9 mm NO3: 14.9 kg/ha NH4: 21.1 kg/ha
 WATER BALANCE      :  IRRIGATE ON REPORTED DATE(S)
 IRRIGATION         :  264 mm IN 16 APPLICATIONS
 NITROGEN BAL.      :  SOIL-N & N-UPTAKE SIMULATION; NO N-FIXATION
 N-FERTILIZER       :  401 kg/ha IN 6 APPLICATIONS
 RESIDUE/MANURE     :  INITIAL: 10 kg/ha; 1000 kg/ha IN 1 APPLICATIONS
 ENVIRONM. OPT.     :  DAYL=0.00; SRAD=0.00; TMAX=0.00; TMIN=0.00; RAIN=0.00; CO2=
                       R330.00; DEW=0.00; WIND=0.00
 SIMULATION OPT     :  WATER : Y; NITROGEN : Y; N-FIX : N; PESTS : N; PHOTO : C; ET : R
 MANAGEMENT OPT     :  PLANTING : R; IRRIG : R; FERT : R; RESIDUE : R; HARVEST : M; WTH : M
```

YR and DOY @DATE	Days after plant CDAY	Leaf num L#SD	Grow stage GSTD	LAI LAID	Dry weight (kg/Ha) Leaf LWAD	Stem SWAD	Grain GWAD	Root RWAD	Crop CWAD	Grain per m² G#AD	Kern. wght mg K GWGD
82057	0	0.0	0	0.00	0	0	0	0	0	0	0.0
82062	5	0.0	0	0.00	0	0	0	0	0	0	0.0
82068	11	2.0	1	0.00	14	14	0	15	29	0	0.0
82074	17	4.0	1	0.06	17	14	0	33	31	0	0.0
82080	23	7.0	1	0.25	113	14	0	121	127	0	0.0
82086	29	9.0	1	0.52	281	14	0	306	295	0	0.0
82092	35	10.0	3	0.91	521	15	0	681	536	0	0.0
82098	41	12.0	3	1.41	886	28	0	1221	913	0	0.0
82104	47	13.0	3	2.00	1369	110	0	1764	1480	0	0.0
82110	53	16.0	3	2.98	2242	531	0	2208	2772	0	0.0
82116	59	18.0	3	3.64	2872	1112	0	2353	3983	0	0.0
82122	65	20.0	3	4.27	3514	2415	0	2568	5929	0	0.0
82128	71	22.0	3	4.60	3947	3707	0	2977	7655	0	0.0
82134	77	23.0	4	4.60	3993	3884	0	3571	9236	0	0.0
82140	83	23.0	4	4.48	3628	4459	0	3842	10884	0	0.0
82146	89	23.0	5	4.35	3538	4556	1326	3899	12458	3506	37.8
82152	95	23.0	5	4.26	3516	4465	2944	3912	13964	3506	84.0
82158	101	23.0	5	4.05	3495	4556	4526	4091	15616	3506	129.1
82164	107	23.0	5	3.72	3474	4556	6104	4243	17173	3506	174.1
82170	113	23.0	5	3.24	3454	4514	7710	4336	18716	3506	219.9
82176	119	23.0	5	2.59	3433	4556	9338	4381	20366	3506	266.3
82182	125	23.0	6	2.02	3419	4556	10417	4408	21431	3506	297.1

System performance. System performance (crop and soil) data are contained in four files designated as FILEP, FILED, FILEA and FILET. The purpose of these files is to store measured performance data for easy comparison with simulated results and for easy use with other analysis and application programs.

FILEP and FILED are the basic performance data files, with information detailed at the replicate level for each treatment, arranged by plots in FILEP and by date of measurement in FILED. FILEA and FILET contain average values derived from the data in FILEP or FILED. Averages are arranged in columns in order of treatment in FILEA, in order of date in the time-course file, FILET. The files will have a variable number of columns, depending on the data available. Each could have as few as one measured variable, or as many variables as measured. Each experiment data file, however, should always have a heading designated by a '*' in the first column, with the institute and site codes and experiment number, the crop group code, and the experiment name. All columns should have at least one leading blank and five spaces for data, and be headed by standard variable abbreviations. Each data column, if appropriate, could have comment lines containing information on the date measurements were made. Examples of 'A' and 'T' files are shown in Table 8.

Data checking and filling

New data-sets almost invariably contain errors, inconsistencies, and gaps despite the best efforts of experimenters. IBSNAT has recognized this and emphasized that data used for model operation must be checked and/or verified for accuracy and have any gaps filled-in using some standard procedures. In particular, IBSNAT has developed software (WeatherMan) that facilitates the checking of weather data and the computation of surrogate values for days when data are missing, and inserts appropriate 'flags' in the files to indicate the nature of any replacement values (Hansen et al., 1994). In similar vein, software has been developed for the computation of missing soil moisture limits from information on texture.

For crop data, IBSNAT has emphasized the need for a manual evaluation of the data. Experience gained by IBSNAT co-operators has highlighted a number of problems, with some discussed in the following sections.

Moisture contents. Grain yield reported without any indication as to whether it has been recorded on a fresh weight (i.e., on an 'as-is' basis) or dry weight basis. Model outputs are generally given in dry-weight terms so comparisons with field data recorded in other than dry weight terms can lead to gross misinterpretation.

Units. Measurements reported without any indication as to the units and basis of measurement used. In some cases this type of problem can be easily spotted but in other cases, for example, when yield data have been reported on a 'per plot' rather than per unit area basis, the problem can be virtually impossible to detect.

Conventions. Data being reported without any indication as to the convention used to define the aspect measured. For example, leaf area index reported

Table 8. Examples of experiment data files.

A. Averages arranged by treatment

*EXP.DATA (A): UFGA8201MZ N X IRRIGATION, GAINESVILLE

!Grain yield (HWAM) expressed as dry weight

@TRNO	HWAM	HWUM	H#AM	H#UM	LAIX	CWAM	BWAH	ADAT	MDAT	GN%M	CNAM	SNAM
1	2929.	0.218	917.	229.	2.26	5532.	3530.	132	185	1.80	69.5	37.8
2	3130.	0.209	1494.	220.	2.84	7201.	4071.	132	185	1.80	104.8	47.8
3	6850.	0.227	3013.	343.	3.26	14581	7729.	132	185	1.80	130.9	38.5
4	11881	0.309	3847.	496.	4.09	22001	10120	132	185	1.60	267.7	74.8
5	6375.	0.234	2722.	356.	2.76	12002	5627.	132	185	1.20	113.4	35.0
6	9344.	0.279	3344.	474.	3.70	17146	7802.	132	185	1.60	211.6	60.2

B. Averages arranged by time within treatments

*EXP.DATA (T): UFGA8201MZ N X IRRIGATION, GAINESVILLE

@TRNO	DATE	CWAD	LAID	GWAD
1	82057	0	0.00	0
1	82089	88	0.17	0
1	82103	341	0.56	0
1	82116	1070	1.43	0
1	82131	3017	2.26	0
1	82137	3263	1.86	0
1	82145	3214	1.18	50
1	82152	3479	1.37	144
1	82158	5688	1.46	994
1	82166	6012	1.33	1044
1	82172	6203	1.00	2182
1	82179	5902	0.66	2009
1	82189	5532	0.18	2002
2	82057	0	0.00	0
2	82089	96	0.19	0
2	82103	626	0.86	0
2	82116	1771	2.22	0
2	82131	4464	2.84	0
2	82137	4576	2.11	0
2	82145	4658	1.56	79
2	82152	5583	1.54	367
2	82158	6627	1.47	1030
2	82166	7019	1.30	1462
2	82172	7551	1.25	2390
2	82179	7525	0.51	2333
2	82189	7201	0.16	3132

without an indication as to whether senescent (yellow or partially yellow) leaves or parts of leaves were included, or whether the area of non-lamina structures (e.g., stems, spikes) was included. Similarly, specific leaf area reported without an indication as to whether mid-ribs, petioles or petiolules were included.

Discrepancies. Yield component data (e.g., grain weight and number) not matching with the overall yield data, canopy component data (e.g., leaf weight, stem weight) not matching with overall canopy weight, or end of season measurements not matching with the final determination of a sequence of measurement made through the growing season. In most cases there may be good reasons for such discrepancies (e.g., different sampling techniques) and if so, this should be explained with a comment in the file, otherwise a user could be at a loss as to which measurement(s) would be best for comparison with model output.

Manipulations. Some of the data in a file have been adjusted without any indication as to the reasons for the adjustment and the data values so adjusted. Such adjustments could occur when a lattice design and analysis was used, but more frequently occur when plots have been damaged in some way (lodging, animal trampling, etc.), or some of the harvest product has been lost or damaged. Such changes should be clearly documented as a comment in the data file; the structures used for data storage were developed, in part, to make this possible.

Data use

Model calibration

Model calibration needs to be undertaken not only to ensure that the constants and response functions being used are appropriate for the species with which the model was developed, but also to ensure that the model works well for the cultivars used in a particular region. Generally, the first of these activities is undertaken by the modeler, whereas the second activity could be undertaken by a user. IBSNAT has recognized these different activities, and has developed 'tools' to help with both of them. A graphics package (Wingraf) was developed to simplify the comparison of model outputs with recorded data, particularly useful when examining outputs and data recorded throughout a growing season. This package is useful for calibration for both 'species' and 'cultivar' traits. For the cultivar characteristics, an additional package (GENCALC) was developed to reduce the need for manual iteration when estimating coefficients (Hunt et al., 1993; Hunt and Pararajasingham, 1994). Both of these packages make use of the standard file structures, and they would thus be applicable with any files or models that conform to the standards. The graphics package, when used for calibration, is used to facilitate a 'visual eye fit' approach to the adjustment of parameters that relate to aspects such as rate of dry matter increase, duration of leaf area growth, rate and duration of pod addition, specific leaf area over time, and rate of single seed growth.

When calibrating, IBSNAT has emphasized the importance of following a systematic approach in which parameters are evaluated in a logical sequence. This approach, relative to what constitutes a logical sequence for a number of different crops, has been captured in a 'rules' file in the GENCALC software, and outlined verbally in many workshop sessions for those using the graphics package to facilitate a 'visual eye fit' approach. The essence of such a sequence, as used for legume crop models, is outlined below.

1. *Crop Life Cycle.* The first step in any calibration exercise should be concentrated on crop development (flowering date and maturity date). A start should be made by selecting the starting parameter values from those available for 'general' cultivars, assuming the critical daylength and photoperiod sensitivity values are correct, and adjusting the duration of the period between germination or emergence and flower appearance until flowering date is simulated correctly. Then, the period between first seed and physiological maturity should be adjusted until maturity date is correct.

2. *Dry Matter Accumulation.* The next step should involve a comparison of simulated and measured biomass values. If simulated dry matter accumulation is too rapid or too slow, parameters that affect leaf and canopy photosynthesis will need to be adjusted.

3. *Leaf Area Index and Specific Leaf Area.* Several 'cultivar' parameters can impact somewhat on dry matter accumulation via their effect on leaf area index and light interception. These include specific leaf area, time to cessation of leaf area expansion, early leaf area expansion, and the timing of pod and seed growth. Of these characteristics, specific leaf area should be adjusted first then the simulated leaf area curve should be compared with real data and the other parameters adjusted to achieve a reasonable match.

4. *Dry Matter Accumulation.* This should now be re-calibrated using correct specific leaf area and leaf area timing aspects.

5. *Seed Size, Seeds per Pod, and Seed Filling Duration.* These should be calibrated next by changing parameters that relate to the maximum weight per seed, the average number of seeds per pod, and the seed filling duration for a pod cohort, until simulated and measured final seed size and seeds per pod are matched.

6. *Initial Rise in Pod and Seed Dry Weight.* Parameters that deal with the duration of periods from flowering to first pod, to first seed, and to the end of pod addition, now should be adjusted until the initial rise in pod and seed dry weight is correct. Durations should be shortened if pod and seed growth needs to start sooner, and the opposite if they need delaying.

7. *Maturity.* The duration of the period from first seed to physiological maturity may need changing once again at this stage, to ensure that the recorded date of physiological maturity is correctly predicted.

8. *Seed Size, Shelling Percentage, and Seed Filling Duration.* Finally, end-point seed size and shelling percentage, which may have changed because of alterations to parameters affecting timing, should be recalibrated by adjusting the parameter dealing with weight per seed.

Model evaluation

For evaluation, an exercise which involves determination of the suitability of a model for an intended use, the approach taken by IBSNAT has been one that simplifies the creation of graphs that take data from a number of simulations and present simulated output on one axis and measured data on the other. To this end, the graphics package, Wingraf, has been set-up so that a user can easily make a comparison of model output and real data for a trait of concern. The latter could be grain yield, for which the graph of interest would be one of simulated vs recorded grain yield, nitrate loss in the drainage water, biomass production, or any one of a number of aspects for which both simulated and measured data are available.

The question of how to quantify the goodness-of-fit of model outputs and recorded data has been taken up at a number of IBSNAT workshops. For comparisons that involve predictions of the time of occurrence of discrete events during the life cycle (e.g., flowering date, maturity date), or of yield or biomass at maturity, the use of the error sum of squares (*SS*) has sometimes been recommended. This would be determined over *i* model runs with *i* different experiments as:

$$SS = (\text{sim } Y_1 - \text{obs. } Y_1)^2 + \cdots + (\text{sim } Y_i - \text{obs } Y_i)^2$$

This sum of squares approach has been used in calibration in an automated optimization procedure that systematically searched for parameters to minimize error (e.g., Grimm et al., 1993). Regression analysis could also be applied to a set of simulated and measured data comparisons to provide the standard error of the estimated intercept, the standard error of the estimated slope, the residual standard error (rse), and an *r* value. Such an approach was taken by Huda (1988) with simulated and recorded yields of sorghum. With such an approach, good predictability is indicated if the regression slope is near 1.0, if the intercept is close to 0, if the rse is low, and if the *r* value is high. If the slope is significantly different from 1.0, then the model is either too sensitive or insufficiently sensitive to the environments in which the experimental data were collected.

Conclusions

One of the overriding impressions gained while working with IBSNAT models is that there is a dearth of data sets for model calibration and evaluation. An associated aspect concerns the lack of attention accorded to the management of data that could be useful in this context, or that has once been used in the same context. This lack of attention has in some cases resulted in the loss of data sets that were aggregated for a particular piece of work, but later were lost after publication or when the scientist concerned moved on to other activities. There is an urgent need, therefore, for the collection of past and future data sets that can be used for calibration/evaluation, for the adoption of standards for recording and transferring such data, and for one or more organizations to take on the task of maintaining a database of such data sets. The widespread use of the data structures developed by IBSNAT, which make possible a 'dispersed' but nonetheless standard database, has proved to be of considerable help in the organization and transfer of data. IBSNAT itself endeavored to maintain a centralized collection of such weather, soil and experimental data files, but now does not have the resources to adequately fill this role. There is thus an urgent need for another organization, preferably the Committee on Data for Science and Technology (CODATA) or under the aegis of CODATA to take on this task (Uhlir and Carter, 1994). Without continued effort in this area, data loss will continue to occur, determination of model precision and accuracy will be difficult, and without knowledge of the outcome of such determinations, model application to decision making in the real world will remain a hope rather than a reality.

A second impression concerns the lack of data that are essential for model operation, which in some cases reflects the inaccessibility rather than the absence of data. In some regions, required weather and soil data have not been collected, or they are archived in such a way as to be virtually inaccessible to a user. For such regions, effort is necessary not only to assemble data, but also to archive it in a way that facilitates transfer to users. For cases where required data have not been collected, effort is needed to develop improved procedures for computing surrogate values. This is important for both weather and soil information.

As with weather and soil data, information on the characteristics of the cultivars used in a region is often not readily available for use in crop modeling. This is so both for the basic characteristics used in simple models (e.g., total duration of the life-cycle, yield under nonstressed conditions), and for the additional coefficients used in some more comprehensive models (e.g., photoperiod sensitivity, heat sensitivity). In many regions, plant breeders and/or germplasm bank curators collect information that could be used for modeling, but present this in ways that render its use by modelers virtually impossible. A final need, therefore, that has been highlighted by the IBSNAT experience, is for enhanced interaction among plant breeders, germplasm custodians, and crop modelers, and for the development of standards for presenting plant breeding and germplasm bank information in ways that can be useful both for traditional decision-making approaches and for model-based approaches.

References

Batchelor W D, Jones J W, Boote K J, Hoogenboom G (1994) Carbon-based model to predict peanut pod detachment. Transactions of the American Society of Agricultural Engineers 37:1639–1646.

Boote K J (1982) Growth stages of peanut (*Arachis hypogaea* L.). Peanut Science 9:35–40.

Davies J A, McKay D C (1988) Estimating solar radiation from incomplete cloud data. Solar Energy 41:153.

Davies J A, McKay D C (1989) Evaluation of selected models for estimating solar radiation on horizontal surfaces. Solar Energy 43:153.

Fehr W R, Caviness C E, Burmood D T, Pennington J S (1971) Stage of development descriptions for soybeans, *Glycine max* (L.) Merrill. Crop Science 11:929–931.

Geng S, Auburn J, Brandsetter E, Li B (1988) A Program to Simulate Meteorological Variables; Documentation for SIMMETEO. Agronomy Progress Report, University of California, Davis, California, USA.

Godwin D C, Thornton P K, Jones J W, Singh U, Jagtap S S, Ritchie J T (1990) Using IBSNAT's DSSAT in strategy evaluation. Pages 59–71 in Proceedings of the IBSNAT Symposium: Decision Support System for Agrotechnology Transfer. Part I: Symposium proceedings. Department of Agronomy and Soil Science, College of Tropical Agriculture and Human Resources, University of Hawaii, Honolulu, Hawaii, USA.

Grimm S S, Jones J W, Boote K J, Hesketh J D. 1993 Parameter estimation for predicting flowering date of soybean cultivars. Crop Science 33:137–144.

Hansen J W, Pickering N B, Jones J W, Wells C, Chan H V K, Godwin D C (1994) WeatherMan: Managing Weather and Climate Data. Pages 137–200 in Tsuji G Y, Uehara G, Balas S (Eds.) DSSAT v3. Vol. 3–3, University of Hawaii, Honolulu, Hawaii, USA.

Huda A K S (1988) Simulating growth and yield responses of sorghum to changes in plant density. Agronomy Journal 80:541–547.

Hunt L A (1994) Data requirements for crop modelling. Pages 15–25 in Uhlir P F, Carter G C (Eds.) Crop modelling and related environmental data. A focus on applications for arid and semiarid regions in developing countries. CODATA, International Council of Scientific Unions, Paris, France.

Hunt L A, Jones J W, Ritchie J T, Teng P S (1990) Genetic coefficients for the IBSNAT crop models. Pages 15–29 in Proceedings of the IBSNAT Symposium: Decision Support System for Agrotechnology Transfer. Part I: Symposium proceedings. Department of Agronomy and Soil Science, College of Tropical Agriculture and Human Resources, University of Hawaii, Honolulu, Hawaii, USA.

Hunt L A, Jones J W, Hoogenboom G, Godwin D C, Singh U, Pickering N B, Thornton P K, Boote K J, Ritchie J T (1994a) Input and output file structures for crop simulation models. Pages 35–72 in Uhlir P F, Carter G C (Eds.) Crop modeling and related environmental data. A focus on applications for arid and semiarid regions in developing countries. CODATA, International Council of Scientific Unions, Paris, France.

Hunt L A, Jones J W, Tsuji G Y, Uehara G (1994b) A minimum data set for field experiments. Pages 27–33 in Uhlir P F, Carter G C (Eds.) Crop modeling and related environmental data. A focus on applications for arid and semiarid regions in developing countries. CODATA, International Council of Scientific Unions, Paris, France.

Hunt L A, Pararajasingham S (1994) GenCalc. Pages 201–234 in Tsuji G Y, Uehara G, Balas S (Eds.) DSSAT v3. Vol. 3–4, University of Hawaii, Honolulu, Hawaii, USA.

Hunt L A, Pararajasingham S, Jones J W, Hoogenboom G, Imamura D T, Ogoshi R M (1993) GENCALC: Software to facilitate the use of crop models for analyzing field experiments. Agronomy Journal 85:1090–1094.

International Benchmark Sites Network for Agrotechnology Transfer (IBSNAT) (1986a) Technical Report 1: Experimental Design and Data Collection Procedures for IBSNAT. Second Edition, Department of Agronomy and Soil Science, College of Tropical Agriculture and Human Resources, University of Hawaii, Honolulu, Hawaii, USA.

International Benchmark Sites Network for Agrotechnology Transfer (IBSNAT) (1986b) Technical Report 5: Documentation for the IBSNAT Crop Model Input and Output Files, Version 1.0, Department of Agronomy and Soil Science, College of Tropical Agriculture and Human Resources, University of Hawaii, Honolulu, Hawaii, USA.

International Benchmark Sites Network for Agrotechnology Transfer (IBSNAT) (1988) Technical Report 1: Experimental Design and Data Collection Procedures for IBSNAT. Third Edition, Department of Agronomy and Soil Science, College of Tropical Agriculture and Human Resources, University of Hawaii, Honolulu, Hawaii, USA.

International Benchmark Sites Network for Agrotechnology Transfer (IBSNAT) (1989a) Technical Report 5: Documentation for the IBSNAT Crop Model Input and Output Files, Version 1.1, Department of Agronomy and Soil Science, College of Tropical Agriculture and Human Resources, University of Hawaii, Honolulu, Hawaii, USA.

International Benchmark Sites Network for Agrotechnology Transfer (IBSNAT) (1989b) DSSAT v2.1 User's Guide, Department of Agronomy and Soil Science, College of Tropical Agriculture and Human Resources, University of Hawaii, Honolulu, Hawaii, USA.

Jintrawet A (1991) A decision support system for rapid appraisal of rice-based agricultural innovations. Ph.D. Thesis, University of Hawaii, Honolulu, Hawaii, USA.

Jones C A (1984) Experimental Design and Data Collection Procedures for IBSNAT. IBSNAT Technical Report 1. Department of Agronomy and Soil Science, College of Tropical Agriculture and Human Resources, University of Hawaii, Honolulu, Hawaii, USA.
Jones J W, Hunt L A, Hoogenboom G, Godwin D C, Singh U, Tsuji G Y, Pickering N B, Thornton P K, Bowen W T, Boote K J, Ritchie J T (1994) Input and output files. Pages 1–94 in Tsuji G Y, Uehara G, Balas S (Eds.) DSSAT v3. Vol. 2-1, University of Hawaii, Honolulu, Hawaii, USA.
Lancashire P D, Bleiholder H, Van den boom T, Langeluddeke P, Strauss R, Weber E, Witzenberger A (1991) A uniform decimal code for growth stages of crops and weeds. Annals Applied Biology 119:561–601.
Lemon E R (1977) Final Report of the USDA Modeling Coordinating Committee. USDA-ARS, USA.
Matthews R B, Hunt L A (1994) GUMCAS: model describing the growth of cassava (Manihot esculenta L. Crantz). Field Crops Research 36:69–84.
Nix H A (1980) Strategies for crop research. Proceedings of the Agronomy Society of New Zealand 10:107–110.
Nix H A (1984) Minimum data sets for agrotechnology transfer. Pages 181–188 in Proceedings of the International Symposium of Minimum Data Sets for Agrotechnology Transfer, 21–26 March 1983, ICRISAT Center, India, Patancheru, India.
Ogoshi R M (1994) Determination of genetic coefficients from field experiments for CERES-maize and SOYGRO crop growth models. Ph.D. Thesis, University of Hawaii, Honolulu, Hawaii, USA.
Richardson C W (1985) Weather simulation for crop management models. Transactions of the ASAE 28(5):1602–1606.
Ritchie J T (1981) Soil water availability. Plant and Soil 58:327–338.
Ritchie J T (1993) Genetic specific data for crop modeling. Pages 77–93 in Penning de Vries F W T, Teng P, Metselaar K (Eds) Systems approaches for agricultural development. Kluwer Academic Publishers, Dordrecht, the Netherlands.
Ritchie J T, Godwin D C, Singh U (1990) Soil and weather inputs for the IBSNAT crop models. Genetic coefficients for the IBSNAT crop models. Pages 31–45 in Proceedings of the IBSNAT Symposium: Decision Support System for Agrotechnology Transfer. Part I: Symposium proceedings. Department of Agronomy and Soil Science, College of Tropical Agriculture and Human Resources, University of Hawaii, Honolulu, Hawaii, USA.
Ritchie S W, Hanway J J (1982) How a Corn Plant Develops. Special Report No. 48 (Revised) Iowa State University, Ames, Iowa, USA.
Schroedter H (1983) Meteorological problems in the practical use of disease-forecasting models, EPPO Bulletin 13:307.
Selirio I S, Brown D M, King K M (1971) Estimation of net and solar radiation. Canadian Journal of Plant Science 51:35.
Sexton P J, White J W, Boote K J (1994) Yield-determining processes in relation to cultivar seed size of common bean. Crop Science 34:84–91.
Soil Survey Staff (1975) Soil Taxonomy: A basic system of soil classification for making and interpreting soil surveys. Soil Conservation Service, U.S. Department of Agriculture Handbook 436, U.S. Government Printing Office, Washington, D.C.
Tichelaar B W, Ruff L J (1989) How good are our best models? Eos, May 16.
Uhlir P F, Carter G C (Eds.) (1994) Crop modeling and related environmental data. A focus on applications for arid and semiarid regions in developing countries. CODATA, International Council of Scientific Unions, Paris, France.
Vanderlip R L, Reeves H E (1972) Growth stages of sorghum (*Sorghum bicolor* (L.) Moench). Agronomy Journal 64:13–16.
World Meteorological Organization (1983) Guide to Agricultural Meteorological Practices. 2nd edition, World Meteorological Organization, Geneva, Switzerland.
Zadoks J C, Chang T T, Konzak C F (1974) A decimal code for the growth stages of cereals. Eucarpia Bulletin 7.

Soil water balance and plant water stress

J.T. RITCHIE
Homer Nowlin Chair, Department of Crop and Soil Sciences, Michigan State University, East Lansing, Michigan 48824, USA

Key words: evaporation, transpiration, water uptake, infiltration, runoff, drainage, simulation

Abstract
The soil water balance is calculated in the DSSAT crop models in order to evaluate the possible yield reduction caused by soil and plant water deficits. The model evaluates the soil water balance of a crop or fallow land on a daily basis as a function of precipitation, irrigation, transpiration, soil evaporation, runoff and drainage from the profile. The soil water is distributed in several layers with depth increments specified by the user. Water content in any soil layer can decrease by soil evaporation, root absorption, or flow to an adjacent layer. The limits to which water can increase or decrease are input for each soil layer as the saturated upper limit. The values used for these limits must be appropriate to the soil in the field, and accurate values are quite important in situations where the water input supply is marginal.

Introduction

Water is essential for plant life. Plants derive water from the soil through their roots in an attempt to maintain a favorable hydraulic balance. Water flow through plants helps to maintain a favorable temperature by evaporative cooling. Water is also a vital substrate for several plant biochemical reactions. Plant roots simultaneously absorb water and nutrients essential for plant growth. The absorbed water transports nutrients to the plant tops. The soil gaseous environment is the complement to soil water. The rate in which gases can move in the soil is inversely related to the soil water content. Since soil oxygen is essential for root function, the well-being of plants is dependent on the water content of the soil being in the right proportions for water uptake and oxygen movement. Fortunately, there is a broad range of soil water contents for sufficient water and nutrient transport to the plant tops and for oxygen transport to the roots.

Thus the water economy of the root-zone soil is a critical determinant of the biophysical activity of plants. The simulation of the soil water balance depends on the capability of water from rainfall or irrigation to enter soil through the surface and be stored in the soil reserve. Other budgetary considerations include drainage of water out of the rootzone by the forces of gravity, runoff of water that does not enter the surface, water lost from the surface by evaporation and water absorbed by plant roots and used in transpiration. Accurate simulation of the soil water content not only depends on getting the rates of each compo-

G.Y. Tsuji et al. *(eds): Understanding Options for Agricultural Production, 41–54*

nent in the water balance determined correctly, but also having sufficient measured biophysical properties of the soil available.

The crop models contained in the DSSAT shell all require the same soil properties and calculate the soil water and soil nitrogen balance with the same procedures. The purpose of this chapter is to summarize the procedures used in the water balance calculations and describe the inputs necessary. As with the philosophy of all the simulations in DSSAT, the water balance is calculated with the minimum input requirements. This minimum data set is sufficient for obtaining goals of the DSSAT models such as yield prediction, farm decision making, risk analysis, strategic planning and policy analysis.

The soil inputs

The functional soil water balance model used in DSSAT requires inputs for establishing how much water the soil will hold by capillarity, how much will drain out by gravity, and how much is available for root uptake. The calculation procedures require knowledge of soil water contents (volumetric fraction) for the lower limit of plant water availability (*LL*), for the limit where capillary forces are greater than gravity forces, the drain upper limit (*DUL*), and for field saturation (*SAT*). The term drained upper limit is not totally appropriate because it is not the upper limit water content from which plant roots can absorb water. If soil drainage is relatively slow, roots absorb water above this limit while drainage is occurring. The simultaneous process of uptake and drainage are simulated in the DSSAT water balance. The *DUL* water content is usually reached in a few days after wetting in soils with good internal drainage. Therefore the value is helpful in describing reservoir of plant water availability of most soils. The *DUL* corresponds closely to field capacity concepts and to water potentials in the range of -0.1 to -0.33 bar. The *LL* water content corresponds to the wilting point and to water potentials of -15 bar.

Since soils are not homogeneous with depth, the water content inputs are needed for several soil layers (*L*) for which they apply. The depth increment of each layer, *DLAYR*(*L*), also needs to be sufficiently small to accommodate the functional simulation procedures needed to reasonably predict plant water status. In general, the values of *DLAYR* should not exceed about 20 cm near the surface and 30 cm for deeper depths. The total layers for the entire potential root-zone should be in the order of 7 to 10 layers, depending on the soil depth and need of the user.

The root weighting factor (*WR*) is an input for each depth increment. This relative variable takes values ranging from 1 – indicating a soil most hospitable to root growth, to near 0 – indicating the soil is inhospitable for root growth. The values of *WR* can be approximated for each depth increment from a default function or it can be supplied by a user. Low values of *WR* make it possible to simulate restricted root growth in layers with unusually poor physical or chemical properties.

Other inputs needed for simulations of the soil water balance are not required for each depth. A single value of the soil surface albedo (*SALB*), the limit of first stage soil evaporation (*U*), the runoff curve number (*CN2*) and the drainage coefficient (*SWCON*) are needed for calculation of various components of the water balance. Those variables will be discussed in the relevant sections that follow.

Procedures for obtaining the soil inputs are discussed in more detail in Ritchie and Crum (1989), and Ritchie et al. (1990). For best accuracy in calculating the soil water balance, the limits *LL* and *DUL* should be determined in the field. Traditional laboratory limits can be biased because of disturbance of the soil required for the measurement. Measurements of the plant water availability limits and possible biases in laboratory measurements are discussed in Ritchie (1981b) and Ratliff et al. (1983).

An initial value of the soil water content (*SW*) for each depth increment is also required as an input. The value can go into the proper file if it has been measured or it can be approximated by logical default values prior to crop sowing. In many cases the soil is either dry to near the *LL* from a previous crop or wet to near *DUL* after a period of fallow. The water balance can be run for any length of time before the crop sowing date when weather data are provided to give a more reasonable initial condition of the soil water contents at sowing. Several initialization procedures contained in a subroutine provide the initial values for several state variables needed for calculation of various components of the water balance.

Each weather input is used in the calculation of the soil water balance. The daily precipitation provides the potential infiltration, the temperature and radiation provide information needed to calculate potential evaporation (*EO*). If irrigation is added, a separate file provides the day of year of the irrigation and the irrigation amount.

Infiltration and runoff

The water balance subroutine calculates runoff by a modification of the USDA-Soil Conservation Service (SCS) curve number method (Williams et al., 1991). The SCS procedure uses the total precipitation from one or more storms which occur in a single day to estimate runoff, and excludes time as an explicit variable, i.e. rainfall intensity is ignored.

The original SCS procedure used antecedent rainfall to determine soil wetness runoff estimations. The procedure developed for layered soils considers the wetness of the soil in the layers near the surface. Figure 1 illustrates the SCS curve number concept with variations which allow for wet or dry conditions near the surface. Note from Figure 1 that little runoff occurs for low rainfall, especially when near surface conditions are relatively dry. Infiltration is assumed to be the difference between precipitation and runoff. The infiltration is assumed

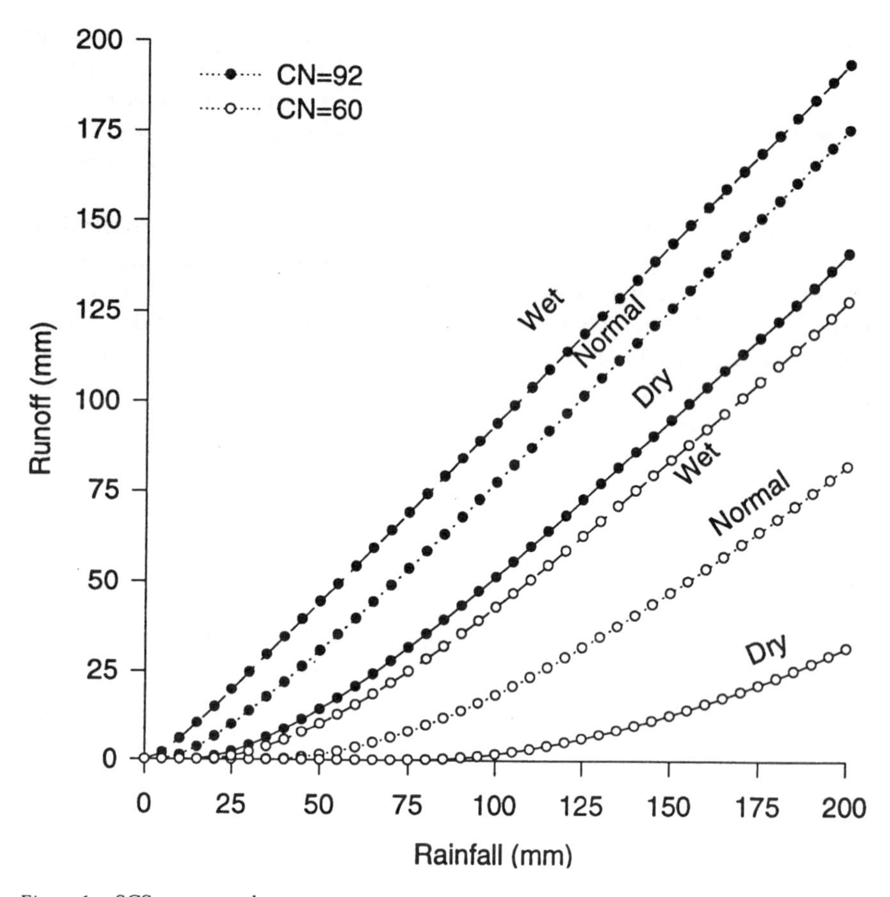

Figure 1. SCS curve number concept.

to be possible because of the presence of macro pores since there is no other more physically based approach to limit infiltration at the surface.

The runoff curve concept is not expected to provide accurate runoff and infiltration information for a specific storm. The curve number concept was empirically derived to approximate the runoff volume when only daily rainfall is known. If a greater accuracy is required, a more physically based approach would be required. Such an approach would require information regarding storm intensities. Because storms vary both spatially and temporally, accurate modeling of infiltration and runoff over large regions would require more frequent rainfall measurements than daily values and would have to be taken in a more dense network. The knowledge of the correct rainfall for a particular site where the model is to be applied is often one of the limiting factors affecting model accuracy.

Drainage

Water can be taken up by plants while drainage out of the root zone is occurring. Thus the drained upper limit soil water content is not always the appropriate upper limit of soil water availability. Many productive agricultural soils drain slowly, providing a potentially significant quantity of water for plant use before gravity affected drainage practically stops. Redistribution of water in the soil profile and drainage out of the rootzone are calculated using a functional model developed from field drainage information.

For soil water redistribution during infiltration water is moved downward from the top soil layer to lower layers in a cascading approach. Drainage from a layer takes place only when the soil water content ($SW(L)$) is between field saturation ($SAT(L)$) and the drained upper limit ($DUL(L)$).

For drainage calculations from each layer the infiltration $PINF$ is converted from mm to cm and a downward flux for each layer calculated ($FLUX(L)$). This water flow rate is also needed for calculating nitrate leaching. When $FLUX(L)$ is not equal to zero, the amount of water that the layer can hold ($HOLD$) between the current volumetric water content ($SW(L)$) and saturation ($SAT(L)$) is calculated.

$$HOLD = (SAT(L) - SW(L)) \times DLAYR(L)$$

If $FLUX(L)$ is less than or equal to $HOLD$, an updated value of $SW(L)$ is calculated prior to drainage.

$$SW(L) = SW(L) + FLUX(L)/DLAYR(L)$$

If this new $SW(L)$ is less than the drained upper limit of volumetric soil water in the layer ($DUL(L)$), no drainage occurs. If this $SW(L)$ is greater than $DUL(L)$, drainage ($DRAIN$) from the layer is calculated from $SW(L)$, $DUL(L)$, $DLAYR(L)$, and $SWCON$, the whole profile drainage rate constant,

$$DRAIN = (SW(L) - DUL(L)) \times SWCON \times DLAYR(L)$$

An updated value of $SW(L)$ is calculated after the drainage

$$SW(L) = SW(L) - DRAIN/DLAYR(L)$$

and a new value of $FLUX(L)$, representing water moving into the layer below, is set equal to $DRAIN$.

If $FLUX(L)$ is greater than $HOLD$, the water in excess of $HOLD$ is passed directly to the layer below. The drainage is then calculated as follows:

$$DRAIN = SWCON \times (SAT(L) - DUL(L)) \times DLAYR(L)$$

An updated value for $FLUX(L)$ is calculated

$$FLUX(L) = FLUX(L) - HOLD + DRAIN$$

After calculating the water movement through all soil layers, drainage from the bottom layer of the profile is equal to *FLUX(L)*. For convenience, this value is converted to mm and set equal to *DRAIN*. *DRAIN* then represents the total outflow from the lowest layer of the soil profile and is an available output variable for those interested in the time course of drainage out of the soil profile.

Evapotranspiration and upward flow

The soil water balance subroutine requires calculations for potential evaporation from the soil and plant surfaces. The equations to predict evaporation are primarily those described in Ritchie (1972). The main difference between this part of the soil water balance subroutine and the Ritchie model is that a Priestley–Taylor (1972) equation for potential evapotranspiration is used instead of the Penman equation. This was done to eliminate the need for vapor pressure and wind inputs while providing sufficiently accurate evapotranspiration information.

Calculation of potential evaporation with a modified Priestley–Taylor equation requires an approximation of daytime temperature (*TD*) and the soil-plant reflection coefficient (*ALBEDO*) for solar radiation. For the approximation of the daytime temperature a weighted mean of the daily maximum (*TEMPMX*) and minimum (*TEMPMN*) air temperatures is used

$$TD = 0.6 \times TEMPMX + 0.4 \times TEMPMN$$

The combined crop and soil albedo (*ALBEDO*) is calculated from the model estimate of leaf area index (LAI) and the input bare soil albedo (*SALB*). Prior to germination, *ALBEDO* is equal to *SALB*. For pre-anthesis conditions the value for *ALBEDO* is

$$ALBEDO = 0.23 - (0.23 - SALB) \times \exp(-0.75 \times LAI)$$

For post-anthesis, the *ALBEDO* is calculated, assuming that the maturing canopy results in an increased albedo,

$$ALBEDO = 0.23 + (LAI - 4) \times 2/160$$

An equilibrium evaporation rate (*EEQ*) defined in Priestley and Taylor (1972) is calculated from *ALBEDO*, *TD*, and the input solar radiation *SOLRAD*. The equation was developed in a simplified mathematical form, but gives quite similar results to the more formal equation in which long wave radiation calculations are made separately. The *EEQ* calculation also estimates daytime net radiation instead of 24-hour net radiation.

$$EEQ = SOLRAD \times (4.88 \times 10^{-3} - 4.37 \times 10^{-3} \times ALBEDO) \times (TD + 29)$$

The units of *EEQ* is mm day^{-1} and *SOLRAD* is MJ m^{-2} day^{-1}. A graphical

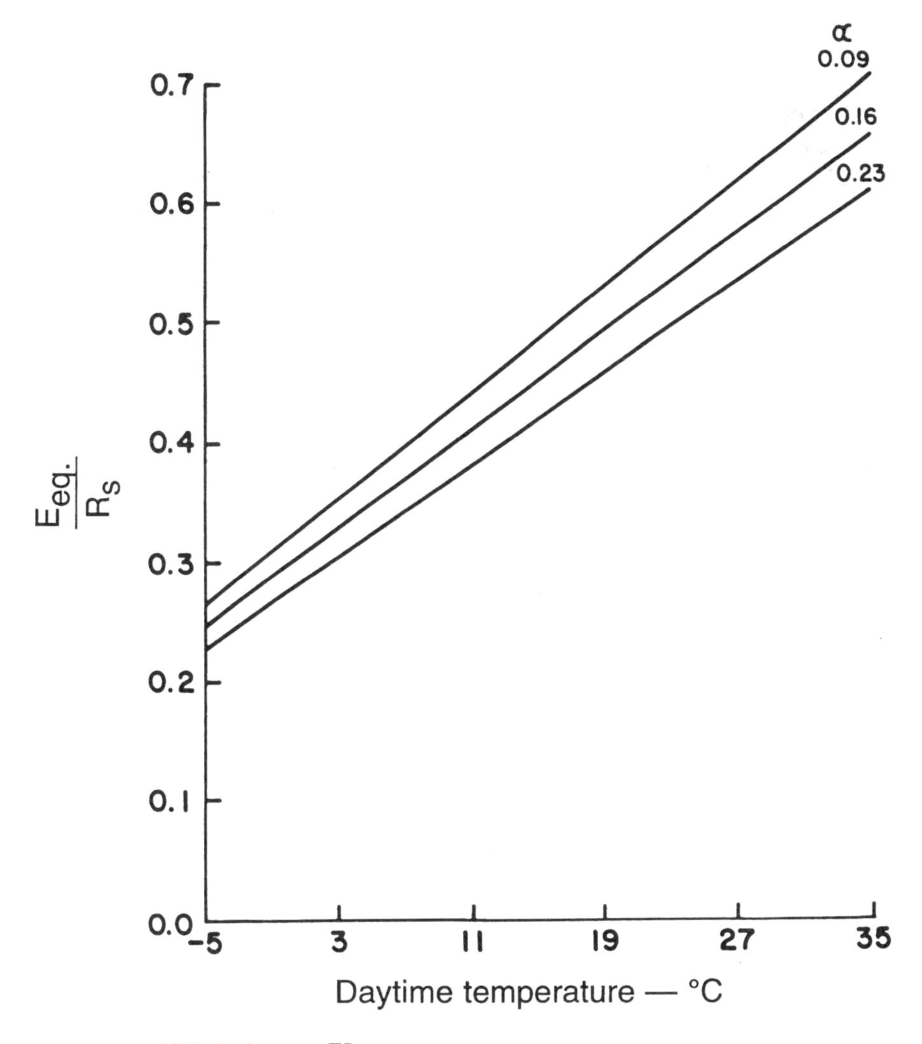

Figure 2. *EEQ/SOLRAD* versus *TD.*

representation of the *EEQ* as a function of *TDAY* and *ALBEDO* is given in Figure 2.

The potential evaporation (*EO*) is calculated as the product of *EEQ* times 1.1 for most conditions. The constant 1.1 increases *EEQ* to a larger value to account for unsaturated air. This is derived from the Priestley–Taylor assumption that the aerodynamic component of evaporation is proportional to the radiant energy into the system instead of being additive as it is in combination equations. The constant 1.1 is a typical value from other studies where daytime net radiation is used. When *TEMPMX* is greater than a high threshold value

($TEMPTH$) the constant 1.1 is increased to account for advection with the equation

$$EO = EEQ \times ((TEMPMX - TEMPTH) \times 0.05 + 1.1)$$

The value of $TEMPTH$ is assumed to be 32°C for all crops except the winter cereals. Winter cereal $TEMPTH$ values are 24°C. When $TEMPMX$ is less than 5°C, the constant is reduced to account for near freezing conditions with the equation

$$EO = EEQ \times 0.01 \times \exp(0.18 \times (TEMPMX + 20))$$

Steiner et al. (1989) found that the calculations for EO used in the DSSAT models, which use only daily radiation and temperature inputs, provided good estimates of grain sorghum ET.

The potential soil evaporation (EOS) for a day is modified by the LAI as follows

$$EOS = EO \times (1 - 0.43 \times LAI); \qquad LAI < 1$$

$$EOS = EEQ \times \exp(-0.4 \times LAI); \qquad LAI \geq 1$$

The calculation for the actual rate of soil evaporation (ES) is based on a two stage evaporation assumption. The first stage limits evaporation by the energy available at the soil surface and continues until a soil-dependent upper limit is reached. This upper limit for stage 1 evaporation is expressed by the input U (mm). After the upper limit of stage 1 is reached, soil evaporation enters stage 2. In stage 2 evaporation, the evaporation decreases proportionally to the time spent in stage 2. Detailed descriptions of the procedures for simulating this two stage soil evaporation process are given in Ritchie (1972).

After ES has been determined, the equivalent water depth is subtracted from the water content of the upper soil layer. If this calculation causes the upper layer soil water content to fall below air dry conditions, ES is reduced to a value that would bring down the upper water content to air dry. Air dry water content is assumed to be approximately half the value of LL of the upper layer, with some variation depending on the upper layer thickness.

The upward flow of water in the top 4 soil layers resulting from evaporation and uptake is calculated next. This upward flow is principally needed to account for upward movement of nitrates in soil. Upward flow is approximated with a normalized soil water diffusion concept operating on a daily time-step.

Variables $THET1$ and $THET2$ represent volumetric water content normalized to values above LL of layers L and $L+1$, as follows

$$THET1 = SW(L) - LL(L)$$

$$THET2 = SW(L+1) - LL(L+1)$$

and values of $THET1$ and $THET2$ are allowed to be no less than 0. Under these conditions the water content is normalized to the lower limit $LL(L)$ where the assumption is made that the diffusivity for all soils is 0.88 $cm^2\ day^{-1}$. The assumed average diffusivity (*DBAR*) above the lower limit is a function of the normalized water content for all soils

$$DBAR = 0.88 \times \exp(35.4 \times (THET1 \times 0.5 + THET2 \times 0.5))$$

If *DBAR* is greater than 100 it is limited to 100 $cm\ day^{-1}$. These diffusivity relationships, requiring no input by users, was fitted from data of Rose (1968). The flow of water is then calculated

$$FLOW = DBAR \times (THET2 - THET1)/((DLAYR(L) + DLAYR(L+1)) \times 0.5)$$

The soil water in layers L and $L+1$ are then changed by the amount of *FLOW*, and the new soil water is calculated.

$$SW(L) = SW(L) + FLOW/DLAYR(L)$$

$$SW(L+1) = SW(L+1) - FLOW/DLAYR(L+1)$$

For user information the cumulative soil evaporation after germination (*CES*) is updated daily,

$$CES = CES + ES$$

The potential plant evaporation (*EOP*) is calculated using procedures approximately the same as those discussed in Ritchie (1972), where *EOP* is related to *EO* and LAI. When LAI is less than 3 then

$$EOP = EO \times LAI/3; \qquad LAI < 3$$

and when LAI is greater than 3 then

$$EOP = EO; \qquad LAI \geq 3$$

If the new value of *EP* added to *ES* is greater than *EO*, then *EP* has to be reduced

$$EOP = EO - ES$$

This reduction in *EP* occurs when the *ES* is high from wet soil and results from the more humid conditions around the plant canopy under these circumstances. Reducing *EOP* from a potential value to an actual one (*EP*) requires the calculation of root water absorption.

Root water absorption

The root water absorption in DSSAT is calculated using a law of the limiting approach whereby the soil resistance, the root resistance, or the atmospheric demand dominate the flow rate of water into the roots. Several details of this approach have been discussed by Ritchie (1981a). The flow rates are calculated using assumptions of water movement to a single root and that the roots are uniformly distributed within a layer.

The maximum daily water uptake by roots in a layer ($RWUMX$) is assumed to be 0.03 cm^3 of water per cm of root. This value, mostly determined by trial and error, sets the upper limit of water absorption for a day by the roots as limited by axial root resistance. First the potential root water uptake ($RWU(L)$) as influenced by soil water flow to roots in a layer is calculated as

$$RWU(L) = 0.00267 \times \exp(62 \times (SW(L) - LL(L)))/(6.68 - \ln(RLV(L)))$$

where $RLV(L)$ is the root length density in the soil layer of natural logarithm. This equation was derived from the theory of radial flow to a single root and assumes that the hydraulic conductivity of all soils are similar when normalized to the lower limit water content $LL(L)$. This assumption is more generally correct when the soil water content is near the lower limit and has larger errors for wetter soils. This equation also assumes that the water potential gradient between the root and the soil remains constant, even when the soil dries out. In reality, the water potential of the roots change considerably throughout the day. However, because we are calculating daily values for water absorption, these less dynamic empiricisms provide sufficient detail for realistic uptake simulations. A graphical representation of the root water uptake function is depicted in Figure 3.

This potential root water uptake is converted from uptake per unit root length to uptake for each layer by the following equation:

$$RWU(L) = RWU(L) \times DLAYR(L) \times RLV(L)$$

and the total potential water uptake from the entire root zone ($TRWU$) is calculated as the sum of $RWU(L)$ for all soil layers with roots.

A water uptake fraction (WUF) is calculated to reduce the potential water uptake if $TRWU$ is greater than the potential plant evaporation (EOP) as follows:

$$WUF = EOP/TRWU$$

If WUF is less than or equal to 1.0, the plants are considered to be free of a water deficit for water uptake calculation. If WUF is greater than 1, then the actual plant evaporation EP is equal to $TRWU$ and the distribution with depth of the absorption is equal the $RWU(L)$ calculation. Actual water uptake distribution in each layer when $WUF > 1$ is calculated

$$RWU(L) = RWU(L) \times WUF$$

Values of $SW(L)$ are then updated,

$$SW(L) = SW(L) - RWU(L)/DLAYR(L)$$

and the total soil water in the profile (TSW) is calculated:

$$TSW = TSW + SW(L) \times DLAYR(L)$$

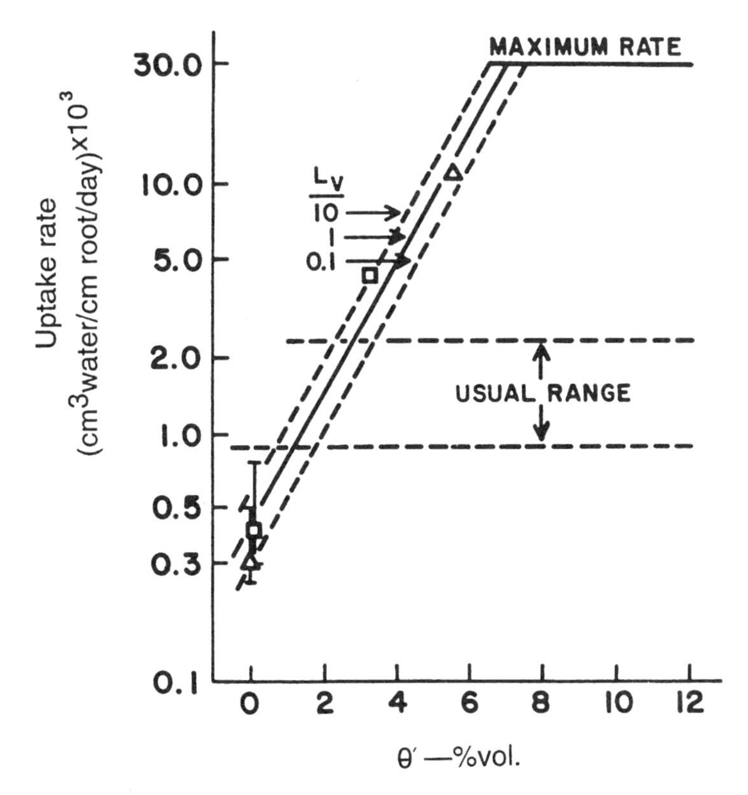

Figure 3. The relationship used to calculate maximum root water absorption as related to θ' (the water content above the lower limit) and root length density (L_v). Also shown is the assumed maximum possible rate and the usual range of absorption when all the soil profile is at an optimum water content.

For user information, the potentially extractable soil water for all soil depths ($PESW$) is then calculated from TSW and the total water content of the profile at the lower limit of the plant-extractable soil water (TLL):

$$PESW = TSW - TLL$$

Plant water deficit factors

When the soil is wet and the roots are plentiful, the potential supply of water by the root system will exceed the potential transpiration rate. As the soil dries, the conductivity of the soil water decreases and the potential root water uptake decreases. When the potential uptake decreases to a value lower than the potential transpiration rate, actual transpiration rate will be reduced by partially closed stomata to the potential root uptake rate. When this happens the potential biomass production rate is assumed to be reduced in the same proportion as the transpiration. The potential transpiration and biomass production rates are reduced by multiplying their potential rates by a soil water deficit factor (*SWDF1*) calculated from the ratio of the potential uptake to the potential transpiration. The value of *SWDF1* is set equal to 1 when the ratio exceeds 1.

A second water deficit factor (*SWDF2*) is calculated to account for water deficit effects on plant physiological processes that are more sensitive than the stomata controlled processes of transpiration and biomass production. Reduced turgor pressure in many crop plants will decrease processes like leaf expansion, branching and tillering before stomata controlled processes are reduced. Values for *SWDF2* are assumed to fall below 1 when the potential root uptake relative to potential transpiration falls below 1.5. Values of *SWDF2* are assumed to be reduced linearly from 1.0 to 0.0 in proportion to this ratio. A graphical representation of the values for *SWDF1* and *SWDF2* are provided in Figure 4.

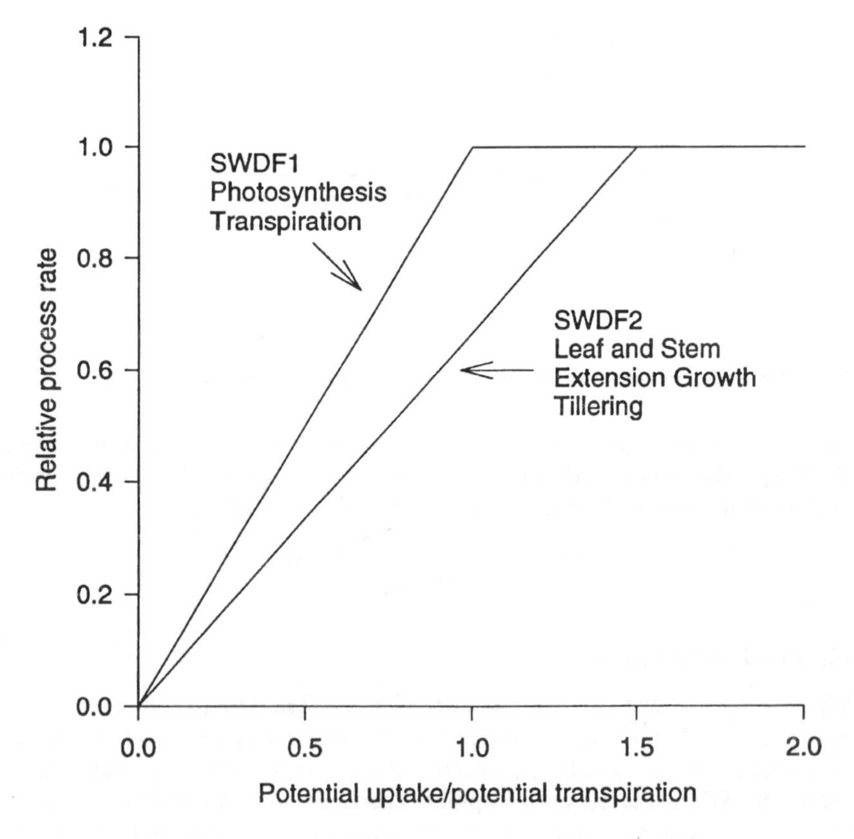

Figure 4. Relationship used to calculate soil water factors, *SWDF1* and *SWDF2*, to incorporate water stresses in the model.

Although the *SWDF2* is difficult to evaluate experimentally by reducing expansive growth more than mass growth, a greater proportion of biomass is partitioned to roots during water deficit conditions. This phenomenon of changing root-shoot partitioning during soil drying is a common experimental observation.

To summarize the information on the soil water deficit influence on plants, two values are calculated: *CSD1* and *CSD2*. These values are an average of the *SWDF1* and *SWDF2* during each growth stage. The values are not used in the model, but are given to provide information to the user for interpretation of the yield responses related to soil water deficits. The values appear on the printouts of the output summary sheet.

Conclusions

Realistic simulation and interpretation of root zone soil water dynamics requires good specifications of the soil properties and the use of reasonably robust functional relationships for expressing various components of the soil water balance. Over the course of the IBSNAT project, the DSSAT models containing the water balance described herein have been tested for a wide range of soil and weather conditions. As the DSSAT system and others like it are being made available to other scientists and engineers and as more data are being collected, testing of crop models is progressing worldwide.

The empiricisms used in calculating each component of the water balance can potentially be improved. Some improvements, such as infiltration and runoff estimation can only come with more input information than contained in daily rainfall. There is definitely a need to have better simulation of the water balance in very poorly drained conditions where oxygen stresses will impact plant growth. We expect that several improvements will be made in various simulation procedures as more testing occurs during the course of validating and using the entire set of DSSAT crop models.

References

Priestley C H B, Taylor R J (1972) On the assessment of surface heat flux and evaporation using large-scale parameters. Monthly Weather Review 100:81–92.

Ratliff L F, Ritchie J T, Cassel D K (1983) Field-measured limits of soil water availability as related to laboratory-measured properties. Soil Science Society of America Proceedings 47:770–775.

Ritchie J T (1972) Model for predicting evaporation from a row crop with incomplete cover. Water Resources Research 8:1204–1213.

Ritchie J T (1981a) Water dynamics in the soil-plant-atmosphere. Plant and Soil 58:81–96.

Ritchie J T (1981b) Soil water availability. Plant and Soil 58:327–338.

Ritchie J T, Crum J (1989) Converting soil survey characterization data into IBSNAT crop model input. Pages 155–167 in Bouma J, Bregt A K (Eds.) Land Qualities in Space and Time. Wageningen, the Netherlands.

Ritchie J T, Godwin D C, Singh U (1990) Soil and weather inputs for the IBSNAT crop models. Pages 31–45 in Proceedings of the IBSNAT Symposium: Decision Support System for

Agrotechnology Transfer. Part I: Symposium proceedings. Department of Agronomy and Soil Science, College of Tropical Agriculture and Human Resources, University of Hawaii, Honolulu, Hawaii, USA.

Rose D A (1968) Water movement in porous materials III. Evaporation of water from soil. British Journal Applied Physics Series 2, Volume 1:1779–1791.

Steiner J L, Howell T A, Schneider A D (1989) Lysimetric evaluation of daily potential evaporation models for grain sorghum. ASAE Paper No. 89–2090. St. Joseph, Michigan, USA.

Williams J R (1991) Runoff and water erosion. Pages 439–455 in Hanks R J and Ritchie J T (Eds.) Modeling plant and soil systems. Agronomy Monograph #31, American Society of Agronomy, Madison, Wisconsin, USA.

Nitrogen balance and crop response to nitrogen in upland and lowland cropping systems

D.C. GODWIN[1], U. SINGH[2]

[1] *Formerly International Fertilizer Development Center; currently 3 Colony Cres, Dubbo NSW, Australia*

[2] *International Fertilizer Development Center, Muscle Shoals, Alabama 35662, USA*

Key words: Nitrogen, modeling, simulation, plant N, ammonia volatilization, N transformations, N uptake, N deficiency

Abstract

CERES N model has two forms–one for upland cereal crops and one for flooded soil rice cropping systems. Both versions simulate the turnover of soil organic matter and the decay of crop residues with the associated mineralization and/or immobilization of N. Nitrification of ammonium and N losses associated with denitrification are estimated by both models. The lowland version adds to this a floodwater chemistry routine which simulates the fluxes of ammoniacal N and urea between floodwater and soil, and calculates ammonia volatilization losses. Both models incorporate a plant N component which simulates N uptake and distribution within the plant and remobilization during grain filling and plant growth responses to plant N status. The models are closely coupled with the CERES water balance and crop growth routines and require few readily obtainable inputs.

Introduction

Nitrogen (N) is the element most abundant in the atmosphere and yet it is the nutrient element most often deficient in agricultural soils. This paradox exists because N is the nutrient element required in largest quantity by plants and because only a small proportion of the N present in soils is in a form amenable to plant uptake. Nitrogen fertilizers are employed to enhance the soil supply of N to crops. Their role in the production of the world's food and fibre crops is now well established and has expanded considerably over the last three decades. Consumption of N fertilizer is now estimated at 74 million tons per annum (Bumb, 1995). The efficiency with which crops use this N is highly variable but often poor. Craswell and Godwin (1984) when examining fertilizer efficiency in a range of cropping systems reported apparent recoveries of fertilizer N by the various crops to vary from 0 to almost 100% with most crops recovering less than 40% of applied N.

The N which is not recovered by the crop may be lost from the soil:plant system through runoff, leaching, denitrification or ammonia volatilization. It may be made unavailable to the plant through immobilization in the soil, or it may become inaccessible to the plant through lack of water. The fraction that is lost from the cropping system is a source of much of the environmental

G.Y. Tsuji et al. *(eds): Understanding Options for Agricultural Production, 55–77*

pollution associated with fertilization. Supply of the N requirement of crops by organic materials, either directly or through the use of green manure crops and crop residues, requires the transformation of N from an organic form to mineral N where it will be subject to the same loss processes as fertilizer N.

Soil N and fertilizer N undergo many transformations involving numerous pathways and states, all of which are influenced by the weather. Given the complexity of the soil N cycle, the myriad of pathways for N transformations and the nuances of weather, determining the most appropriate way to manage N in a cropping system is difficult. Simulation models which describe the effects of weather on the major soil N transformations, the availability of water to the crop, and crop growth and development can provide valuable insights into the management of N.

The crop models contained within the Decision Support System for Agrotechnology Transfer (DSSAT) attempt to meet these criteria and are designed to have widespread applicability. They have common water balance and nitrogen balance routines and have crop growth and development routines that are specific to each crop. The N component of these models is closely linked with the water and growth components and is consistent with the philosophy of DSSAT that only a minimum amount of data inputs is required to run the crop models. The models for each of the upland crops share a common suite of procedures for simulating the soil N balance. The rice model as well as the aroid model adds to these procedures additional routines for the simulation of N transformations in flooded soils and in floodwater. Extra procedures in the water balance to cope with flooded soil hydrology are also present in the rice model. Figure 1 indicates the processes simulated in the flooded soil N model. The N model used for upland crops does not have the ammonia volatilization and floodwater chemistry and transport routines depicted in this figure. This chapter summarizes procedures used in the N balance calculations of the DSSAT crop simulation. Where the N component of the model requires inputs from the soil water balance, water balance components have been referred to with the same terminology as used in the chapter that discusses the soil water balance and plant water stress.

Nitrogen transformations in upland soils

Mineralization and immobilization

Mineralization refers to the decay of crop residues and soil organic matter, which can lead to the net release of mineral N. Immobilization is also associated with the decay of residues and occurs when inorganic N compounds are transformed to the organic state and rendered temporarily unavailable to the crop. Immobilization occurs during the growth of cells of the organisms involved in the decay process.

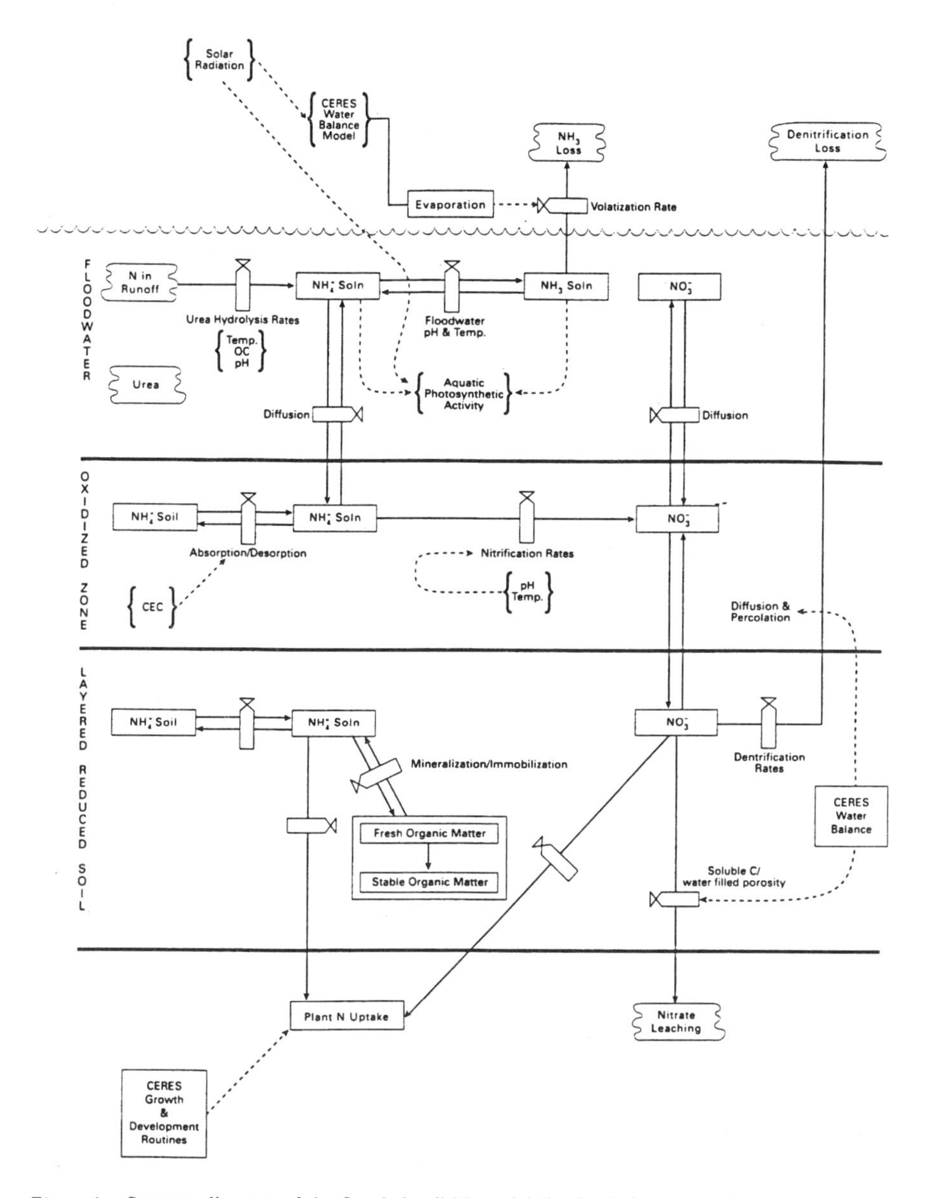

Figure 1. Systems diagram of the flooded soil N model (lowland rice systems).

The perceived applications for the DSSAT models in studies examining crop growth in response to soil fertility and fertilizer management require that a model of mineralization and immobilization be simple, require few inputs, be capable of working on a diversity of soils, and be suitable for examining the fate of residues of widely differing composition. We originally chose the approach described by Seligman and van Keulen (1981) in PAPRAN but have modified several important components as other components of our N model have developed.

The model simulates the decay of 'fresh organic matter' (FOM) and the associated turnover of the fresh organic N (FON). FOM comprises crop residues and/or green manure crop material. The amounts of this material, its C:N ratio, and its depth of incorporation into the soil, if any, are inputs to the model. From these inputs, and an estimate of the root residue remaining from the previous crop, FOM and the corresponding FON amounts are ascribed to the various soil layers. In practice the estimates for root residue can be quite crude. Usually a value of 20% of the past crop's yield is satisfactory. If pastures or long fallows or other conditions occur before the current cropping seasons more suitable estimates will be needed. The C:N ratio of this root material is internally fixed at 40.

There also exists a more slowly decaying fraction of organic matter in the soil, which can be estimated from a soil organic carbon determination. This more stable fraction or humic fraction is termed HUM and the N associated with it NHUM.

Mineralization of fresh organic matter. Three pools comprise the FOM pool in each layer (L); they are :

$$FPOOL\ (L, 1) = \text{carbohydrate}$$

$$FPOOL\ (L, 2) = \text{cellulose}$$

$$FPOOL\ (L, 3) = \text{lignin}$$

Each of the three FOM pools ($FPOOL\ (L, 1$ to $3)$) has a different decay rate ($RDECR$ (1 to 3)). Under non-limiting conditions the decay constants for each of these pools are 0.20, 0.05 and 0.0095, respectively. For carbohydrate the decay constant implies 20% of $FPOOL\ (L, 1)$ would decay in one day under optimal conditions. Optimal conditions seldom prevail however, and limitations due to moisture, temperature or the nature of the residue itself generally occur.

A water factor (MF) is first determined from the volumetric soil water content ($SW(L)$) relative to the lower limit (LL), drained upper limit (DUL) and saturated water content (SAT). When the soil is drier than DUL, MF is calculated as:

$$AD = LL(L) \times 0.5$$

$$MF = (SW(L) - AD(L))/(DUL(L) - AD)$$

where:

$$AD = \text{air dry moisture content (volume fraction)}$$

When the soil is wetter than DUL, MF is calculated as:

$$MF = 1.0 - (SW(L) - DUL(L))/(SAT(L) - DUL(L)) \times 0.5$$

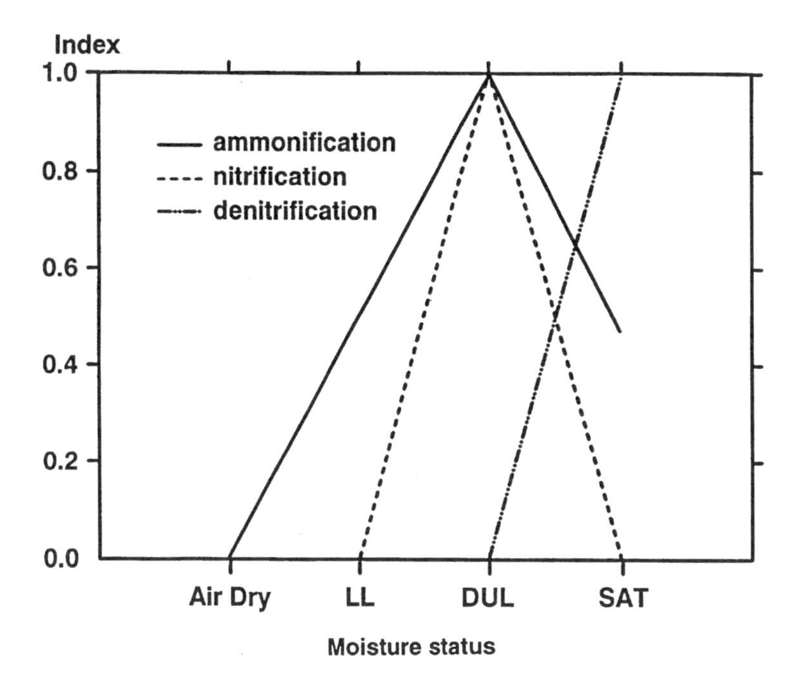

Figure 2. Moisture indices used in the dryland soil N model to modify rates of ammonification, nitrification, and denitrification.

These functions follow the observations reported by Myers et al. (1982) and Linn and Doran (1983) of moisture effects on ammonification. The first function allows ammonification to proceed at slow rates when soil moisture content is below the lower limit of plant water availability. Reichman et al. (1966) have reported ammonification occurring in soils drier than 1.5 mPa and this would approximately correspond to *LL*. Under paddy conditions, 100% of water filled porosity, ammonification proceeds at approximately half the rate of ammonification as at field capacity (Linn and Doran, 1983). The moisture indices used in the upland model to modify the rates of ammonification, nitrification and denitrification are shown in Figure 2.

The zero to unity temperature factor (TF) is calculated directly from the soil temperature ($ST(L)$):

$$TF = (ST(L) - 5.0)/30.0$$

The third factor limiting the decay rate is the effect of the composition of the residue itself as reflected by its C:N ratio. In this case the C:N ratio is calculated as the C contained in the FOM divided by the N 'available' for the decay process. This N available for decay is the sum of the N contained in the *FOM*, which is *FON*, and the extractable mineral N present in the layer

($TOTN$). Thus,

$$CNR = (0.4 \times FOM(L))/(FON(L) + TOTN)$$

From CNR the zero to unity index ($CNRF$) is calculated:

$$CNRF = \exp(-0.693 \times (CNR - 25))/25.0$$

For each of the FOM pools, a decay rate appropriate for that pool (JP) can be calculated by multiplying the rate constant by the three indices as:

$$G1 = TF \times MF \times CNRF \times RDECR(J)$$

where $G1$ is the proportion of the pool that decays in 1 day. The amount of material that has decayed is then the product of $G1$ and the pool size. The gross mineralization of N associated with this decay ($GRNOM$) is then calculated according to the proportion of the pool that is decaying as:

$$GRNOM = G1 \times FPOOL(L, JP)/FOM(L) \times FON(L)$$

A revised value of $GRNOM$ is obtained for each layer by summing the $GRNOM$ values for each of the three FOM pools ($JP = 1, 2, 3$).

Mineralization of humic fraction.

The more stable fraction of soil organic matter (as distinct from crop residues) turns over at a much slower rate. To estimate this rate ($RHMIN$), TF and MF as determined above are used but the C:N ratio factor is not used. A very small rate constant ($DMINR$) (8.3×10^{-5}) is used in this procedure as shown below. It was found that this procedure could not be universally applied to all soils without modification. A modifier ($DMOD$) was added to the equation to enable users to modify predicted mineralization rates. This was found necessary on recently cultivated virgin soils where $DMOD$ was usually given an arbitrary value of 2.0 and on andepts where the organic matter present decayed at very slow rates ($DMOD = 0.2$). In virtually all other circumstances $DMOD$ has the value of unity. The procedure for calculating $RHMIN$ then, is the product of the various indices and the N contained within the humus ($NHUM(L)$).

$$RHMIN = NHUM(L) \times DMINR \times TF \times MF * DMOD$$

Immobilization. As organic matter decomposes, some N is required by the decay process and may be incorporated into microbial biomass. The amount of N that is required is dependent on both the amount of fresh organic matter decaying and its N content. The procedure used is as described by Seligman and van Kuelen (1981). If net mobilization occurs it is firstly drawn from the soil ammonium pool and if there is not sufficient N to retain this pool with a concentration of 1 ppm, then withdrawals are made from the nitrate pool.

Nitrification

Nitrification refers to the process of oxidation of ammonium to nitrate. This occurs under aerobic conditions and is facilitated by various species of nitrifying organisms. The process is limited by the availability of ammonium, oxygen, soil pH, and temperature and to a lesser extent soil water.

Many different approaches have been taken in soil/crop models to simulate the process of nitrification. Some models, e.g., McGill et al. (1981), van Veen (1977) and van Veen and Frissel (1981), simulate the effects of the environment on the populations of nitrifying bacteria and then calculate a rate of nitrification based on this population. Most models ignore microbial population aspects and describe nitrification as a simple first order kinetic process, (e.g., Davidson et al., 1978, Tillotson et al., 1980), with ammonium concentration as a rate limiting variable. The simplest approach has been to assume it is either instantaneous or occurs so rapidly that it can be deemed independent of soil type, for example as in PAPRAN (Seligman and van Keulen, 1981). This enables that soil ammonium can be ignored and only a mineral N pool consisting of nitrate N needs to be considered. This approach is probably appropriate where the soil does not become periodically inundated, lags to nitrification due to prolonged drying or to other factors do not occur, or where the presence of certain clay minerals does not limit nitrification rates. For more widespread applicability and to examine the potential of nitrification inhibitors we have chosen to include a description of the process in nitrogen model that is part of the DSSAT crop models, albeit at a very simple level. We have found that this approach described below works satisfactory for upland soils but that a modified version described by Gilmour (1984) worked better for flooded soils.

The approach used in the DSSAT upland N model has been to calculate a potential nitrification rate and a series of zero to unity environmental indices to reduce this rate. This potential nitrification rate is a Michaelis–Menten kinetic function dependent only on ammonium concentration and is thus independent of soil type. A further index termed a 'nitrification capacity' index is introduced which is designed to introduce a lag effect on nitrification if conditions in the immediate past (last 2 days) have been unfavorably for nitrification. Actual nitrification capacity is calculated by reducing the potential rate by the most limiting of the environmental indices and the capacity index. The capacity index is an arbitrary term introduced to accommodate an apparent lag in nitrification observed in some conditions. The functions described below have been found to be appropriate across a wide range of data sets.

Firstly a soil ammonium concentration factor (*SANC*) is calculated:

$$SANC = 1.0 - \exp(-0.01363 \times SNH4(L))$$

This is a zero to unity index which has approximately zero values when there is less than 1 ppm of ammonium present and has a value of 0.75 at 100 ppm.

Zero to unity indices for temperature (TF) and water (WFD) as well $SANC$ are used to determine the environmental limit on nitrification capacity ($ELNC$):

$$ELNC = AMIN1(TF, WFD, SANC)$$

The water factor has a zero value for moisture contents below the lower limit and increases linearly from the lower limit to the drained upper limit.

To accommodate lags that occur in nitrifier populations, $ELNC$ and the previous day's relative microbial nitrification potential in the layer ($CNI(L)$) are used to calculate the interim variable $RP2$ which represents the relative nitrification potential for the day.

$$RP2 = CNI(L) \times \exp(2.302 \times ELNC)$$

$RP2$ is constrained between 0.01 and 1.0.

Today's value of the nitrification potential ($CNI(L)$) is then set equal to $RP2$. Since $\exp(2.303 \times ELNC)$ varies from 1.0 to 10.0 when $ELNC$ varies from 0.0 to 1.0, relative nitrification potential can increase rapidly, up to ten fold per day. An interim variable A is then determined from these indices and a zero to unity effect of soil pH on nitrification. This pH index is similar to that reported by Myers (1975).

$$A = AMIN1(RP2, WFD, TF, PHN(L))$$

This interim variable A is used together with the ammonium concentration ($NH4(L)$) in a Michaelis–Menten function described by McLaren (1970) to estimate the rate of nitrification. The function has been modified to estimate the proportion of the pool of ammonium ($SNH4(L)$) that is nitrified on a day.

$$B = (A \times 40.0 \times NH4(L)/(NH4(L) + 90.0)) \times SNH4(L)$$

A check is made to ensure some ammonium is retained in the layer and thus the daily rate of nitrification ($RNTRF$) is:

$$RNTRF = AMIN1(B, SNH4(L))$$

Following this calculation soil nitrate and ammonium pools can be updated. Finally, the soil temperature, moisture and the new $NH4$ concentration are used to update ($CNI(L)$) which is used in the subsequent day's calculations. In these calculations the least limiting of today's and yesterday's temperature and water factors are used. This prevents a single day of low soil temperature or severe water shortage from substantially reducing $CNI(L)$.

In the rice model nitrification has been simulated quite differently. Here a temperature factor ($TFACTOR$) as well as a water factor ($WF2$) dependent on the water filled porosity are calculated. A first order kinetic function adapted from Gilmour (1984) is then used with these modifiers to determine nitrification rate in the oxidized layer of flooded soils. The model does not attempt to simulate nitrification occurring in the rhizosphere under flooded condition. Predicting nitrification rate accurately on rice soils immediately following drainage when the soil is close to saturation but oxidized continues to be a challenge. Given the importance of nitrate contamination in groundwater and loss of nitrate due to denitrification, additional testing and refinements to the nitrification routine are required.

Dentrification

Denitrification represents a substantial N loss pathway in soils and flooded conditions where anaerobic conditions persist. Denitrifying organisms require an energy source and this is usually derived from soil carbon. In the DSSAT models an estimate of the soil soluble carbon (CW) is made using a modification of Rolston's (1980) procedure.

$$CW = (HUM(L) \times FAC(L)) \times 0.0031 + 24.5 + 0.4 \times FPOOL1(L, 1)$$

where $FPOOL(L, 1)$ is the mass of carbohydrate fraction of organic matter in layer L as described previously, and

$FAC(L)$ = conversion factor to allow for bulk density and layer thickness

$HUM(L)$ = mass of stable or 'humus' organic matter in layer L

A water factor (FW) for denitrification and a temperature factor (TF) as described previously are calculated. In this case the water factor acts as a surrogate for oxygen availability. As soil layers fill beyond the drained upper limit soil oxygen supply is depleted, thus FW approaches zero as SW approaches SAT.

The denitrification rate can then be calculated as:

$$DNRATE = 6.0 \times 1.0E-05 \times CW \times NO3(L) \times BD(L) \times FW \times FT \times DLAYR(L)$$

If floodwater is present, it is assumed that all nitrate present, except for a residue of 0.5 ppm, is immediately lost.

Nitrate and urea movement

In the DSSAT upland crop models only nitrate and urea are deemed capable of moving across layer boundaries and the movement of ammonium is not considered. The same procedures are used for the simulation of both urea and nitrate movement.

Nitrate movement in the soil profile is highly dependent upon water movement. In the DSSAT models the water balance routine calculates the volume of water moving through each layer in the profile ($FLUX(L)$). The volume of water present in each layer ($SW(L) \times DLAYR(L)$) and the water draining from each layer ($FLUX(L)$) in the profile are used to calculate the nitrate lost from each layer as follows:

$$NOUT = SNO3(L) \times FLUX(L)/(SW(L) \times DLAYR(L) + FLUX(L))$$

A fraction of the mass of nitrate, $SNO3(L)$, present in each layer thus moves with each drainage event. A simple cascading approach is used where the nitrate lost from one layer is added to the layer below. When the concentration of nitrate in the layer falls below 1 μg NO_3 (g^{-1} soil) then no further leaching occurs. In the approach described here the implicit assumption is that all nitrate present in a particular layer is uniformly and instantaneously in solution in all of the water in that layer. Similar procedures are used to model the rate of upward movement of nitrate and urea with evaporation of water from the surface layers. The water balance model again provides an estimate of the flux of water and this is used as the driving force to move solute upward. This upward movement of N is confined to the upper layers and is generally small compared to rates of downward movement. No upward movement from the surface layer occurs.

The model's assumptions will not hold in soils with large cracks or in variable-charged soils with nitrate retention properties.

Nitrogen transformations in rice paddies

The presence of floodwater in rice paddies leads to large differences in nitrogen behavior in rice cropping systems compared to upland systems. In these systems the presence of a shallow layer of floodwater limits oxygen transfer to deeper layers of soil. At the same time it provides a very biologically active environment for many organisms at the soil/atmosphere interface. This limited flux of oxygen, combined with demand for oxygen for soil processes leads to reduced soil conditions in the soil profile. Under these reduced conditions nitrate is readily denitrified and nitrification all but ceases. Mineral N will exist mostly in the form of ammoniacal N. As such any ammoniacal N present in floodwater will be vulnerable to loss through ammonia volatilization if the ammonium/ammonia equilibrium that exists in floodwater is shifted toward ammonia. This will occur under conditions of where the floodwater has a high pH.

Compared to the reaction rates of the soil N transformations, reaction rates occurring in the floodwater and at the floodwater/soil interface in rice paddies are quite rapid. This has necessitated a departure from the daily time-step as used in other components of the nitrogen model. During the period following application of fertilizer or when floodwater N concentrations remain high, twelve timesteps are used per day for the calculation of floodwater transformations and transport processes. This arrangement of timesteps represents a reasonable compromise between the demands of accuracy in rapidly occurring processes and the requirements for data and computation speed. Part of the compromise also requires that hourly floodwater temperature, floodwater evaporation and incident radiation have to be interpolated from daily values. Interpolation procedures to accomplish this are built into the code. For example floodwater temperature (*FTEMP*) is given by:

$$FTEMP = TMIN + HTMFAC \times (TMAX + 2.0 - TMIN)$$

where *HTMFAC*, hourly floodwater temperature factor increases from near zero in the beginning of the day to almost one at the 6th timestep at midday to zero at the end of the diurnal cycle. Likewise a sinusoidal function is used to interpolate hourly floodwater evaporation from daily values.

Hydrolysis of urea in floodwater is generally quite rapid, but sensitive to floodwater temperature (*FTEMP*). Urease enzymes are often associated with algae. In the absence of algae, hydrolysis rates are slower and are determined more by soil properties as described for the upland model.

The temperature effect (*TEMPFU*) on floodwater hydrolysis rates (*FUHYDR*) is given by:

$$TEMPFU = 0.04 \times FTEMP - 0.2$$

where *TEMPFU* has a maximum value of 0.9. *FUHYDR* is a first order rate process with the rate determined by the maximum of *ALGACT* and *AK*.

$$FUHYDR = AMAX1(AK, ALGACT) \times TEMPFU \times FLDU$$

where

$FLDU$ = floodwater urea (kg N ha^{-1} as urea)

AK = soil determined hydrolysis rate, and

$ALGACT$ = floodwater photosynthetic activity index (0.0–1 value)

Following hydrolysis, floodwater urea and ammoniacal N pools can be updated.

$$FLDU = FLDU - FUHYDR$$

$$FLDH4 = FLDH4 + FUHYDR$$

The ammoniacal N mass (kg N ha^{-1}) is then converted to a concentration basis (*FLDH4C*, μg ml^{-1}), by using the floodwater depth (*FLOOD*, mm).

$$FLDH4C = FLDH4 \times 100.0/FLOOD$$

Floodwater pH, FPH, is calculated as a function of time of day (I) in 2-hour intervals and PHSHIFT.

$$FPH = 7.0 + PHSHIFT \times \sin(3.1412 \times FLOAT(I)/12.0)$$

The main factor influencing *PHSHIFT* and hence *FPH* is the *ALGACT*, causing *FPH* increments of up to 2.5 units. It has been shown (Fillery et al.,

1984) that floodwater pH is subject to marked diurnal fluctuations due to photosynthetic activity of algae.

$$PHSHIFT = 0.5 + 2.0 \times ALGACT$$

Algal growth

Rather than attempting to simulate the complex processes of algal growth and photosynthesis, we developed an index of floodwater photosynthetic activity (*ALGACT*). This index varies between zero and unity and is derived by determining the most limiting of a series of factors which limit algal growth. These factors attempt to describe the limitations imposed by temperature, radiant energy supply, floodwater N and P as well as recent history of algal growth. The index is described in more detail in Godwin and Singh (1991). These factors are also described as zero to unity indices. The first of these indices relates the relative propensity for algal growth to floodwater N concentration. This function is rather arbitrary in design, but it is asymptotic in nature with the asymptote reached at a floodwater N concentration of about 20 mg N l^{-1} floodwater.

A second index describing the effects of floodwater temperature on algal activity was also constructed. This index increases linearly from a zero value at 15°C to unity at 30°C. Above 30°C it decreases linearly to a zero value at 45°C. The third index is used to describe the effect of radiant energy supply passing through the crop canopy to the reach the floodwater surface. This 'light index' is calculated as a function of the crop leaf area index and an extinction coefficient using a formulation of Beer's law. The relationship used was also rather arbitrary but is asymptotic with an asymptote reached at 15 MJ m^2 day^{-1} to account for light saturation. No attempts were made to account for variability in radiant energy transmission due to differences in crop architecture or plant arrangements or due to the turbidity of the water. The index is merely a pragmatic attempt at describing in relative terms how much energy will be available for photosynthesis.

Phosphorus (P) has been implicated in algal blooms and is known to have large effects on algal activity. The model does not simulate floodwater P dynamics but in an attempt to simply accommodate the effects of P, a P index with only two values is used. These two values are 'low' (0.5) in the absence of phosphatic fertilizer additions and 'high' (1.0) in the presence of phosphatic fertilizer.

To initialize the floodwater model, an initial value of algal activity is determined from organic carbon in the soil surface layer. A potential daily increase in algal activity is calculated using a sigmoidal growth function. This potential is in turn reduced according to the most limiting of the environmental factors (light, temperature, N and P) as described by the zero to unity indices.

During the hydrolysis of urea, the temporary formation of ammonium carbamate in floodwater will buffer floodwater pH at about 8.4 (Vleck and Carter,

1983; Vleck and Stumpe, 1978). In the model described here, maximum floodwater pH is largely determined by algal activity. However when urea hydrolysis is occurring, maximum pH is set to 8.4.

Ammonia volatilization

When floodwater pH and floodwater temperature have been calculated, the equilibrium between $NH_4^+ - N$ and $NH_3 - N$ can be determined using the procedures described by Denmead et al. (1977).

$$FLDH3C = FLDH4C/(1.0 + 10.0^{**}(0.009018 + 2729.92/TK - FPH))$$

where:

TK = floodwater temperature (°K)

$FLDH3C$ = floodwater ammonia-N concentration

The floodwater ammonia-N will exist as both aqueous ammonia and as gaseous ammonia in equilibrium. The partial pressure of the ammonia-N (*FNH3P*) can be determined from the aqueous concentration when the absolute temperature is known as reported by Freney et al. (1981). *FNH3P* provides the potential for ammonia volatilization loss. The actual NH_3 loss is dictated by wind velocity. An alternative is to relate ammonia loss to evaporation of water as has been proposed by Bouldin and Alimagno (1976). From a reworking of the data of Fillery et al. (1984, 1986) and using an estimated hourly floodwater evaporation rate as the driving force, Godwin et al. (1990) derived a function to estimate ammonia loss as:

$$AMLOS = 0.036 \times FNH3P + 0.05863 \times HEF + 0.000257 \times FNH3P \times 2.0 \times HEF \times FLOOD$$

where:

HEF = hourly floodwater evaporation rate (mm hr^{-1})

$FNH3P$ = floodwater partial pressure of ammonia, and

$AMLOS$ = ammonia volatilization loss (kg N ha^{-1} hr^{-1})

Following ammonia loss, floodwater ammoniacal N can be updated.

$$FLDH4 = FLDH4 - AMLOS$$

Calculations can then proceed which would enable diffusion of more ammonium and urea from soil surface layer pools into the floodwater subsequent to the next time step.

Floodwater transport processes

Diffusion between soil and floodwater for the nitrate, ammonium and urea pools is calculated in the model. For ammonium, an adsorption/desorption routine centered upon a Langmuir isotherm is used to determine the amount of soil ammonium in solution 'available' for diffusion. The key equation for this is:

$$SOLN = \exp(B \times alog(C) - A)$$

where:

$SOLN$ = the solution concentration of ammonium

A = a constant (0.83)

B = a coefficient derived from the soil surface layer cation exchange capacity (CEC) as:

$$B = 4.1 - 0.07 \times CEC$$

C = soil ammonium concentration

When floodwater is present an effective diffusion coefficient (DE) for ammonium is calculated from the aqueous diffusion coefficient ($AQDC$) and a tortuosity factor calculated as the square root of the saturation moisture content (Nye and Tinker, 1977), and the soil buffering power for ammonium (BPS). The soil buffering power (Nye and Tinker, 1977) for ammonium is determined from the soil cation exchange capacity and the prevailing KCl–extractable ammonium concentration.

$$DE = AQDC \times 1.0\text{E}{-5} \times SQRT(SW(1)) \times SW(1)/BPS$$

To calculate the actual diffusion rate ($DIFFN$) for the timestep $DELT$, DE is used together with the diffusion distance (half the thickness of the surface layer, $DELX$) and the concentration gradient (difference between floodwater concentration and soil solution concentration, $DELC$)

$$DIFFN = DELC/DELX \times DE \times DELT$$

Plant vegetative tissue nitrogen concentrations

Plant growth is greatly affected by the supply of N. N is required for the synthesis of chlorophyll, proteins and enzymes and is essential for the utilization of carbohydrates. Typically when plants are young, tissue concentrations of N are high as result of the synthesis of organic N compounds required for photosynthesis and growth. As plants age less of this material is required and export from the old tissue to new tissue occurs lowering the whole plant N concentration. Critical concentrations of N required for optimum growth thus

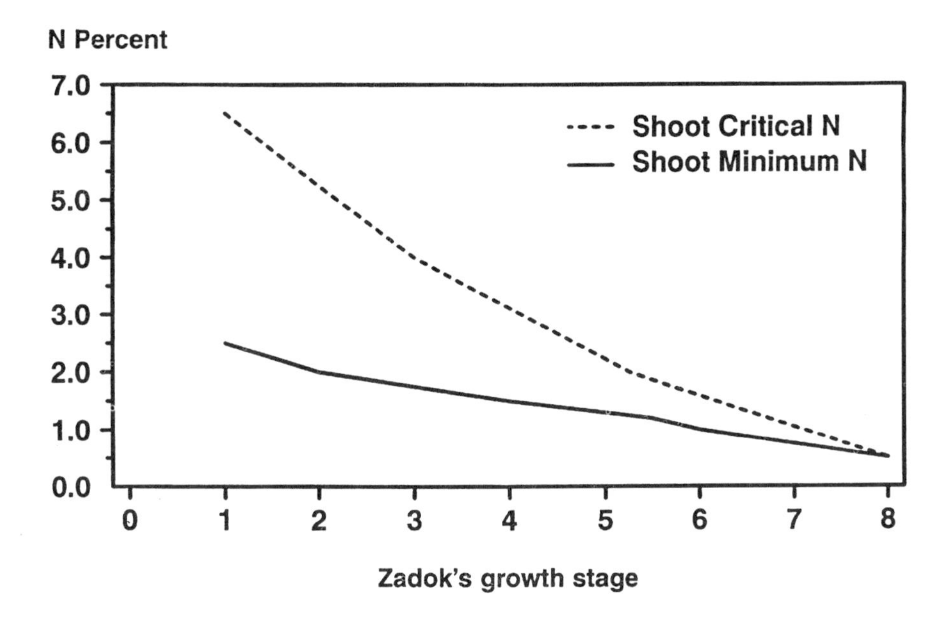

Figure 3. Relationship between phenological age as indicated by the estimated growth stage using the Zadok's system and shoot critical and minimum N concentrations for spring wheat.

change with plant physiological age. The critical N concentration is the lowest concentration at which maximum growth occurs. If tissue N concentrations fall below this critical concentration growth processes are affected. If the N concentration rises above this no further growth occurs and luxury consumption of N is deemed to have occurred. The minimum concentration below which all growth ceases can also be identified. The functions used for the critical (*TCNP*) and minimum concentrations (*TMNC*) used in the models for wheat and rice are depicted in Figure 3. Usually plant N concentrations will range between *TCNP* and *TMNC*. The actual concentration *TANC* relative to these concentrations can be used to define an N factor (*NFAC*) which ranges from zero when *TANC* is at its minimum value *TMNC* and unity when *TANC* is at *TCNP* or above. *NFAC* is thus an index of N deficiency which is used in the simulation of the effect of N shortages on plant growth processes. It is calculated as:

$$NFAC = 1.0 - (TCNP - TANC)/(TCNP - TMNC)$$

All concentrations are expressed as $g\ N\ g^{-1}$ dry weight of plant tissue.

All plant growth processes are not affected equally by N stress. Several zero to unity stress indices have been derived for different processes. These indices together with similarly derived indices for stresses for water deficit and effects of other nutrients are used within the growth component of the crop simulation models to reduce potential rates of growth. The indices are used to affect

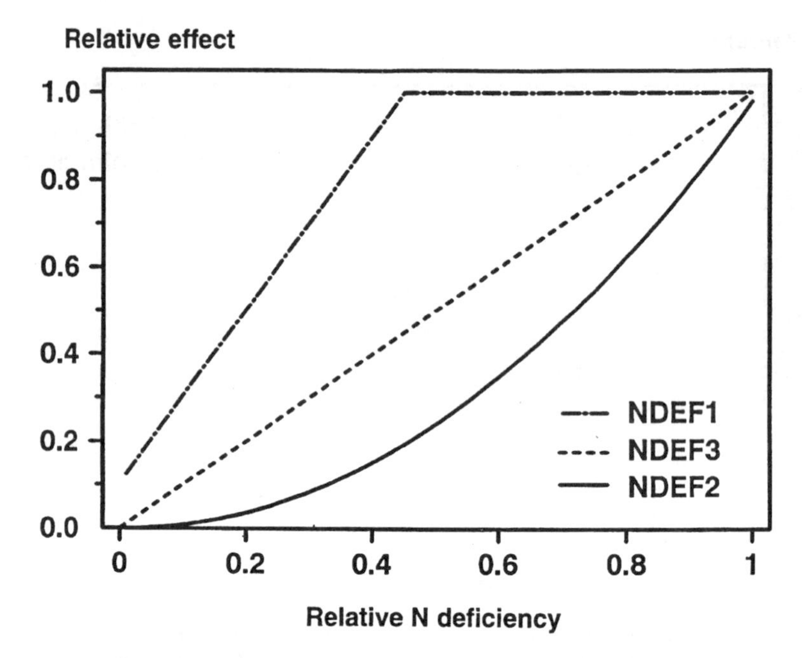

Figure 4. Relationship between the relative N deficiency indicator (*NFAC*) and indices of N deficiency for photosynthesis per unit area (*NDEF1*), expansion growth processes (*NDEF2*), and relative tillering rate (*NDEF3*).

photosynthesis per unit area (*NDEF1*), leaf expansion growth, senescence and partitioning (*NDEF2*), and grain nitrogen concentration (*NDEF3*). Typically with the onset of N stress, expansion growth processes are more severely affected at first than photosynthesis per unit area. This is reflected in the shape of these indices (Figure 4).

Nitrogen uptake

In the crop models used within the DSSAT, crop N uptake is simulated by examining the potential supply of N to the crop from the soil and the capacity or demand of the crop for N. Crop N demand has two components. Firstly there is 'deficiency' demand which represents the amount of N required to restore the actual concentration of N to the critical concentration. Using the definitions made above for critical and actual concentrations, deficiency demand for tops (*TNDEM*) in units of kg N ha^{-1} can be calculated as this difference in concentration multiplied by the existing mass of shoots (*TOPWT*) in units of kg dry weight ha^{-1}:

$$TNDEM = TOPWT \times (TCNP - TANC)$$

Similarly root N demand (*RNDEM*) can be calculated from the discrepancy in root N concentrations, root critical (*RCNP*), and root actual (*RANC*) concentrations, and root biomass (*RTWT*) in kg ha^{-1}:

$$RNDEM = RTWT \times (RCNP - RANC)$$

If luxury consumption of N has occurred, the actual N concentration will be greater than the critical and demand will be negative. This can occur under conditions of high N supply when some constraint reduces growth. When this does occur the negative demand will ensure uptake does not proceed. Subsequent growth or redistribution will lower N concentrations again such that *TANC* will generally only ever exceed *TCNP* for a few days.

The second and usually smaller component of N demand is that required for new growth. Newly synthesized tissue tends to have a high concentration of N, but as the plant becomes larger the mass of new growth compared to the existing mass becomes smaller. This new growth demand can thus be a large component of demand when the plants are small and young, but diminishes in importance as the crop grows. The growth models used within the DSSAT estimate a potential amount of new growth before any of the stresses are applied. This potential new growth increment, together with the prevailing critical concentration are used to determine this new growth demand. When actual growth is checked by some stress, the N demand for new growth provides a driving force for luxury consumption to occur provided there is an adequate supply of N.

During grain filling, N required by the grain is removed from the vegetative and root pools to form a grain N pool. The resultant lowering of concentration in these pools may lead to an increase in N demand. The total plant N demand (*ANDEM*) is the sum of the deficiency demand and the demand for new growth.

N supply

Plants may take up N as either nitrate or ammonium ions. The concentration of each is determined by the various transformations described previously. In general in upland soils nitrate predominates and in flooded rice soils ammonium is present in larger amounts. To calculate the potential supply of N to the crop, zero to unity availability factors for nitrate (*FNO3*) and ammonium (*FNH4*) are calculated from the soil concentrations of the respective ions as below:

$$FNO3 = 1.0 - \exp(-0.030 \times NO3(L))$$

$$FNH4 = 1.0 - \exp(-0.030 \times (NH4(L) - 0.5)$$

In the model for flooded rice, similar availability indices are also calculated from floodwater concentrations of these ions.

When the soil is not flooded a zero to unity soil water factor (*SMDFR*), which reduces potential uptake, is calculated as a function of the relative availability of soil water:

$$SMDFR = (SW(L) - LL(L))/ESW(L)$$

where $ESW(L)$ is the extractable soil water in layer L. To account for increased anaerobiosis and declining root function at moisture contents above the drained upper limit, $SMDFR$ is reduced as saturation is approached. In flooded rice, when floodwater is present, this index is not used.

The maximum potential N uptake from a layer is calculated as a function dependent on a coefficient describing the maximum daily uptake of N per unit length of root and the total amount of root present in the layer. Initial estimates for this coefficient were derived from the work of Warncke and Barber (1974) for maize, but have since been modified with experience.

The simulated actual uptake of each of ammonium ($RNH4U(L)$) and nitrate ($RNO3U(L)$) from a layer may then be determined by reducing this potential uptake by the factors describing the concentration effects on ion availability and the soil water availability index described previously:

$$RNO3U(L) = RLV(L) \times 0.008 \times SMDFR \times 2 \times DLAYR(L) \times 100 \times FNO3$$

Note here that the layer thickness ($DLAYR$) and the coefficient 100 are necessary unit conversions to allow the uptake to be expressed in units of kg N ha^{-1}.

Potential N uptake from the whole profile ($TRNU$) is the sum of $RNO3U(L)$ and $RNH4U(L)$ from all soil layers where roots occur. Thus $TRNU$ represents an integrated value that is sensitive to rooting density, concentrations of both ammonium and nitrate and soil water effects on uptake. If the potential N supply from the whole profile ($TRNU$) is greater than crop N demand ($ANDEM$), an N uptake factor (NUF) is calculated:

$$NUF = ANDEM/RNU$$

NUF is used to reduce the N uptake from each layer to the level of demand. When demand is greater than supply NUF has a value of 1.0 and potential uptake is not reduced. When supply is high and demand is small, NUF will be less than 1.0 and uptake from each layer is reduced. This latter situation often arises when plants are young and are exposed to a high N supply.

After daily N uptake has been determined the partitioning of this N between roots and shoots is calculated on the basis of the proportions of the total plant N demand arising from each. When the increments of shoot N and root N have been added to the appropriate pools, plant N dry weight concentrations are updated for use in the following days plant N stress calculations.

Grain nitrogen concentration

Under most systems of crop husbandry, the crop has utilized most of the soil N available to it by the time grain filling commences. Crop plants must satisfy the requirements of the developing grains by remobilization of protein N from vegetative organs as well as from any uptake which may occur. When the soil

N supply is increased, the proportion of grain N arising from remobilization declines and the proportion from uptake increases (Vos 1981).

In the DSSAT crop models, the N accumulation rate of single grains is simulated as a function driven by the prevailing temperatures and is quite similar to the functions for the rate of grain mass accumulation. Grain N accumulation is more sensitive to temperature than grain mass accumulation, thus at high temperatures grain N concentrations will tend to be higher. Moisture stress affects the supply of assimilate to developing grains. When moisture is limiting, simulated grain N concentrations can rise because the dilution of N in the grain by assimilate is lowered. The temperature driven function for the rate of N accumulation in single grains (*RGNFIL*) for wheat is given by:

$$RGNFIL = 4.83 - 3.25 \times DTT + 0.25 \times (TMAX - TMIN) + 4.306 \times TEMPM$$

where: *DTT* is the daily thermal time (°C days), *TMAX* and *TMIN* are ambient maximum and minimum temperatures, respectively, and *TEMPM* is the mean ambient temperature. For the other crops adjustments are made for differing temperature optima and for different grain sizes. From this potential rate of grain N accumulation a daily grain N sink is determined based on this daily concentration of N in new grain growth and the number of grains filling.

Two pools of N within the plant are potentially available for translocation: a shoot pool (*NPOOL1*) and a root pool (*NPOOL2*). These pools are determined from the N concentration (*TANC* or *RANC*) relative to the critical concentration (*TMNC* or *RMNC*) and the dry weight of the pool (*RTWT* or *TOPWT*):

$$NPOOL1 = TOPWT \times (TANC - TMNC)$$

$$NPOOL2 = RTWT \times (RANC - RMNC)$$

These pools of N are however not immediately available for translocation. N contained within these would have to undergo various biochemical transformations before it would become available for grain protein synthesis. The proportion (*XNF*) which is labile on any day is affected by the N status of the plant.

$$XNF = 0.15 + 0.2 \times NFAC$$

The total N supply available for remobilization (*NPOOL*) is determined by summing the shoot and root labile pools, *TNLAB* and *RNLAB*, respectively.

$$TNLAB = XNF \times NPOOL1$$

$$RNLAB = XNF \times NPOOL2$$

$$NPOOL = TNLAB + RNLAB$$

In the rice model provision is made here for separation of N sinks and sources arising from tillers versus the main stem.

When the *NPOOL* is not sufficient to supply the grain N demand (*NSINK*), *NSINK* is reduced to *NPOOL*. As the supply of N for remobilization becomes limiting, shoot N is utilized in preference to root N. This is accomplished as follows: if *NSINK* is greater than that which can be supplied by the tops (*NPOOL1*), then *NPOOL1* is set to zero and the tops N concentration is set to its minimum value (*VMNC*). The remaining *NSINK* is then satisfied from the root N pool and the N pool is updated accordingly. After these adjustments to *NSINK* the total amount of N contained in the grain can then be accumulated.

$$GRAINN = GRAINN + NSINK$$

These procedures together with the N deficiency indices and the growth routines provide several pathways by which N stress during grain filling can affect grain yield and grain protein content.

Plant N stress indices

The DSSAT crop models provide summary outputs describing the average N stress prevailing during each of the growth stages of the crop. These are calculated from the daily values of *NDEF1* and *NDEF2* described previously. These values are accumulated and divided by the duration of each growth stage to produce average stress values (*CNSD1* and *CNSD2* corresponding to *NDEF1* and *NDEF2*, respectively). They are not used in the simulation but provide a useful summary of crop N status during each of the growth stages for the user.

Model inputs

Since the soil nitrogen model used within the DSSAT is intrinsically linked to the water balance model, those parameters used by the water balance model that define ranges of soil water availability, soil drainage and deep percolation characteristics and layer depth increments are also required by the N model. The inputs for the soil water balance model is described by Ritchie (1985). The N model itself requires input data that describe the initial amount of mineral N present in the soil profile and information that will enable the estimation of how much N will be mineralized from soil organic matter, the potassium chloride extractable nitrate (NO_3^-) and ammonium (NH_4^+) present in each of the layers. The soil bulk density is used in the calculations of concentrations of N from mass. It is also used within the rice model to determine the effect of puddling on the effective bulk density of the surface layer.

Rather than use soil and situation specific constants for N mineralization rates derived from incubation experiments, the DSSAT models use inputs more likely to be available to a wider group of model users. These are soil organic carbon in each of the layers.

The amount of crop residue added and its C/N ratio are also required inputs. Root residue in practice is estimated by assuming a reasonable harvest index of the previous crop and a shoot/root ratio of 10 at harvest. Since data for the root C/N ratio would not normally be available this is internally fixed at 45.

The rice model also requires the cation exchange capacity of the surface layer of soil for use in ammonium diffusion calculations.

Both models require data specifying the dates, amounts and types of N fertilizer used. The depth of incorporation and the degree of incorporation are also required inputs. Since the method of incorporation can have a profound effect on fate of fertilizer in rice cropping systems, a system of codes for each is used.

Conclusions

The model described here represents a pragmatic approach to describing the complexities of the N cycle. We have strived to capture the major influences of weather and soil properties on determining N availability to a crop and subsequently how that crop responds to N.

The model does not consider the processes leading to volatilization of ammonia from upland soils nor does it consider ammonium fixation, leaching of ammonium, or direct losses of N from plants to the atmosphere. Where these losses are substantial, the model will be inaccurate. In the upland model fertilizer placement is considered.

Many of the procedures used in the simulation of each of the major N transformations can potentially be improved. Quemada and Cabrera (1995) reported improvements to the simulation of N mineralization for differing conditions. The lack of an ammonia loss routine for upland soils may prove to be a serious omission for some conditions or circumstances. The challenge is to conduct any nitrogen model changes or improvements reliably while at the same time not require any additional or least only minor readily attainable inputs data.

References

Bouldin D R, Aligmagno B V (1976) NH_3 volatilization from IRRI paddies following broadcast applications of fertilizer nitrogen. Terminal Report as Visiting Scientist, International Rice Research Institute, Manila, Philippines.

Bumb B L (1995) Global fertilizer perspective, 1980–2000: the challenges in structural transformation. Technical Bulletin T42, International Fertilizer Development Center, Muscle Shoals, Alabama, USA.

Craswell E T, Godwin D C (1984) In Tinker P B, Lauchli A (Eds.) The efficiency of nitrogen fertilizers applied to cereals grown in different climates. Advances in Plant Nutrition, Praeger Scientific, New York, New York, USA.

Davidson J B, Graetz D A, Rao P S C, Sclim M (1978) Simulation of nitrogen movement, transformation, and uptake in the plant root zone. U.S. Environmental Protection Agency EPA-600/3-78-029. Office of Research and Development, Athens, Georgia, USA.

Denmead O T, Simpson J R, Freney J R (1977) The direct measurement of ammonia emission after injection of anhydrous ammonia. Soil Science Society of America Journal 41:1001–1004.

Fillery I R P, Simpson J R, de Datta S K (1984) Influence of field environment and fertilizer management on ammonia loss from flooded rice. Soil Science Society of America Journal 48:914–920.

Fillery I R P, Roger P A, de Datta S K (1986) Ammonia volatilization from nitrogen sources applied to rice fields. II. Floodwater properties and submerged photosynthetic biomass. Soil Science Society of America Journal 50:86–91.

Freney J R, Denmead O T, Watanabe I, Craswell E T (1981) Ammonia and nitrous oxide losses following application of ammonium sulfate to flooded rice. Australian Journal of Agricultural Research 32:37–45.

Gilmour J T (1984) The effects of soil properties on nitrification and nitrification inhibition. Soil Science Society of America Journal 48:1262–1266.

Godwin D C, Singh U, Buresh R J, DeDatta S K (1990) Pages 320–325 in Transactions of the 14th International Congress of Soil Science, Modeling of nitrogen dynamics in relation to rice growth and yield. vol IV, Commission IV, Kyoto, Japan.

Godwin D C, Jones C A (1991) Nitrogen Dynamics in soil-crop systems. Pages 287–321 in Hanks R J and Ritchie J T (Eds.) Modeling plant and soil systems. Agronomy Monograph #31, American Society of Agronomy, Madison, Wisconsin, USA.

Godwin D C, Singh U (1991) Modelling nitrogen dynamics in rice cropping systems. Paper presented at International Symposium on Rice Production on Acid Soils of the Tropics, Kandy, Sri Lanka, June 26–30, 1989.

Linn D M, Doran J W (1984) Effect of water filled pore space on carbon dioxide and nitrous oxide production in tilled and non-tilled soils. Soil Science Society of America Journal 48:1267–1272.

McGill W B, Hunt H W, Woodmansee R G, Reuss J O (1981) Phoenix – a model of the dynamics of carbon and nitrogen in grass-land soils. In Clark F E, Rosswall T (Eds) Terrestrial nitrogen cycles. Eological Bulletins (Stockholm) 33:49–115.

McLaren A D (1970) Temporal and vectorial reactions of nitrogen in soil: A review. Canadian Journal of Soil Science 50:97– 109.

Myers R J K (1975) Temperature effects on ammonification and nitrification in a tropical soil. Soil Biol Biochemistry 7:83–86.

Myers R J K, Campbell C A, Weier R L (1982) Quantitative relationship between net nitrogen mineralization and moisture content of soils. Canadian Journal of Soil Science 62:111–124.

Nye P H, Tinker P B (1977) Solute movement in the soil-root system. Blackwell, UK.

Quemada M, Cabrera M L (1995) CERES-N model predictions of nitrogen mineralized from cover crop residues. Soil Science Society America Journal 59:1059–1065.

Reichman G A, Grunes D L, and Viets F G Jr (1966) Effect of soil moisture on ammonification and nitrification in two northern great plains soils. Soil Science Society of America Proceedings 30:363–366.

Ritchie J T (1985) A. User-oriented model of the soil water balance in wheat. Pages 293–305 in Day W, Atkin R K (Eds.) Wheat growth and modelling, Plenum Press, New York, New York, USA.

Rolston D E, Sharpley A N, Toy D W, Hoffman D L, Broadbent F E (1980) Denitrification as affected by irrigation frequency of a field soil. EPA -600/2-80-06 U.S. Environmental Protection Agency, ADA, Oklahoma, USA.

Seligman N C, van Keulen H (1981) PAPRAN: A simulation model of annual pasture production limited by rainfall and nitrogen. Pages 192–221 in Frissel M J, van Veen J A (Eds.) Simulation of nitrogen behavior of soil plant systems, PUDOC, Wageningen, the Netherlands.

Tillotson W R, Robbins C W, Wagenet R J, and Hanks R J (1980) Soil water, solute, and plant growth simulation. Utah Agric. Exp. Stn. Bull 502, Utah State University, Logan, Utah, USA.

Van Veen J A (1977) The behavior of nitrogen in soils. A computer simulation model. PhD Thesis, Vrije Universiteit, Amsterdam, the Netherlands.

Van Veen J A, Frissel, M J (1981) Simulation model of the behavior of N in soil. Pages 126–144 in Frissel M J, van Veen J A (Eds.) Simulation of nitrogen behavior of soil plant systems, PUDOC, Wageningen, the Netherlands.

Vleck R G L, Carter M F (1983) The effect of soil environment and fertilizer modifications on the rate of urea hydrolysis. Soil Science 136:56–63.

Vleck R G L, Stumpe J M (1978) Effects of solution chemistry and environmental conditions on ammonia volatilization losses from aqueous systems. Soil Science Society of America Journal 42:416–421.

Vos J (1981) Effects of temperature and nitrogen supply on post-floral growth of wheat; measurements and simulations. Agricultural Research Report 911, PUDOC, Wageningen, the Netherlands.

Warncke D D, Barber S A (1974) Root development and nutrient uptake by corn grown in nutrient solution. Agronomy Journal 66: 514–516.

Cereal growth, development and yield

J.T. RITCHIE[1], U. SINGH[2], D.C. GODWIN[3], W.T. BOWEN[4]
[1]*Homer Nowlin Chair, Department of Crop and Soil Sciences, Michigan State University, East Lansing, Michigan 48824, USA*
[2]*International Fertilizer Development Center, Muscle Shoals, Alabama 35662, USA*
[3]*3 Colony Cres, Dubbo NSW, Australia*
[4]*International Fertilizer Development Center/Centro Internacional de la Papa, Apartado 1558, Lima, Peru*

Key words: simulation, biomass, partitioning, roots, shoots, ears, grains, maize, wheat, barley, sorghum, rice, millet

Abstract
The objective of the CERES crop simulation models is to predict the duration of growth, the average growth rates, and the amount of assimilate partitioned to the economic yield components of the plant. With such a simulation system, optimizing the use of resources and quantifying risk related to weather variation is possible. The cereal crops included in the DSSAT v3 models are maize, wheat, barley, sorghum, millet and rice. A feature of each model is its capability to include cultivar specific information that make possible prediction of the cultivar variations in plant ontogeny and yield component characteristics and their interactions with weather. Biomass growth is calculated using the radiation use efficiency approach; biomass produced is partitioned between leaves, stems, roots, ears and grains. The proportion partitioned to each growing organ is determined by the stage of development and general growing conditions. The partitioning principles are based on a sink source concept and are modified when deficiencies of water and nutrient supplies occur. Crop yields in the CERES models are determined as a product of the grain numbers per plant times the average kernel weight at physiological maturity. The grain numbers are calculated from the above ground biomass growth during a critical stage in the plant growth cycle for a fixed thermal time before anthesis. The grain weight in all the CERES models is calculated as a function of cultivar specific optimum growth rate multiplied by the duration of grain filling. Grain filling is reduced below the optimum value when there is an insufficient supply of assimilate from either the daily biomass production or from stored mobile biomass in the stem. The CERES models have been tested over a wide range of environments. Although there are improvements that can be made in the simulation procedures, results have shown that when the weather, cultivar and management information is reasonably quantified, the yield results are usually within acceptable limits of $\pm 5\%$ to 15% of measured yields.

Introduction

The simulation of crop development, growth, and yield is accomplished through evaluating the stage of crop development, the growth rate and the partitioning of biomass into growing organs. All of these processes are dynamic and are affected by environmental and cultivar specific factors. The environmental factors can be separated into the aerial environment and the soil environment. The aerial environment information comes from a record of weather conditions

G.Y. Tsuji et al. *(eds): Understanding Options for Agricultural Production, 79–98*

such as temperature and solar radiation. The soil environment information must come from synthesized approximations of the soil water status, the availability of soil nitrogen and other nutrients, the aeration status of the soil and the spatial distribution of the root system.

The potential biomass yield of a crop can be thought of as the product of the rate of biomass accumulation times the duration of growth. The rate of biomass accumulation is principally influenced by the amount of light intercepted by plants over an optimum temperature range. The duration of growth for a particular cultivar, however, is highly dependent on its thermal environment and to some extent the photoperiod during floral induction. Thus, accurate modeling of crop duration and crop growth rates are necessary for a reasonable yield estimation.

Variation between modern annual cultivars within a species is usually most evident in the duration of growth and least evident in the rate of growth. Older cultivars with approximately the same duration of growth as modern ones differ from the modern ones principally in partitioning less assimilate to the reproductive organs and more to the vegetative organs, causing a lower harvest index. Many modern cereal cultivars have shorter, stronger stems, helping to prevent lodging when high soil fertility levels are used. Little cultivar variation exists within a species for the processes of photosynthesis and respiration.

The objective of this chapter is to present the procedures used in the CERES crop models to estimate crop growth, development and yield. The CERES models include simulation procedures for wheat (*Triticum aestivum* L.), maize (*Zea mays* L.), rice (*Oryza sativa* L.), barley (*Hordeum vulgare* L.), grain sorghum (*Sorghum bicolor* L.), and pearl millet (*Pennisetum americanum* L.)

Basic principles of crop simulation with CERES models

In its simplest form, the total biomass (B_T) of a crop is the product of the average growth rate (g) and the growth duration (d):

$$B_T = g \times d$$

The simulation of yields at the process level must involve the prediction of these two important processes. The economic yield of a crop is the fraction of B_T that is partitioned to grain. This fraction can range from practically 0.0 for crops with severe stresses at critical times to more than 0.5 for crops that are grown under optimum conditions. Thus, the partitioning of B_T into yield takes into consideration both the effects of growth and development.

Table 1 presents several aspects of the processes of growth rates and duration that come from experimental observations of several grain crops. The CERES models attempt to take these factors into consideration for estimation of crop yield. Growth and development each have two distinctly different aspects to consider. For growth, the mass accumulation deals with the absorption of

Table 1. Information regarding factors of plant growth and development processes and their sensitivity to stresses (From Ritchie, 1991).

	Growth		Development	
	Mass	Expansion	Phasic	Morphological
Principal environmental factor	Solar radiation	Temperature	Temperature Photoperiod	Temperature
Degree of variation among cultivars	Low	Low	High	Low
Sensitivity to plant water deficit	Low – Stomata Moderate – leaf wilting and rolling	High – vegetative stage Low – grain filling stage	Low – delay in vegetative stage	Low
Sensitivity to nitrogen deficiency	Low	High	Low	Low – main stem High – tillers and branches

photosynthetically active radiation (PAR) for the assimilation of plant biomass. Thus mass growth is dependent on the amount of PAR received from the sun and the amount of leaf surface available for the absorption of PAR for photosynthesis. Only a portion of the carbon fixed by photosynthesis eventually appears in the harvestable dry matter because the plant loses some of the carbon through respiration. The other dimension of growth, that of expansion of the area or volume of plant parts, is not controlled by the net amount of carbon fixed. Environmental factors and stresses affecting the cell expansion are different from those that affect mass growth. This is why plants growing vegetatively at different temperatures and at similar radiation levels, will, with time, have a different size and mass. These differences in mass and expansion suggest that the partitioning patterns of the plant are altered some way to accommodate the differences in plant size when the net assimilation rate would be approximately the same when normalized to the area of leaves absorbing the PAR.

The duration of plant growth also has two distinctly different features, namely phasic and morphological development. Phasic development involves changes in stages of growth and is almost always associated with major changes in biomass partitioning patterns. Note from Table 1 that phasic development is one of the principal plant processes with a high degree of cultivar diversity. This cultivar diversity provides the opportunity for selecting cultivars that have the growing season length synchronized with the supply of water or length of warm season for a particular location. This feature is of major importance in the analysis of risks associated with crop production.

Morphological development refers to the beginning and ending of development of various plant organs within the plant life cycle. Modeling morphologi-

Table 2. Growth stages of crops in CERES models and the organs growing during those stages (Barley = BA; Maize = MZ; Pearl-Millet = ML; Rice = RI; Sorghum = SG; Wheat = WH).

Stage	Duration	Crop	Organs growing
7	Fallow	All	
8	Sowing to germination	All	
9	Germination to emergence	All	Root
1	Emergence – end juvenile Emergence – terminal spikelet	MZ, SG, ML, RI, BA, WH	Leaf, root
2	End juvenile – floral or panicle initiation Terminal spikelet – end leaf growth	MZ, SG, ML, RI, BA, WH	Leaf, stem, root
3	Floral or panicle initiation – end leaf growth Begin ear growth – end ear growth	MZ, SG, ML, RI, BA, WH	Leaf, stem, root Stem, ear
4	End leaf growth – begin grain fill End ear growth – begin grain fill	MZ, SG, ML, RI, BA, WH	Stem, ear/panicle, root
5	Begin grain fill – physiological maturity	All	Grain, root
6	Physiological maturity – harvest	All	

cal features is attempted to provide an estimate of the number of leaves, tillers, and grains that will be produced by a plant. The principal environmental factor affecting both phasic and morphological development rates is the temperature of the growing part of the plant. The temperature response functions of the two developmental processes may be different, however. The cultivar characteristics affecting plant response to photoperiod is also an important determinant of the duration of growth in addition to the temperature influence.

A logical reason for separating the growth and development processes is to determine how differences in water or nutrients alter each process. Information in Table 1 indicates that vegetative expansion growth is more sensitive to water deficits than the other three processes. In an environment where water supply is uncertain, these growth and development features should be considered separately for accurate yield simulation, especially in situations where the water deficit condition may occur at any stage of the crop ontogeny.

Plant development

Phasic development in the CERES models quantifies the physiological age of the plant and describes the duration of nine growth stages (Table 2). With the exception of fallow duration (stage 7), which is user specified (the start of simulation and sowing dates), the model simulates the duration of the growth stages.

Temperature influence on development

The time scale of cereal plants is closely linked to the temperature of the growing parts of the plant. Thus, simulating how some development characteristics are affected by temperature, independent of photoperiod or other environmental constraints, was thought important. Rameur first suggested in 1735 that the duration of particular stages of growth was directly related to temperature and that this duration for a particular species could be predicted using the sum of mean daily air temperature (Wang, 1960). This procedure for normalizing time with temperature to predict plant development rates has been widely used. A term believed to be most appropriate to describe the duration of plant development is 'thermal time' (Ritchie and NeSmith, 1991). Because the time scale of plants is closely coupled with its thermal environment, thinking of thermal time as a plant's view of time is appropriate.

Thermal time has units of °C day. The simplest and most useful definition of thermal time t_d is

$$t_d = \sum_{i=1}^{n} (\bar{T}_a - T_b)$$

where $\bar{T}_a$ is daily mean air temperature, T_b is the base temperature at which development stops and n is the number of days of temperature observations used in the summation. The calculation of T_a is accomplished in the CERES models by averaging the daily maximum and minimum temperatures under most circumstances. This calculation of thermal time is appropriate for predicting plant development if: (a) the temperature response of the development rate is linear over the range of temperatures experienced; (b) the daily temperature does not fall below T_b for a significant part of the day; (c) the daily temperature does not exceed an upper threshold temperature for a significant part of the day; and (d) the growing region of the plant has the same mean temperature as T_a.

When one or more of the above assumptions are not correct, alternative methods for calculating thermal time are required. For such conditions, the diurnal temperature values are approximated from the daily maximum and minimum temperature because the system becomes non-linear when outside the range of the normal linear function. If the temperature is below T_b, the thermal time is assumed to be zero. If it is above an upper temperature threshold, the thermal time is assumed to be equal to the upper threshold value or some value lower than the threshold (Covell et al., 1986; Ritchie and NeSmith, 1991). More precise models should use plant or canopy temperature rather than air temperature where appropriate.

Temperature and development

Time between developmental events can be measured and expressed as an equivalent development rate by taking the inverse of time as the rate. Developmental events such as time between the appearance of leaf tips or time between emergence and flowering provided the information needed to derive

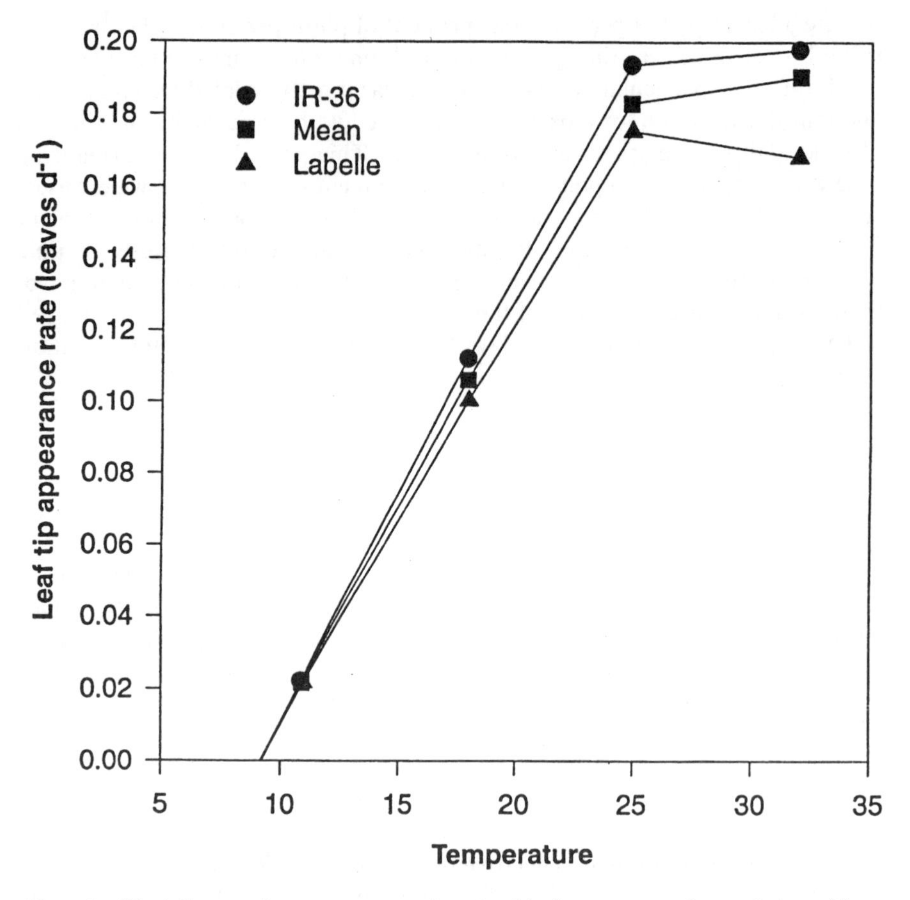

Figure 1. The influence of temperature on the rate of leaf appearance of several rice cultivars. The curve labeled mean is the average for 12 cultivars. The varieties IR36 and Labelle represent the fastest and slowest development rates of the 12 cultivars studied (Ritchie, 1993).

temperature response functions for cereal development processes. Often when such development rates are measured at various temperatures, the resulting relationship is linear over a relatively large range of temperatures. Figure 1 provides information for rice of the type used to develop temperature response functions for vegetative development. The rate of appearance of leaf tips were measured on the main stems of plants that were growing at four contrasting growth cabinet temperatures (Ritchie, 1993). There were differences in individual varieties in leaf appearance rates of 12 diverse varieties. This type of information demonstrates that the base temperature for vegetative development of rice is approximately 9°C, that a plateau, termed supra-optimal by Summerfield et al. (1992), begins at about 26°C, and that there is little difference between varieties in this relationship although the 12 varieties in the study consisted of Indica, Japonica and mixed types.

This type of experimentation was used for all the cereal crops to determine the relevant vegetative developmental functions used in CERES. They are reported in various sources (Warrington and Kanemasu, 1983; Ong and Monteith, 1985; Alocilja and Ritchie, 1988 and 1991; Ritchie and Alagarswamy, 1989; Alagarswamy and Ritchie, 1991; Ritchie, 1991; Ritchie and NeSmith, 1991).

Grain filling duration was more difficult to quantify than visual developmental events such as leaf appearance or time to flowering. For many crops, there is a lag period after anthesis before rapid grain filling begins. The grain filling rate is almost constant if average temperatures are relatively constant until the grain is almost filled. If a shortage of assimilate, nitrogen, or stored carbohydrate ends before grain filling, the filling rate decreases. The most accurate determination of the beginning and ending of grain filling requires destructive sampling, thus creating a higher level of variability in the data than for non-destructive type observations. Determining the timing of grain filling from visual observations such as flowering to physiological maturity has a great deal of uncertainty because of the lack of a clearly observable plant feature at the beginning and end of grain filling. Black layer formation has been used in recent years for defining maize maturity. It has not proven to be a highly accurate indication of end of grain filling, although it does provide qualitative evidence of differences between cultivars needed in modeling (Daynard, 1972). The temperature functions for grain filling duration in CERES have been estimated from field derived information where possible.

The primary difference in development between cultivars within a species occurs in the length of their vegetative phase. Besides the temperature influence on duration of stages, variation in duration can result from cultivar differences in photoperiod sensitivity, maturity type, or vernalization sensitivity. Maturity type variations are exhibited in the length of a juvenile stage of the crop species rice, maize, sorghum and millet in which young plants do not begin photo-induction to flowering until they reach a certain leaf number. During this juvenile stage the plants are insensitive to photoperiod.

Cultivar variation in response to photoperiod

The length of the day (or night) can influence the rate at which plants change from vegetative growth to reproductive growth. These differences in photoperiod response are evident by the total number of leaves developed on plants grown during various photoperiods. Maize, sorghum, pearl millet and rice are termed short day plants because they have minimum vegetative development in long days. In contrast, wheat and barley are long day plants and minimize their development during short days.

To quantify the photoperiod development functions, growth cabinets are used that provide constant environmental variables except for photoperiod. The extra time taken for floral induction to occur for photoperiods longer or shorter than an optimum is used to derive a cultivar specific rate of photo-induction. The optimum photoperiod for short day plants is a threshold photoperiod below which there is no further photoperiod sensitivity. This optimum for maize is about 12.5 hours. For rice, sorghum, and millet, the optimum (P2O) varies with cultivars, ranging between 11 and 15 hours.

The rate of change of development with photoperiod is expressed as a cultivar specific coefficient in the models. In the maize model a coefficient (P1D) for delaying reproductive growth is expressed in units of days hr^{-1}. In sorghum, pearl millet, and rice, it is thermal time hr^{-1} (Ritchie and Alagarswamy, 1989; Alagarswamy and Ritchie, 1991). In wheat and barley it is expressed as a relative change in thermal time (Ritchie, 1991). Hopefully we can learn more about the photoperiod response functions, and express them for various crop species in a more uniform manner than those described herein.

The photoperiod influence on leaf appearance and leaf number within or between maize cultivars was demonstrated for maize using field experiments at three contrasting locations. A photoperiod sensitive tropical maize cultivar, X304C, was grown in Michigan, Texas, and Hawaii. These locations have large contrasts in daylength during development. However, they had almost the same rate of leaf development when expressed in terms of thermal time, but quite different leaf numbers as can be seen in Figure 2 (Ritchie and NeSmith, 1991).

At the Hawaii site, the photoperiod (daylength and civil twilight) at tassel initiation was about 13.5 hours, at the Texas site it was about 14.2 hours and at the Michigan site it was about 16.3 hours. These data clearly demonstrate that for photoperiod sensitive cultivars, thermal time to flowering is not a constant but also a function of the photoperiod at each site. Thermal time would only be constant for plants grown with common photoperiods during floral induction (Kiniry et al., 1983a).

Cultivar variation in juvenile stage

Maize, rice, millet, and sorghum have a distinctive juvenile phase that varies between cultivars. The juvenile stage can be defined as the pre-induction stage when the plant is not sensitive to variations in photoperiod. The cultivar variation in the length of a juvenile stage gives rise to the classification of plants in terms of maturity types. The maturity type concept, however, is only qualitative and may be only regionally transferrable if the maturity type classification is the result of both the juvenile stage duration and the responsiveness to photoperiod. Thus a cultivar that is long season in one region can be short season in another region if the photoperiod is quite different at the time of floral initiation.

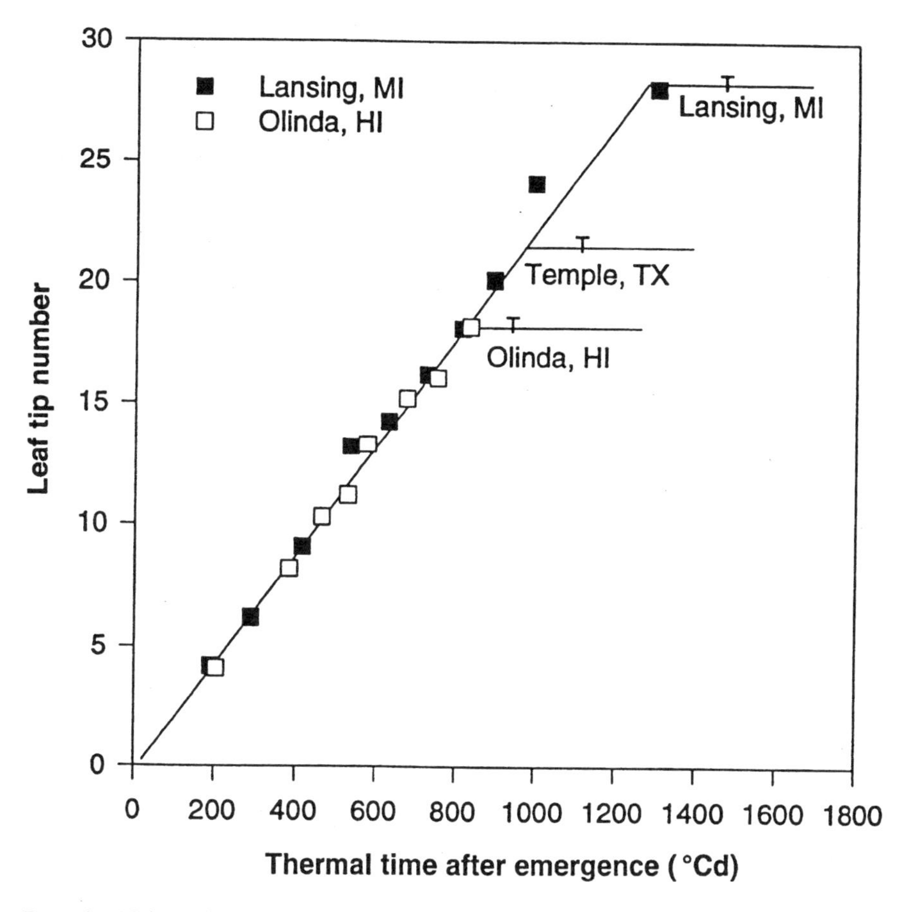

Figure 2. Maize leaf tip number as a function of thermal time after emergence at Olinda, HI; Temple, TX; and East Lansing, MI. Final leaf number is indicated by horizontal lines and time of tasseling is indicated by T (Ritchie and NeSmith, 1991).

To simulate the juvenile stage it is necessary to know the variation among cultivars. In short day plants, this stage can be observed by growing the plants in short day conditions (usually 12 hours or less) and observing, by plant dissection, the time to floral initiation. The end of the juvenile stage is usually about five days before floral induction in maize, a phenomenon that can be determined with special experiments in which plants are interchanged between long days and short days during the late juvenile and floral induction phase (Kiniry et al., 1983b). Photoperiod exchange experiments with all the CERES crops have demonstrated a similar pattern to maize with a similar minimum induction period. The duration of the juvenile stage is almost totally controlled by temperature. Thus a single cultivar specific coefficient (P1) can be determined to describe the duration of the juvenile period in thermal time. During the juvenile and floral induction phase, leaf primordia are being developed. When floral induction ends, leaf initiation no longer occurs and the final leaf number has been determined. Because the rate of appearance of the final leaves that have differentiated, but not yet appeared, is controlled by temperature, the thermal time required for the remainder of the vegetative stage is a consequence of what has happened in the juvenile and floral induction phase.

Cultivar variation in response to vernalization of winter cereals

Winter wheat and barley varieties usually require exposure to relatively low temperatures before spikelet formation can begin. This low temperature requirement for flowering, called vernalization, begins at germination. Vernalization is assumed to occur at temperatures between 0 and 18°C (Ahrens and Loomis, 1963; Tirone and Metzer, 1970). The optimum temperature for vernalization is assumed to be in the range of 0 to 7°C, with temperatures between 7 and 18°C having a decreasing influence on the process. Minimum and maximum daily temperatures are used to calculate a daily vernalization effectiveness factor with a value ranging between 0 and 1. The daily relative vernalization effectiveness factor is totaled to determine what is termed vernalization days. Although there is cultivar variability in sensitivity to vernalization between cultivars, 50 vernalization days are assumed to be sufficient to completely vernalize all cultivars. This variability is considered by the use of a cultivar coefficient (P1V) to calculate the influence of vernalization on stage 1 growth. The relative development rates as influenced by vernalization for various cultivars is determined from comparisons made of plant development in growth cabinet experiments. In the experiments, plants are exposed to effective vernalizing temperatures (≈ 2°C) for variable lengths of time and then grown in warmer, more typical temperatures. The ratio of days to terminal spikelet for a 50 vernalization day treatment to the days of terminal spikelet for treatments < 50 d provides a relative development rate for use in the wheat and barley model (Ritchie, 1991). A relative development of 0.5 indicates that it takes twice as long for a plant to develop to the terminal spikelet stage with less than full vernalization than it did for one with 50 vernalization days.

Spring wheat varieties have a low sensitivity to vernalization. They are incorporated in the model in the same way as winter wheat varieties by expressing the differences in vernalization through the cultivar coefficient P1V. Spring wheats should have P1V values < 0.5.

Devernalization can occur when young seedlings are exposed to high temperature. In the model, if the number of vernalization days is < 10 and the maximum temperature > 30°C, then the number of vernalization days decreases by 0.5 days per degree above 30°C. If the number of vernalization days is > 10, no devernalization is assumed to occur.

Table 3. Phenological and growth variables used to simulate cereal crops.

Coefficient		Barley	Millet	Maize	Rice	Sorghum	Wheat
Radiation use efficiency (RUE) g MJ^{-1}		3.5	5.0	5.0	2.6–4.0*	4.0	2.6–4.0
Extinction coefficient (K)		−0.85	−0.85	−0.65	−0.625	−0.85	−0.85
Photosynthesis optimum temperature, °C		18	26.0	26.0	14.0–32.0	20–40	18.0
Root growth minimum fraction of daily biomass	1	0.35	0.25	0.25	0.35	0.25	0.35
	2	0.20	0.20	0.25	0.20	0.25	0.20
	3	0.15	0.20	0.08	0.15	0.25	0.15
	4	0.10	0.20	0.08	0.15	0.08	0.10
Phyllochron (PHINT) °C day $leaf^{-1}$		77.5**	43	75	83	49	95***
Base temperature, °C	Veg.	0.0	8.0	8.0	9.0	8.0	0.0
	Grain Fill	1.0	8.0	8.0	9.0	8.0	1.0

*These vary with the value of PAR. Lower RUE at high PAR; High RUE at low PAR.
**The phyllochron varies somewhat from this value based on the rate of change of photoperiod at the time of seedling emergence.
***The number 95, is the default value. Users can change the value in the cultivar file, if necessary.

Biomass growth

Monteith (1977) demonstrated that cumulative seasonal light interception for several crops grown with adequate water and nutrients was linearly related to biomass production. Hesketh and Baker (1967) reported that 15 minute averaged net photosynthesis rates of maize and cotton canopies had a reasonably linear relationship to light interception. Since those reports were published, many other researchers have found similar results specifically for cereal crops (Norman and Arkebauer, 1991). Because this concept of radiation use efficiency (RUE) is simple and conservative, it was chosen for use in calculating the total biomass production in CERES. The calculated PAR is the light quantity available for crop interception in CERES and is assumed to be half the input daily solar radiation (Monteith, 1977). When PAR values for a day are expressed in units of MJ m^{-2}, RUE units are g MJ^{-1}.

A major difficulty associated with verifying the RUE value to use for the various crops has been the lack of sufficiently precise information on the fraction of biomass partitioned to the root system. Some studies have reported that as much as half the seasonal total biomass production is partitioned below ground. However, root respiration, exudation and death can result in short term loss of measurable below ground biomass. The measurement of root mass should yield considerably lower biomass than was partitioned to the roots. Because a goal of the CERES models is to develop a reasonably realistic root system for water and nutrient capture, root system biomass and length of roots per unit soil volume needs to be as accurate as possible. To compensate for this loss of biomass in the root system, the values of RUE used in CERES are larger than most field studies have indicated. Table 3 provides RUE values used for the various crops in the DSSAT v3.

The equation used for potential biomass production ($PCARB$) for a day is

$$PCARB = RUE \times IPAR$$

where IPAR is the fraction PAR intercepted by the plants. The units of $PCARB$ are $g\ m^{-2}$. Values of IPAR are obtained from

$$IPAR = PAR \times [1 - \exp(-k \times LAI)]$$

where k is an extinction coefficient and LAI is the green leaf area index of the plant canopy. Values of k used in each crop in DSSAT v3 are given in Table 3. Porter et al. (1993) showed that the value of k (0.85) used in CERES-Wheat was too large, but the error had little influence on biomass production. The values of LAI use only the blade area but do not include exposed sheath area as done in some models. As a result, values of k in CERES are somewhat larger than other investigators have reported.

The actual daily biomass production ($CARBO$) may be less than $PCARB$ because of non-optimal temperature or deficits of water or nitrogen. A temperature weighted to approximate the daytime temperature is calculated for a temperature reduction factor ($PRFT$). The weighted equation is

$$TDAY = 0.75 \times TMAX + 0.25 \times TMIN$$

and $PRFT$ is approximated in most of the models with the equation

$$PRFT = 1 - T_c \times (TDAY - T_o)^2$$

where T_c is an empirically fit constant and T_o is the optimum temperature for photosynthesis. Values of T_o used in the models for the various crops and are given in Table 3. The values used to decrease $PCARB$ for water deficits ($SWDF1$) and nitrogen deficits ($NDEF1$) are discussed in other chapters on those subjects in this book. The equation to calculate $CARBO$, the daily biomass production uses a law of the limiting concept to reduce $PCARB$;

$$CARBO = PCARB \times \min(PRFT, SWDF1, NDEF1, 1)$$

where min indicates that the minimum value of the list in parenthesis is used. Most of the factors in the minimum list are fractions that vary between 0 and 1, where 1 is non-limiting and 0 is the maximum deficit. This value of $CARBO$, like $PCARB$, is in units of $g\ m^{-2}$. It is converted to a per plant basis when divided by the plant population ($PLANTS$). In this way a single plant growing in competition with other plants is logically represented. The models assume that all plants are the same and equally spaced within rows.

Leaf expansion growth and tillering

The CERES models attempt to evaluate the sink capacity of the above ground biomass to determine if the crop is sink-limited or source-limited for each day.

Estimating the sink-limitation of plant growth during early vegetative stages is critical, especially for obtaining LAI values needed for the biomass production calculations. The appearance of leaves has already been discussed as an important and predictable developmental process. It then remains to calculate the expansion rate of the leaves that have appeared to determine the LAI. Whole plant leaf expansion, however, is the result of leaf growth on the main stem as well as on tillers. Leaf area expansion is first calculated on the main stem and then additions are made to include tiller expansion. Tillering and tiller leaf area expansion have been difficult to simulate accurately, partially because there is usually large spatial variability in field plant tiller numbers and leaf area when making measurements for use in model development and testing.

There is no uniformity in procedures for calculating main stem leaf area expansion growth between the various CERES crop models. A typical one as used in CERES-Sorghum calculates the potential leaf blade area on the main stem ($PLAM$) as a function of the leaf tip number (LN) using a Gompertz equation

$$PLAM = 6000 \times \exp[-10.34 \times \exp(-PLC \times LN)]$$

where PLC is an empirically fitted constant. The units of PLA are cm^2. Daily potential main stem leaf expansion growth rate ($PLAMO$) is calculated as the difference in PLA between two consecutive days obtained using LN on the two days. Actual leaf expansion of the main stem ($PLAGM$) is obtained by multiplying the potential expansion rate by a fraction varying between 0 and 1 related to non-optimal temperature ($TEMF$) or to water ($SWDF2$) or nitrogen ($NDEF2$) deficits. The water and nitrogen deficit factors used for expansion growth are more sensitive than the biomass growth deficit factors and reduce growth as follows:

$$PLAGM = PLAMO \times \min(TEMF, SWDF2, NDEF2)$$

where min indicates the minimum value of the factors in parenthesis used to modify leaf expansion growth.

Other CERES models calculate cumulative leaf area growth with equations other than the Gompertz equation used for sorghum. Improvements in the leaf area calculations of CERES-Maize have been the subject of two papers written in recent years (Carberry et al., 1989; Hodges and Evans, 1992). Their suggested improvements likely indicate the need to include cultivar variations in leaf area in the model.

Tiller appearance is estimated in a similar manner to leaf appearance. Most tillers begin to appear between the appearance of the second and third main stem leaf. However, in crops like maize tillers may never appear. In sorghum, tillers appear in some regions, usually the cooler climates, while not appearing in others. Tiller appearance and tiller leaf expansion growth are highly sensitive to water and nutrient deficits. When the assimilate supply is limited, tillers almost always have less priority for the assimilate. As a result, they discontinue their development. Thus, the challenge to model tiller development and growth for field conditions remains a major one. The CERES models attempt to deal with this highly interactive feedback system of tiller growth using a sink-source comparison. When the source is adequate, a potential tiller leaf expansion rate is defined. When the source is limited, the source defines the expansion rate. In all the CERES models, the tiller leaf area is calculated separately from the main stem in order to use a priority system for assimilate partitioning. This procedure is helpful when plant populations vary because plants that tiller can usually compensate for low populations. Total new leaf area is determined by adding the main stem and tiller leaf areas. Tillers usually stop appearing when the stems begin to grow because of the high assimilate requirement for both stem growth and leaf growth. In most of the tillering crops, some tillers begin to abort after rapid stem expansion growth.

Leaf senescence and leaf area index calculations

In all the CERES models, leaf senescence is linked to plant leaf development. In wheat and barley, leaf senescence occurs when there are more than four green leaves on the main stem or tiller. Thus, only the four most recently developed leaves are maintained green while the others senesce. For the other CERES crops, an attempt is made to define a natural senescence pattern coupled with plant development. This natural pattern is then accelerated if there is a high plant population and LAI where the lower leaves are shaded, thus causing a lack of leaf activity that would result in senescence. Deficits of nitrogen and water also accelerate senescence. After senescence is determined, the green LAI can be calculated from the total plant leaf area (PLA) developed and the senescent leaf area *SENLA* as follows:

$$LAI = (PLA - SENLA) \times PLANTS/10\,000$$

where *PLANTS* is the plant population (plants m^{-2}). The 10 000 converts the units of PLA and *SENLA* from cm^2 to m^2.

Stem growth and storage of mobile assimilates

All plant species in the CERES models have a period in their early development in which leaves are the only above ground sink. After a certain development time, defined in the phenology sub-model, the stems begin to grow and become a major sink for assimilates. In the case of wheat, barley and rice, stem internode elongation does not occur until after floral induction. In maize, sorghum, and millet, stem elongation begins after a well defined developmental stage regardless of whether floral induction has occurred. When the stems begin to elongate internodes, the partitioning pattern of the assimilate changes rapidly. In the models, the assumption is made that while the leaves are expanding, the leaf mass growth and the stem mass growth are proportional to one another and that there is no priority for either growing organ caused by stresses. The definition of the leaf growth rate thus determines the stem growth rate. The proportionality between leaf and stem partitioning gradually changes in most crops, starting with a relatively small fraction for stems when stem elongation begins to be a large fraction when leaf growth is naturally ending.

The stems are assumed to be the primary organ for storage of assimilates to be used during grain filling. The amount of mobile assimilates stored for the various crops was difficult to quantify and had to be approximated from literature sources where stem growth had been evaluated closely around the time of anthesis and during grain filling. The stem weight at the end of active stem elongation for wheat and barley is assumed to be structural stem weight, or the minimum stem weight ($SWMIN$) if all mobile assimilates are removed from the stem. For the other crops, $SWMIN$ is assumed to be a fraction of the stem weight at the end of ear or panicle expansion growth. After stem elongation or ear expansion is complete, assimilates are partitioned to the stem as reserves that may be totally or partly used later for grain filling during the grain filling stage. Because there is no active sink in growing organs except roots near the time of anthesis, it is necessary to constrain the rates in which stored assimilates are deposited in the stem. Although each model has somewhat different approaches to the problem, the wheat model uses the following equation to determine the daily fraction of the total above ground biomass production (PTF) that will be partitioned to the stem for storage:

$$PTF = SWMIN/STMWT \times 0.35 + 0.65$$

where $STMWT$ is the total stem weight. The stem weight increases daily ($CARBO \times PTF$) as long as there are excess carbohydrates. When the grain begins to grow the stored assimilates usually stop accumulating because the grains are a large sink and have priority for all the above ground assimilate production. During grain filling the stored assimilates are used on days when the biomass production is inadequate for the sink controlled grain growth as long as the stem weight does not fall below $SWMIN$.

Root growth

Root growth can potentially occur through all of the plant life cycle (Table 1). There is ordinarily little growth or even net root loss during grain filling. The amount of root growth has been difficult to quantify because of problems in recovery of roots in natural soil environments. Also the loss of root material by natural turnover of root hairs and exudation of assimilates directly out of roots makes the determination and simulation of root growth highly uncertain. The equations used for water uptake of all the DSSAT v3 models use root length density to determine the potential uptake rates. The partitioning part of the model provides a specific amount of mass for the root system each day. That mass must be converted to root length for the uptake calculations. A constant is used for that conversion, although, in reality, experimental evaluations of the root length to mass ratio vary more than an order of magnitude. Thus any calculation made with the combined root length model and the water uptake model must be considered as approximations.

A unique feature of the CERES models is the dynamic nature of their root-shoot partitioning. The models were developed to work under a range of environmental conditions, many of which give rise to variations in root-shoot partitioning. It is generally accepted that the light, nutrient and soil water environment affect the root-shoot partitioning in cereal crops (Brouwer, 1965). The general principles are that when light is lower than normal because of low radiation or high plant to plant competition for light, the tops have priority for assimilates. When soil derived resources of water and nutrients are low, the roots have priority for assimilates. This dynamic feedback system helps plants maintain a balanced system. When light is low, the plants produce more leaf area to enhance their capability for future growth. When nutrients or water are short, the plant top growth is reduced in favor of roots so that the plants have time to develop a larger root system needed for uptake. Also, relatively more assimilates are partitioned to the root system when the plants are in the seedling stage and their younger vegetative phase. During this time the root system is establishing itself for the adequate uptake of nutrients and water. Most cereal plants are sink limited when there is small because they have little competition for light. Thus there is usually enough assimilate for rapid root and top growth during this period.

The definition of minimum root requirement as defined in the model (Table 3) was determined by trial and error. For the most part, experiments in which plant population was variable were chosen to approximate when above ground biomass production was reduced on a per plant basis because of the competition. The time when the competition was evident was assumed to represent the time in which the plant top growth became source limited. In this case, the root partitioning was assumed to be a fixed proportion of the assimilate production (Table 3).

The depth distribution of the root system is accomplished by using the root weighting factors for each soil depth defined in the soil input file. This root weighting factor is a fraction between 0 and 1. It is intended to express the fraction of root growth that will be distributed in each depth increment. The depth of rooting is assumed to increase each day by an amount proportional to the crop development rate. Thus, rooting depth is predicted by multiplying a constant (0.22 cm deg^{-1} day^{-1} for maize) times the thermal time for the day to get the new rooting depth. This rooting depth is used with the weighting factors to constrain the root growth only to depths where roots have penetrated. The root weighting factor from the soil file is multiplied by a soil water factor to obtain the actual root depth distribution. In this way if soils are quite dry, root growth is constrained to zones where the soil water environment is better.

The quantification of this dynamic partitioning system is quite empirical, but the source-sink type models were developed to have this capability. The assumption is made that the root sink is infinite, thus there is no need to model potential root growth. This essentially means that secondary and tertiary roots can develop off primary roots without limit.

Ear and panicle growth

In most cereal crops, when the leaf growth ends the ear or panicle begins to expand rather rapidly. The reproductive part of the plant grows slowly after floral induction, but the sink capacity for assimilates is quite small and can be neglected until leaf growth practically stops. Then the stem and ear or panicle are the active above ground sinks for assimilates. The details of how this is accomplished varies between the crop models. In wheat, barley and maize the pre-anthesis ear growth and stem growth are not separated. In sorghum, millet and rice the panicle growth is separated from stem growth. Growth assessment of the stem and ear or panicle is important during this stage because grain numbers per ear or panicle that will become mature kernels are assumed to be closely related to the ear and stem growth immediately prior to anthesis.

A sink capacity is defined for panicle or ear growth that limits growth when the source is adequate. The sink capacity of the growing reproductive structures is calculated much like the leaf expansion sink capacity, being closely related to the temperature, with constraints from relative plant size and deficiencies in water and nitrogen supply.

Grain numbers determination

Grain numbers per unit area that will make mature kernels are usually the most critical determinant of crop yield. Exact quantification of the factors determining grain numbers are not well understood, but there is qualitative knowledge that plant size relative to optimum size around the time of ear or panicle growth, the biomass production rate and duration during ear or panicle expansion, and good conditions for pollination are the primary essentials. It is well known that there is cultivar variability in final grain weight and grain numbers. To accommodate this variability, a cultivar coefficient G2 is used to convert from the weight of all or part of a plant organ, such as the ear or stem, to grain number.

In CERES-Wheat, the stem weight at anthesis is assumed to be proportional to the grain number, following the concepts of Fischer (1985). His work demonstrated that there was a good correlation between kernel number and incident solar radiation in the 30 days preceding anthesis. Since most of the stem weight is developed during this 30 day period, the CERES-Wheat routines should approximate the form of the relationship developed by Fisher. The relationship employed in the CERES-Wheat model uses a fixed thermal time for stem development which averages about 30 days when the temperature averages about 20°C. Fischer (1985) also reported that the ratio of intercepted daily solar to mean temperature above 4.5°C in the 30 days preceding anthesis had an even better correlation with grain number than incident radiation alone. The CERES routines indirectly take this phenomenon into account.

In CERES-Maize, the average rate of photosynthesis per plant during a period around silking is used along with a grain number cultivar coefficient to determine grain number. Edmeades and Daynard (1979) developed the concepts for this relationship from field data in Canada.

For rice, sorghum and millet, a panicle weight is calculated through the grain filling stage without estimating grain number before grain filling. Grain numbers are approximated from the panicle weight at maturity, using a weight per kernel cultivar coefficient (G3) to estimate grain numbers. This is a more conservative approach to estimating grain numbers and yield, but it is generally accepted that grain numbers are determined just before grain filling begins.

Grain filling

For the crops models that calculate grain numbers before grain filling, the filling rate per kernel is calculated daily using a source-sink-reserves procedure. A sink capacity is calculated based on temperature and a cultivar coefficient (G3) which is the potential daily single kernel growth rate at optimal temperature. The temperature dependent function has a broad optimal such that only unusually cool or hot temperatures cause a reduction in the rate of kernel growth. The source for grain filling is the daily biomass production plus the stored assimilates in the stem. Stored assimilates are exhausted when the stem weight reaches the minimum stem weight *(SWMIN)*. In most instances of grain growth simulation, the grain filling rate near the end of grain filling is reduced because of a low supply of assimilates due to leaf senescence and the depletion of the reserves.

As indicated in Table 1, deficits of water and nutrients do not reduce grain filling rates. Stresses do not alter partitioning and have little influence on moving of assimilates from storage in the stem into the grains. The reduction in yield and grain weight from stresses occurs because of the reduction in assimilate production during grain filling. Final grain yield is the product of the grain numbers per plant, the individual kernel grain weight and the number of plants per unit area. Adjustments are made if there are severe stresses during grain fill to abort some kernels so that some will fill at near normal rates.

Conclusions

A primary contribution that crop simulation can make to crop improvement is the ability to predict the environmental modulation of phenological events with the view of optimizing the use of resources for increased crop production and profitability. The ability to predict phenology, using cultivar specific coefficients, in different environmental circumstances as they express themselves over space and time should provide a more rapid method for determining where cultivars should be adapted and the climate related risks associated with their production. The ability to simulate crop production should also facilitate screening of cultivars to select those best adapted to specific target environments.

Although there will surely be advances made in our ability to simulate cereal crop growth and development, our present knowledge should dictate that for most field trials where cultivars and management factors are being evaluated, it is important that the times of major phenological events be recorded, that a quality weather station be nearby and that the soil properties are known. Our testing of the CERES model in the DSSAT system has provided information and ideas for model improvement. Cultivar specific variations within a species that are known to exist and could be quantified include partitioning related phenomenon such as leaf size, stem dimensions, grain numbers and grain growth rates. Quantitative cultivar variations in response to pest damage, excesses or deficiencies of soil water, and nutrient deficiencies are also needed in cases when those stress related phenomenon exist. For prediction of the crop response to these stresses, however, good simulation models of the pest occurrence and soil related stresses are needed.

References

Ahrens J F, Loomis W E (1963) Floral induction and development in winter wheat. Crop Science 3:463–466.

Alagarswamy G, Ritchie, J T (1991) Phasic development in CERES-Sorghum model. Pages 143–152 In Hodges T (Ed.) Predicting crop phenology. CRC Press, Boca Raton, Florida, USA.

Alocilja E C, Ritchie J T (1988) Upland rice simulation and its use in multicriteria optimization. Research Report Series 01. IBSNAT, Dept. of Agronomy and Soil Science, College of Tropical Agriculture and Human Resources, University of Hawaii, Honolulu, Hawaii, USA.

Alocilja E C, Ritchie J T (1991) A model for the phenology of rice. Pages. 181–189 In Hodges T (ed.) Predicting Crop Phenology. CRC Press, Boca Raton, Florida, USA.

Brouwer, R. (1965) Plant productivity and environment. Science 218:443–338.

Carberry P S, Muchow R C, McCown R L (1989) Testing the CERES-Maize simulation model in a semi-arid tropical environment. Field Crops Research 20:297–315.

Covell S, Ellis R H, Roberts E H, Summerfield R J (1986) The influence of temperature on seed germination rate in grain legumes. Journal of Experimental Botany 37(178):705–715.

Daynard T B (1972) Relationships among black layer formation, grain moisture percentage, and heat unit accumulation in corn. Agronomy Journal 64:716–719.

Edmeades G O, Daynard T B (1979) The relationship between final yield and photosynthesis at flowering in individual maize plants. Canadian Journal of Plant Science 59:585–601.

Fischer R A (1985) Number of kernels in wheat crops and the influence of solar radiation and temperature. Journal of Agricultural Sciences 105:447–461.

Hesketh J, Baker D (1967) Light and carbon assimilation by plant communities. Crop Science 7:285–293.

Hodges T, Evans D W (1992) Leaf emergence and leaf duration related to thermal time calculations in CERES-Maize. Agronomy Journal 84(4):724–730.

Kiniry J R, Ritchie J T, Musser R L, Flint E P, Iwig W C (1983a) The photoperiod sensitive interval in maize. Agronomy Journal 75:687–690.
Kiniry J R, Ritchie J T, Musser R L (1983b). Dynamic nature of the photoperiod response in maize. Agronomy Journal 75:700–703.
Monteith, J L (1977) Climate and the efficiency of crop production in Britain. Philosophical Transactions of the Royal Society London Series B 281:277–294.
Norman J M, Arkebauer T J (1991) Predicting canopy light-use efficiency from leaf characteristics. Pages 125–143 in Hanks R J and Ritchie J T (Eds.) Modeling plant and soil systems. Agronomy 31, American Society of Agronomy, Madison, Wisconsin, USA.
Ong C K, Monteith J L (1985) Response of pearl millet to light and temperature. Field Crops Research 11(2/3):141–160.
Porter J R, Jamieson P D, Wilson D R (1993) Comparison of the wheat simulation models AFRCWHEAT2, CERES-Wheat and SWHEAT for non-limiting conditions of crop growth. Field Crops Research 33:131–157.
Ritchie J T (1991) Wheat phasic development. Pages 31–54 in Hanks R J and Ritchie J T (Eds.) Modeling plant and soil systems. Agronomy Monograph #31, American Society of Agronomy, Madison, Wisconsin, USA.
Ritchie J T (1991) Specifications of the ideal model for predicting crop yields. Pages 97–122 in Muchow R C and Bellamy J A (Eds.) Climatic risk in crop production: models and management for the semi-arid tropics and subtropics. Proceedings International Symposium, St. Lucia, Brisbane, Queensland, Australia. July 2–6, 1990. C.A.B. International, Wallingford, UK.
Ritchie J T (1993) Genetic specific data for crop modeling. Pages 77–93 in Penning de Vries F, Teng P S, Metselaar K (Eds.) Systems Approaches for Agricultural Development. Kluwer Academic Publishers, Dordrecht, the Netherlands.
Ritchie J T, Alagarswamy G (1989) Genetic coefficients for CERES models. Pages 27–34 in Virmani S M, Tandon H L S, Alagarswamy G (Eds.) Modeling the growth and development of sorghum and pearl millet. ICRISAT Research Bulletin no. 12. Patancheru, India.
Ritchie J T, NeSmith D S (1991) Temperature and crop development. Pages 5–29 in Hanks R J and Ritchie J T (Eds.) Modeling plant and soil systems. Agronomy 31, American Society of Agronomy, Madison, Wisconsin, USA.
Summerfield R J, Collinson S T, Ellis R H (1992). Photothermal responses of flowering in rice (Oryza sativa). Annals of Botany 69(2):101–112.
Trione E J, Metzger R J (1970) Wheat and barley vernalization in a precise temperature gradient. Crop Science 10:390–392.
Wang J Y (1960) A critique of the heat unit approach to plant response studies. Ecology 41:785–790.
Warrington I J, Kanemasu E T (1983) Corn growth response to temperature and photoperiod. III. Leaf number. Agronomy Journal 75:762–766.

The CROPGRO model for grain legumes

K.J. BOOTE[1], J.W. JONES[2], G. HOOGENBOOM[3]
and N.B. PICKERING[4]

[1] *Department of Agronomy, University of Florida, Gainesville, Florida 32611, USA*
[2] *Department of Agricultural and Biological Engineering, University of Florida, Gainesville, Florida 32611, USA*
[3] *Department of Biological and Agricultural Engineering, Georgia Station, The University of Georgia, Griffin, Georgia 30223-1797, USA*
[4] *Formerly Post-Doctoral Associate, Department of Agricultural and Biological Engineering, University of Florida, Gainesville, Florida 32611, USA*

Key words: CROPGRO, crop growth model, soybean, peanut, drybean, photosynthesis, evapotranspiration, N balance, N_2-fixation, crop development, growth processes, climatic factors, cultural management.

Abstract
The CROPGRO model is a generic crop model based on the SOYGRO, PNUTGRO, and BEANGRO models. In these earlier crop models, many species attributes were specified within the FORTRAN code. CROPGRO has one set of FORTRAN code and all species attributes related to soybean, peanut, or drybean are input from external 'species' files. As before, there are also cultivar attribute files. The CROPGRO model is a new generation model in several other ways. It computes canopy photosynthesis at hourly time steps using leaf-level photosynthesis parameters and hedge-row light interception calculations. This hedgerow approach gives more realistic response to row spacing and plant density. The hourly leaf-level photosynthesis calculations allow more mechanistic response to climatic factors as well as facilitating model analysis with respect to plant physiological factors. There are several evapotranspiration options including the Priestley-Taylor and FAO-Penman. An important new feature is the inclusion of complete soil-plant N balance, with N uptake and N_2-fixation, as well as N deficiency effects on photosynthetic, vegetative and seed growth processes. The N_2-fixation option also interacts with the modeled carbohydrate dynamics of the plant. CROPGRO has improved phenology prediction based on newly-optimized coefficients, and a more flexible approach that allows crop development during various growth phases to be differentially sensitive to temperature, photoperiod, water deficit, and N stresses. The model has improved graphics and sensitivity analysis options to evaluate management, climate, genotypic, and pest damage factors. Sensitivity of growth processes and seed yield to climatic factors (temperature, CO_2, irradiance, and water supply) and cultural management (planting date and row spacing) are illustrated.

Background of CROPGRO

The background history leading to CROPGRO started with soybean crop model development between 1980 and 1983 at the University of Florida and release of the original SOYGRO V4.2 in 1983 (Wilkerson et al., 1983). In 1982, the IBSNAT project was initiated. As part of this effort, the Ritchie (1985) soil water balance and a preliminary phenology model were added to the model

G.Y. Tsuji et al. *(eds): Understanding Options for Agricultural Production, 99–128*

and SOYGRO V5.0 was released (Wilkerson et al., 1985). Also during this time, the peanut crop growth model was developed based on modification of SOYGRO (Boote et al., 1986). As part of the IBSNAT project, standard input and output formats for climate, soil, and crop data files were developed. Subsequently, SOYGRO V5.4 and PNUTGRO V1.0 were released, with code improvements and compatibility with the IBSNAT standard input-output structure (Boote et al., 1987; Jones et al., 1987). SOYGRO V5.42 and PNUTGRO V1.02 were developed to fit within Version 2.1 of the DSSAT (Decision Support System for Agrotechnology Transfer) (Boote et al., 1989c; IBSNAT, 1989; Jones et al., 1989). These were the last official releases of SOYGRO and PNUTGRO, although unofficial improved versions of these two models with leaf-level, hedgerow canopy photosynthesis were used by the developers (Boote et al., 1991; Boote and Tollenaar, 1994). Starting in 1985, a model for common bean (BEANGRO V1.0) was developed from SOYGRO and PNUTGRO code (Hoogenboom et al., 1990), and later released as BEANGRO V1.01 (Hoogenboom et al., 1994c). BEANGRO was the first of the models to use leaf-level, hedgerow photosynthesis. During the past six years, efforts were initiated to add soil N balance and N uptake features as well as N_2-fixation to SOYGRO. The decision was made to use one set of FORTRAN code for all three legume models, in order to eliminate the need to make parallel changes in code of all three models. Early prototype versions of CROPGRO that simulated all three legumes with added N balance were described by Hoogenboom et al. (1991, 1992, 1993). The approach and model changes resulting in CROPGRO V3.0, released in August 1994, are reviewed in this chapter.

Overview of processes simulated by CROPGRO

CROPGRO is process-oriented and considers crop carbon balance, crop and soil N balance, and soil water balance. In this approach, state variables are the amounts, masses, and numbers of tissues whereas rate variables are the rates of inputs, transformations, and losses from state variable pools. For example, the crop carbon balance includes daily inputs from photosynthesis, conversion and condensation of C into crop tissues, C losses due to abscised parts, and C losses due to growth and maintenance respiration. The carbon balance processes also include leaf area expansion, pod addition, seed addition, shell growth rate, seed growth rate, nodule growth rate, senescence, and carbohydrate mobilization. Addition of pods and seeds and their growth rates actually determine partitioning during the seed filling phase. Prior to the seed growth phase, the growth rates of leaves, stems, and roots are determined by the current partitioning to respective tissue types multiplied by the rate of total growth. Important ancillary processes include rate of leaf appearance, rate of reproductive development, rate of height and width increase, and rate of root depth increase. The crop N balance processes include daily N uptake, N_2-fixation, mobilization from vegetative tissues, rate of N use for new tissue growth, and rate of N loss in abscised parts. Soil water balance processes include infiltration of rainfall and irrigation, soil evaporation, distribution of root water uptake, drainage of water through the root zone, and crop transpiration.

Species differences

With the CROPGRO model, there is one common set of FORTRAN code and all species attributes associated with dry bean, peanut, and soybean are input from species files, plus information contained in ecotype and cultivar files. There are no hardcoded, crop-specific subroutines in CROPGRO; rather, all species or cultivar differences are handled externally through input parameters and relationships described in these three input files. The species file for each crop describes the basic tissue compositions, photosynthetic, respiratory, N-assimilation, partitioning, senescence, phenological, and growth processes, as well as the sensitivity of those processes to environmental factors. Cultivars are described in the cultivar file using 15 traits per cultivar. The ecotype file contains genetic attributes that are descriptive of broad categories of cultivars (determinate versus indeterminate) and change less frequently.

Crop development

Crop development in CROPGRO uses a flexible approach that allows development during various growth phases to be differentially sensitive to temperature, photoperiod, water deficit, and N stresses. There are up to thirteen phases, each having its own unique developmental accumulator starting at a unique prior endpoint growth stage. The physiological time development rate during any one day in a phase is typically a function of temperature, photoperiod, and water deficit. If conditions are optimal, one physiological day is accumulated per calendar day. The number of physiological days required for a phase to be completed is equal to calendar days if temperature, photoperiod, water status are optimal, thus allowing the plant to develop at the maximum rate possible.

The crop development subroutine allows the use of different equation shapes for each function as well as different cardinal temperatures. The species file for each crop defines those equation shapes and cardinal temperatures (base temperature, first optimum, second (highest) optimum, and maximum temperature). It also defines the thirteen phases, lists the starting point for accumulators for each phase, and indicates whether or not a given phase is sensitive to temperature, photoperiod, and water deficit. The threshold accumu-

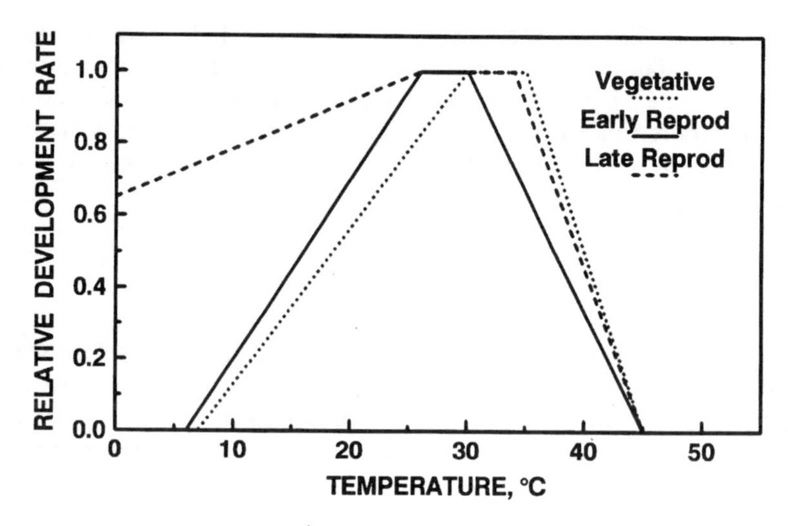

Figure 1. Rate of vegetative node appearance, reproductive development rate before beginning seed stage, and reproductive development rate after beginning seed stage, for soybean in response to temperature.

lation values for the phases (in physiological days) and the critical photoperiod parameters are given in the cultivar and ecotype files.

For soybean in particular, improved phenological parameter values were derived from field data sets by Grimm et al. (1993, 1994). These authors used a simplex optimization algorithm to solve for base temperature, optimum temperature, critical minimum and maximum daylength, and cumulative threshold durations, based on extensive data available on time to flowering, beginning seed, and physiological maturity of many soybean cultivars grown over a wide range of climatic sites. It is important that such data sets for each cultivar come from sites varying widely in temperature and daylength. With field data sets, there is commonly a natural confounding of high temperature with long daylengths.

CROPGRO's flexibility in using different temperature functions before and after flowering is very helpful for correctly predicting reproductive phenology of soybean. Cardinal temperatures for reproductive growth in soybean differ for pre-flowering versus post-flowering phases (particularly after beginning seed). Both of these processes have cardinal temperatures that differ from those for rate of vegetative node (V stage) development (Figure 1). Vegetative node development has a higher base and a higher optimum temperature than does rate of progress toward flowering. Of particular significance is the fact that rate of development after beginning seed had a much lower derived base temperature (−48°C). Of course, soybean does not survive freezing temperatures, but its rate of progress from beginning seed (R5) to physiological maturity (R7) is scarcely slowed by cooler temperatures. This change (use of different temperature functions before and after seedset) has greatly improved our ability to predict maturity of soybean cultivars grown farther north in the USA or in cooler conditions. The single temperature function in SOYGRO V5.42 was adequate for predicting flowering and maturity dates at warm sites, but gave consistently late predictions of maturity when used in cool regions. We have less information on differential phase sensitivity to temperature in the peanut and dry bean models, but we believe the basic concept is correct and the approach gives the needed flexibility.

Crop photosynthesis: daily canopy versus hourly leaf-level hedgerow

The CROPGRO model is considerably improved over SOYGRO V5.42 in its photosynthesis calculations. It now computes hourly values of canopy photosynthesis using a hedge-row light interception model and leaf-level photosynthesis parameters. The hedgerow approach gives more realistic response to row spacing and plant density. Leaf-level photosynthesis in the hourly canopy assimilation loop allows a more mechanistic response to climate change factors as well as facilitating model analysis with respect to plant physiological factors. Users have the option to use either hourly leaf level photosynthesis or the daily canopy photosynthesis approach.

Daily canopy

The daily canopy photosynthesis approach is similar to SOYGRO V5.42, computing daily photosynthesis as a function of daily irradiance for an optimum canopy, then multiplying by factors ranging from 0 to 1 for light interception, temperature, leaf N status, and water deficit. There are also adjustments for CO_2 concentration, specific leaf weight (SLW), row spacing, and variation in cultivar leaf photosynthesis rate. These functions have been improved over those in SOYGRO V5.42. The approach is similar to that of radiation use efficiency (RUE) models, except that the RUE (slope of gross photosynthesis versus irradiance) is not constant and respiration is predicted separately.

Hourly leaf-level, hedgerow photosynthesis

The hourly leaf-level hedgerow photosynthesis and light interception approach is described more completely by Boote and Pickering (1994). Briefly, the procedure calculates absorption of direct and diffuse irradiance by complete or incomplete hedgerow canopies as a function of canopy height, canopy width, leaf area index (LAI), leaf angle, row direction, latitude, day-of-year, and time-of-day. Absorption of photosynthetic photon flux by both sunlit and shaded classes of leaves is computed. Canopy assimilation is the sum of sunlit and shaded leaf photosynthesis rates over their respective LAI classes.

Canopy height and width

In the hourly leaf-level photosynthesis and light interception model, individual plants are assumed to be shaped like ellipsoids. On an hourly basis, the shadow projected by individual plants is computed as a function of canopy height, canopy width, time-of-day, day-of-year, latitude, and row azimuth. The shadow projections account for both row spacing and spacing between plants in the row to allow computation of the fraction of the soil surface shaded by plant canopy at least once. Direct beam light absorption by LAI is thus confined within this fraction.

A new subroutine in CROPGRO predicts canopy height and width as a function of rate of vegetative node formation (V stage) and internode length which in turn is dependent on temperature, irradiance, daylength, water deficit, and V stage. Parameters in the species file for each crop describe the potential internode length relative to node position, as well as relative effects of temperature, irradiance, daylength, and water deficit. This approach works well for peanut and soybean (Boote et al., 1989b; Boote et al., 1991). Figure 2 illustrates concurrent width and height increase as well as V stage for Florunner peanut. When the height and width functions are calibrated to the V stage, the model predicts the correct light interception for both 46- and 92-cm row spacings of a peanut crop grown with the same land-area plant population (Figure 3).

Calculation of hourly weather data

A function from Spitters (1986) is used to compute the hourly distribution of solar radiation and photosynthetic photon flux density (PFD) from daily inputs. Hourly temperatures are calculated from daily maximum and minimum air temperatures using a combined sine-exponential curve versus time-of-day (Parton and Logan, 1981; Kimball and Bellamy, 1986). Hourly total and photosynthetic irradiance are split into direct and diffuse components using a fraction diffuse versus atmospheric transmission algorithm (Erbs et al., 1982; Spitters et al., 1986).

Calculation of radiation absorption

Direct and diffuse beam irradiance absorbed by sunlit and shaded LAI is computed as described by Spitters (1986) with modifications by Boote and Pickering (1994) to address effects of incomplete hedgerow canopies and regular ellipsoidal plant envelopes within hedgerow canopies. Leaf angle distribution follows the procedure proposed by Campbell and Norman (1989). Canopy light absorption is calculated as the sum of direct-beam PFD absorbed by sunlit leaves, plus three categories of diffuse PFD absorbed by sunlit and shaded leaves. The three types of diffuse PFD include direct-beam converted to diffuse within the canopy by scattering processes, skylight originating from

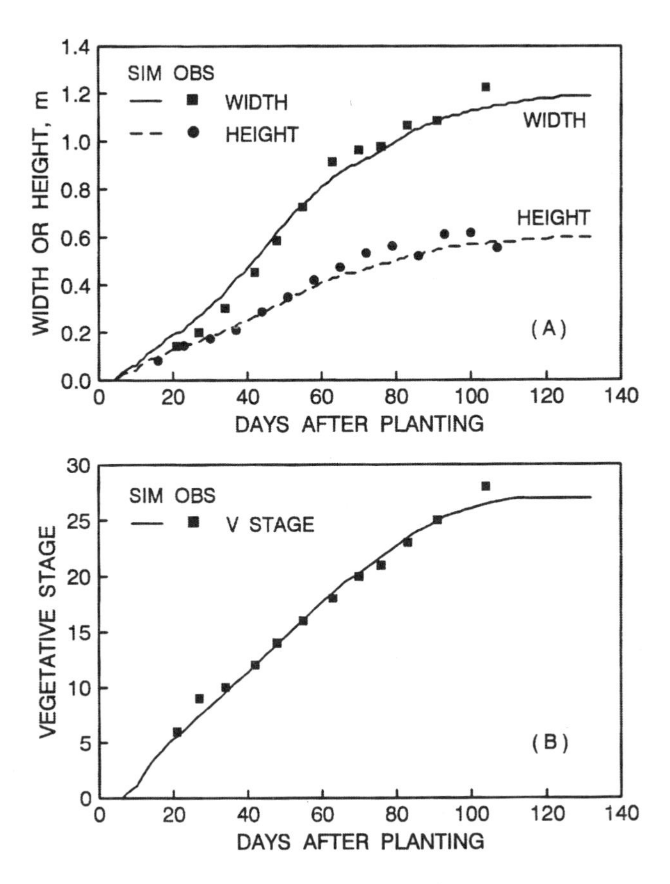

Figure 2. (A) Height and width and (B) V-stage progress of Florunner peanut grown in 1990 at Gainesville, Florida.

a uniformly-lit sky, and PFD reflected from the soil. Absorption of diffuse skylight and absorption of diffuse PFD reflected from the soil is based on an approach from Goudriaan (1977) that uses the path width between plants within rows and between rows in order to compute the view factor (fraction of sky seen). For absorption of all three categories of diffuse PFD, sunlit LAI is assumed to be positioned above the shaded LAI. Boote and Pickering (1994) suggested that canopy photosynthesis models should have multi-layering of LAI for absorption of diffuse light. They concluded that assumption of sunlit LAI over shaded LAI was the most acceptable for single layer sunlit-shaded approach, giving between 3 and 4% underprediction of assimilation compared with a multi-layered approach. Canopy light interception includes not only light absorption but also reflectance of direct and diffuse irradiance from the canopy.

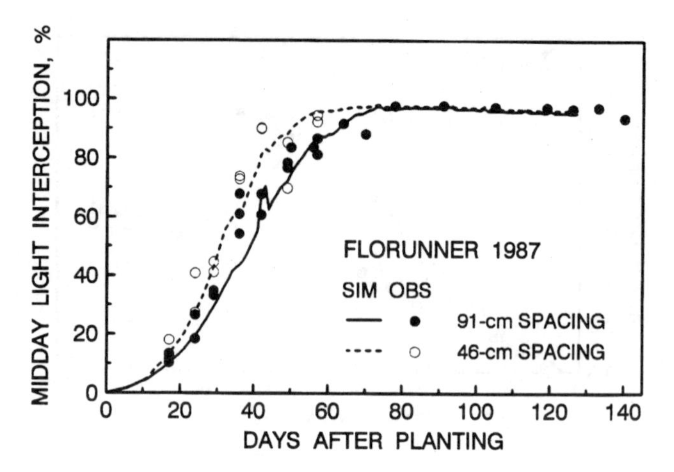

Figure 3. Percent light interception of Florunner peanut grown in 46- and 92-cm row spacings in 1987 at Gainesville, Florida.

Leaf photosynthesis

Sunlit leaves receive direct-beam plus part of the diffuse light, whereas shaded leaves absorb only diffuse irradiance. Photosynthesis of sunlit and shaded leaves is computed with the asymptotic exponential light response equation, where quantum efficiency (Q_E) and light-saturated photosynthesis rate (*Pmax*) variables are functions of CO_2, oxygen, and temperature. The Farquhar and von Caemmerer (1982) equations for the ribulose-1,5-bis-phosphate regeneration-limited region are used to model the basic kinetics of the rubisco enzyme and to compute the efficiency of electron use for CO_2 fixation. This includes temperature effects on specificity factor of rubisco and on CO_2 compensation point in the absence of mitochondrial respiration (see Boote and Pickering (1994) and Pickering et al. (1995) for a full description of simplified algorithms for these computations). Single leaf *Pmax* is modeled as a linear function of specific leaf weight (*SLW*) and as a quadratic function of leaf N concentration. Leaf carbohydrate accumulation is excluded from the *SLW* and N terms used in these functions.

Photosynthetic response to temperature, CO_2 and light

Simulated daily canopy photosynthesis for soybean has a broad temperature optimum, for both the hourly and the daily options (Figure 4A). The simulations were conducted with an 8°C spread between *Tmax* and *Tmin*. With the hourly version, the predicted canopy rate was within 5% of maximum for mean temperature from 21 to 36°C (*Tmax* from 25 to 40°C). The leaf-level hourly version gives a broad temperature optimum for two reasons: (1) in a canopy many leaves operate at low light flux which increases the effect of Q_E,

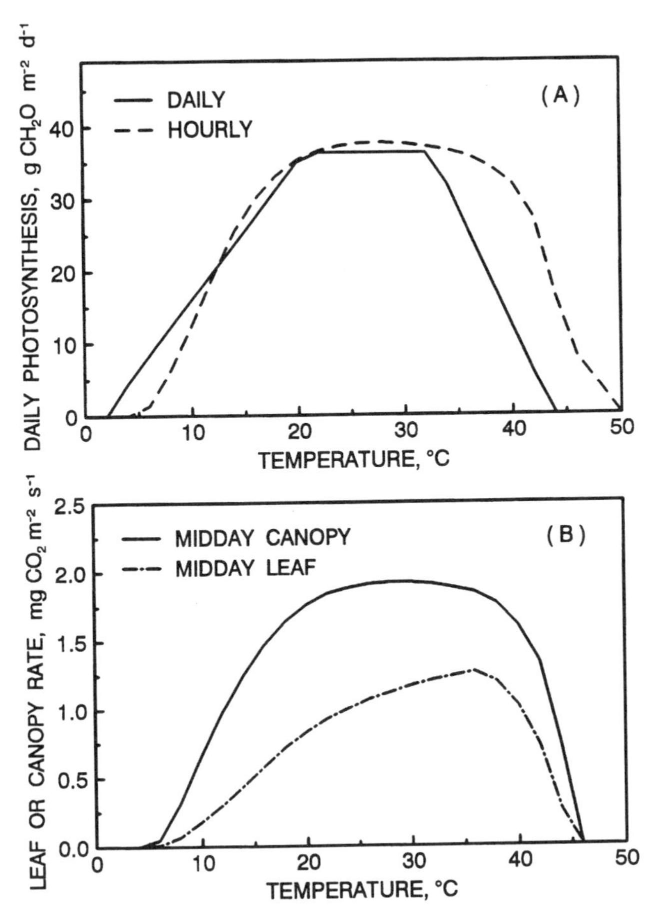

Figure 4. (A) Daily canopy assimilation response to temperature for the daily option and the hourly-leaf option. (B) Midday canopy and leaf assimilation response to temperature (hourly-leaf option). Simulated with 8°C spread between maximum and minimum temperature, and plotted versus mean temperature. Standard simulations for Figures 4–7 used soybean species parameters, 350 ppm CO_2, 30°C maximum and 22°C minimum temperature, 22 MJ m^{-2} d^{-1}, $LAI = 5.4$, simulated at 65 days after planting, except where the given factor was the variable of interest.

and (2) leaf photosynthesis depends on temperature functions influencing both Q_E and *Pmax*, and these two temperature functions operate in opposite directions. Figure 4B illustrates this contrast for instantaneous soybean leaf and canopy photosynthesis simulated at midday with the hourly version. Single leaf rate has an apparent optimum at 36°C, but the resulting midday canopy optimum is very broad. The temperature parameters for the daily version were initially taken from SOYGRO V5.42, based on relative growth rate data of Hofstra and Hesketh (1975), but we decreased the base temperature from 5 to 3°C and decreased the first (lowest) optimum temperature from 24 to 22°C,

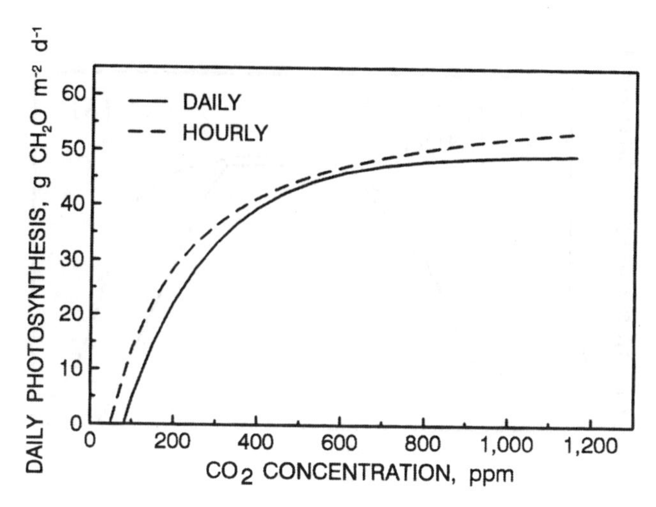

Figure 5. Predicted canopy assimilation response to carbon dioxide concentration (ppm) for the daily and hourly-leaf options, using soybean parameters and conditions described in Figure 4.

based on experience with SOYGRO V5.42 in the northern USA and in France. The hourly photosynthesis model has higher rates at high air temperature than does the daily version; this was substantiated by canopy assimilation measurements on soybean at 40 to 44°C air temperatures in controlled environment studies (Pan et al., 1994). For leaf photosynthesis, the base temperature of 8°C and linear response up to 40°C for rate of electron transport were derived from Harley et al. (1985).

Daily canopy assimilation shows the expected asymptotic response to CO_2 (Figure 5), but the response differs somewhat between the hourly leaf-level model and the daily version. The relative shape of the daily model response to CO_2 was derived by simulating daily canopy photosynthesis for a range of CO_2 levels using a stand-alone version of the hourly leaf-level model and fitting a normalized exponential function to the results. The two responses should be closer but differences are attributed to the following: (1) the daily model is shifted too far to the right because the CO_2 compensation parameter in the released species files is 80 ppm CO_2; CO_2 compensation is temperature-dependent and should really be between 50 and 60 ppm for the 30/22°C temperatures simulated. (2) The hourly model is too responsive at low CO_2, a problem that will be solved if we allow the light-saturated electron transport rate to be limited by low CO_2 concentration as done by Boote and Pickering (1994). Daily assimilation response for soybean is increased 32 and 29% for the daily and hourly models with a doubling of CO_2 from 350 to 700 ppm, and seed yield is increased 33 and 30% by the same doubling.

Response to daily solar irradiance shows a gradual saturation of response beginning at 20 MJ m^{-2} d^{-1}, particularly for the hourly model (Figure 6). The apparent saturation of the hourly version is associated with the algorithm for

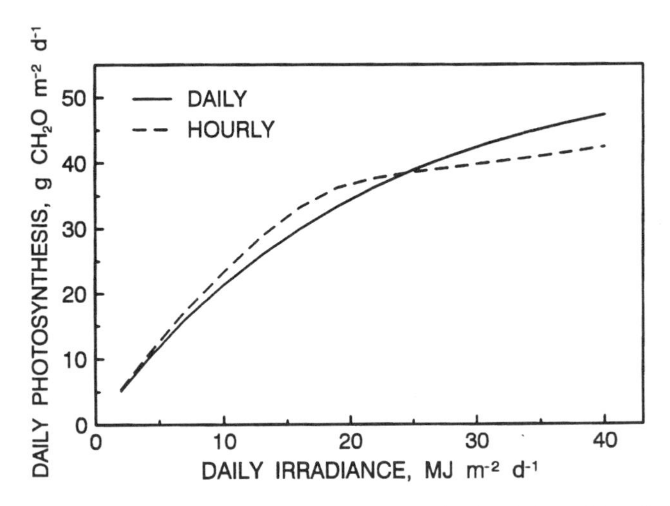

Figure 6. Daily assimilation response to daily irradiance for the daily and hourly-leaf options, using soybean parameters and conditions described in Figure 4.

computing fraction diffuse, the approach used for diffuse light absorption, and the fact that the solar constant remained the same for the day of simulation even through the 'input' irradiance was varied. Fraction diffuse decreased as input irradiance increased, and light was less optimally distributed among sunlit and shaded leaves. Irradiance during mid-summer months is typically between 14 to 24 MJ m^{-2} d^{-1}, and may average 17 to 22 MJ m^{-2} d^{-1} over the growing season.

Response to LAI (Figure 7) is similar for the two assimilation options, although the daily version has an steeper initial response and flattens sooner. The response to LAI was computed at 65 days after planting for a closed canopy. The reason for the lower apparent extinction coefficient for photosynthesis of the hourly versus the daily version is not clear, because the canopy was closed on the date simulated. The calculated light extinction coefficient for the daily version is 0.652 for 0.76 m row spacing.

Evapotranspiration options

The CROPGRO model has several evapotranspiration (*Et*) options that may be selected, including the Priestley-Taylor and the FAO-Penman. The Priestley-Taylor (1972) method is the default option, as in SOYGRO V5.42, and is described by Ritchie (1985). It requires temperature and solar radiation to compute potential evapotranspiration (E_o). When the FAO-Penman function is specified, the FAO version of the Penman *Et* equation, as described by Jensen et al. (1990), is used to compute potential evapotranspiration. Because the Penman method (1948) requires windspeed and humidity data in addition

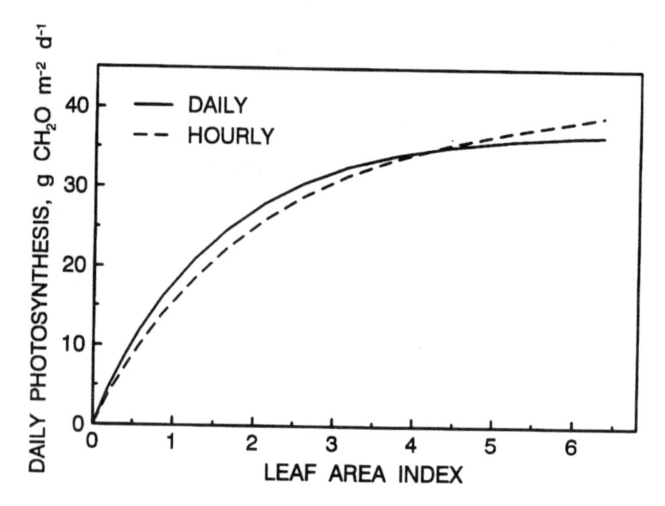

Figure 7. Daily assimilation response to leaf area index for the daily and hourly-leaf options at a closed canopy stage, using soybean parameters and conditions described in Figure 4.

to the other weather data, the weather input subroutine was modified to read daily wind movement (km d^{-1}) and dew point temperature (°C). We recommend using the FAO-Penman option only when these measurements are available. Nevertheless, there are defaults to allow the option to be used even when data are unavailable. A third *Et* option is presently being tested, which would provide for hourly energy balance and prediction of foliage temperatures in conjunction with the hourly leaf-level photosynthesis option. For more information on this approach, see papers by Pickering et al. (1990, 1993, 1995).

Growth conversion costs, growth and maintenance respiration

Growth and maintenance respiration are handled similar to SOYGRO V5.42 (Jones et al., 1989; Wilkerson et al., 1983) and PNUTGRO V1.01 (Boote et al., 1986). Maintenance respiration depends on temperature, crop photosynthesis rate, and current crop biomass (minus oil and protein stored in the seed). The relative sensitivity of maintenance respiration is similar to that reported by McCree (1974). Growth respiration and efficiency of conversion of glucose to plant tissue are computed using the approach of Penning de Vries and van Laar (1982, pp 114–136, also Penning de Vries et al., 1974). This approach requires approximate estimates of tissue composition in six types of compounds: protein, lipid, lignin, carbohydrate-cellulose, organic acids, and minerals (summarized for peanut by Boote et al., (1986) and for soybean by Wilkerson et al., (1983) and Jones et al., (1989)). Because the N balance allows variable N concentration in newly-produced tissue, CROPGRO calculates composition for new tissues prior to computing growth conversion efficiencies for each day's growth. Composition values are entered in the species file, except that protein concentration of newly-synthesized vegetative tissue depends on N balance/deficit. Any change in protein concentration is offset by an opposite change in the carbohydrate-cellulose fraction.

Carbon balance and assimilate partitioning algorithm

The approach for C balance in CROPGRO is modified from SOYGRO V5.42, PNUTGRO V1.02, and BEANGRO 1.0. The algorithm is similar except that N deficit is now allowed to limit vegetative and reproductive growth and to alter partitioning between root and shoot (described in the N balance section). CROPGRO is essentially a source-driven model except under three circumstances: (1) during early V stage development, potential leaf area expansion and sink strength can be limiting, thus reducing growth and photosynthesis; (2) when severe N deficit limits growth of vegetative or seed components, then carbohydrates accumulate in vegetative tissue; and (3) after a full seed-load has been added and if the crop is past a critical seed addition period, then it is possible for reproductive sink strength to be limiting. The latter is rare, but can occur if conditions are adverse during seed addition and improve dramatically after the seed addition period is past.

During vegetative growth, partitioning among leaf, stem, and root is dependent on V-stage progression, water deficit, and N deficit. As reproductive development progresses, new sinks (podwalls, seeds) are formed, and assimilate is increasingly partitioned to these tissues rather than to vegetative growth. At the beginning pod stage, CROPGRO starts adding new classes or cohorts of reproductive sinks on a daily basis. Fruits of each cohort increase in physiological age and pass through slow and then rapid shell-growth phases. Once part way through the rapid shell-growth phase, seeds in each fruit begin their rapid growth phase. Thus, reproductive 'sink' demand consists of many individual reproductive tissues, all of different ages, each having a genetic potential assimilate demand as limited by temperature. The priority order for assimilate use is seeds, podwalls, addition of new pods, and vegetative tissues. Once a full complement of reproductive sites has been added and is growing rapidly, there will be little assimilate remaining for vegetative growth. A genetic limit of fraction partitioning to pods and seeds ($XFRUIT$) is defined to account for the fact that some species or cultivars are indeterminate and continue to grow vegetatively even during rapid seed-filling. With this approach, a fraction ($1.0 - XFRUIT$) of the assimilate is 'reserved' for vegetative growth only.

CROPGRO explicitly defines sink strength only for the reproductive tissues: podwalls and seeds. Leaves, stems (including petioles), roots, and nodules can each be viewed as having one aggregated or lumped class with no age or positional structure. The approach used to compute sink strength of seeds and podwalls has changed slightly with CROPGRO and now requires initial estimates of potential seed size, potential seed filling duration, and threshing percentage under optimum conditions. From these parameters, the potential growth rate per seed and per shell (podwall) under optimum conditions are internally computed. Actual growth rate per seed or shell is subsequently modified by actual temperature, C supply, and N supply. The approach used to compute pod addition rate has also changed and depends on the actual daily assimilate supply divided by actual growth requirement per seed (or shell) and a 'normal' pod addition duration. This approach automatically adjusts for large- or small-seeded cultivars and for long or short seed-fill duration types. Pod addition duration can be estimated experimentally from the physiological time elapsed from addition of the first pod to the point of maximum pod number.

Nitrogen uptake, balance, and fixation with respect to carbon balance

Soil N balance and root N uptake processes are included in CROPGRO, using the code from CERES-Wheat (Godwin et al., 1989). SOYGRO V5.42 does not simulate soil N balance or soil N uptake. The added soil N balance processes (N uptake, mineralization, immobilization, nitrification, denitrification, leaching, etc.) are the same as for the CERES models and are described by Godwin and Jones (1991). Uptake of NO_3^- and NH_4^+ are Michaelis-Menten functions of NO_3^- and NH_4^+ concentration, soil water availability, and the root length density in each layer. The daily crop N uptake cannot be larger than N demand that is computed from the dry matter increase in each organ type multiplied by the maximum N concentration allowed for each tissue type. The N uptake is also allowed to 're-fill' part of N demand contributed from previously-synthesized tissue if those tissues contained less than the maximum N concentration allowed. The upper N concentrations for each tissue are specified in the species file. This upper limit declines as the crop progresses through its reproductive cycle from beginning seed to maturity.

A nitrogen fixation component has been incorporated into CROPGRO. When N uptake is deficient (less than N demand) for growth of new tissues, carbohydrates can be used for N_2-fixation to the extent of the nodule mass and the species-defined nodule specific activity. If nodules are not present or the nodule mass is insufficient, then the assimilates are used for nodule growth at a rate less than the species-defined nodule relative growth rate. When N uptake and N_2-fixation are less than N demand, vegetative tissue will be grown at lower N concentration (see Figure 8A, at 30 to 40 days of age). There is a minimum N concentration, specified in the species file, below which tissue growth rate is progressively decreased to hold a constant minimum N concentration. Under this situation, excess carbohydrates will accumulate in leaves and stems (see Figure 8B, at 30 to 40 days of age). These carbohydrates are subject to a normal mobilization rate. The N_2-fixation rate is also influenced by temperature, soil water deficit, soil aeration (flooding), and plant repro-

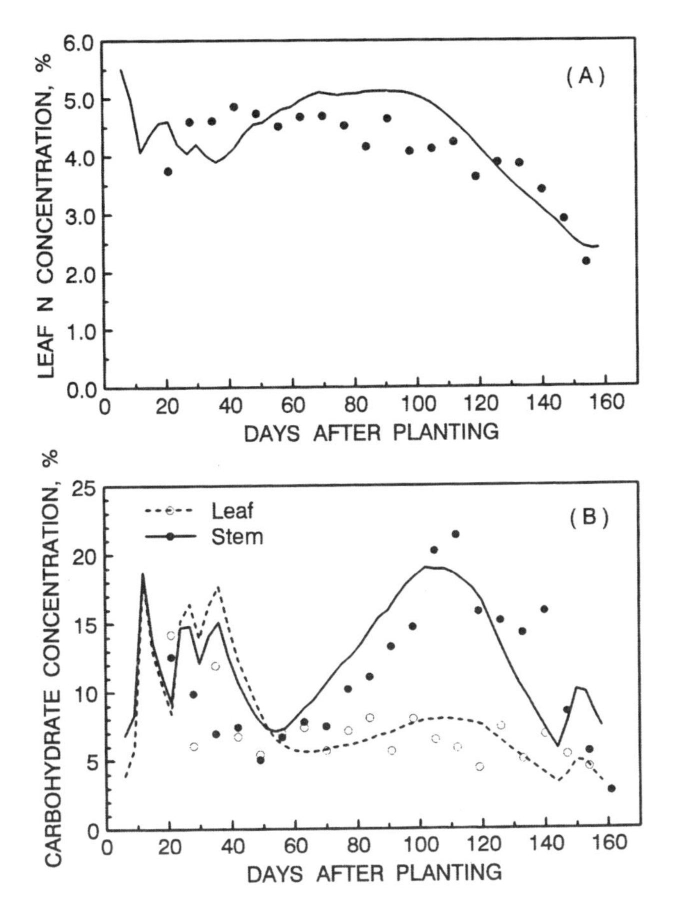

Figure 8. (A) Leaf N concentration and (B) Total available carbohydrate concentration of leaves and stems of Bragg soybean planted 5 May, 1976 at Gainesville, Florida. Lines are CROPGRO simulations and symbols are measured values.

ductive age. Costs for N assimilation (carbohydrate cost for N uptake, N_2-fixation, and reduction to amide form) are subtracted from the available assimilate pool after each process is simulated.

A primary consequence of N deficiency is decreased leaf and canopy photosynthesis as leaf N concentration declines, following a quadratic function from maximum to minimum N concentration. If N is deficient, photosynthesis will be decreased (using the above leaf N effect) and new vegetative tissue will be produced at a lower N concentration. If tissue concentration approaches the lower N limit, as described above, then vegetative growth is reduced and carbohydrates begin to accumulate in vegetative tissues. Below a certain N deficit (if the ratio of N supply to N demand is less than 0.5), assimilate partitioning to roots is increased. Presently, seed growth is presumed to occur at constant N concentration, with growth being limited if either N or carbohydrate supply is deficient. This approach will be changed in a future version of the model to allow seed protein to vary.

Mobilization of nitrogen and carbohydrates, senescence, and maturity

In CROPGRO, carbohydrates accumulate in leaf and stem tissues under several situations: N deficiency that limits growth, insufficient sink during early sink-limited growth or after a full seed load is set, or 'programmed' accumulation during the reproductive transition from flowering until rapid seed growth begins (Figure 8B). During this 'programmed' accumulation, up to 30% of the vegetative growth increment can be allocated to carbohydrate storage in stems (primarily) and leaves (secondarily), ending when a full seed load occurs and no assimilate is left for vegetative growth. Carbohydrate mobilization occurs continuously, but is allowed to accelerate during seed growth (days 100 to 160 in Figure 8B). The simulated increase in carbohydrates after 140 days is caused by insufficient N for seed growth.

Mobilization of protein (C and N) from old to new vegetative tissue occurs slowly during vegetative growth, but mobilization from vegetative tissue occurs up to twice as fast when seeds begin to grow (after 100 days in Figure 8A). The maximum rate of protein mobilization depends on the rate of reproductive development. To the extent that mobilized protein is available, more seed mass can be produced per day than when all amide-N must come from current N assimilation or N_2-fixation. As protein is mobilized, leaf N declines, leaf photosynthesis and maintenance respiration are reduced, and some leaves are abscised. Loss of non-protein leaf and stem mass also occurs when protein mass is mobilized, so leaf area, leaf mass, leaf protein, stem mass, and stem protein all decline.

Senescence of leaves and petioles is dependent on protein mobilization and is enhanced by drought stress (Figure 9A). All three legumes have gradual senescence of foliage and loss of leaf protein as seed fill progresses, thus reducing seed growth rate; however, soybean and dry bean additionally have a grand senescence phase that starts at physiological maturity (R7 stage) and causes almost all remaining leaves to abscise. The grand senescence of soybean and bean foliage causes all seed growth to cease even if seeds have not filled the pods. Peanut, by contrast, is indeterminate and foliage normally remains green to maturity even though growth of individual seeds will cease once they reach the limits of their individual pod cavities (maximum shelling percentage). With peanut, there is an optimum harvest date at which harvestable pod yield will be maximum, i.e., there is a maximum mass of mature pods relative to continued growth of younger pods and abscission of mature pods.

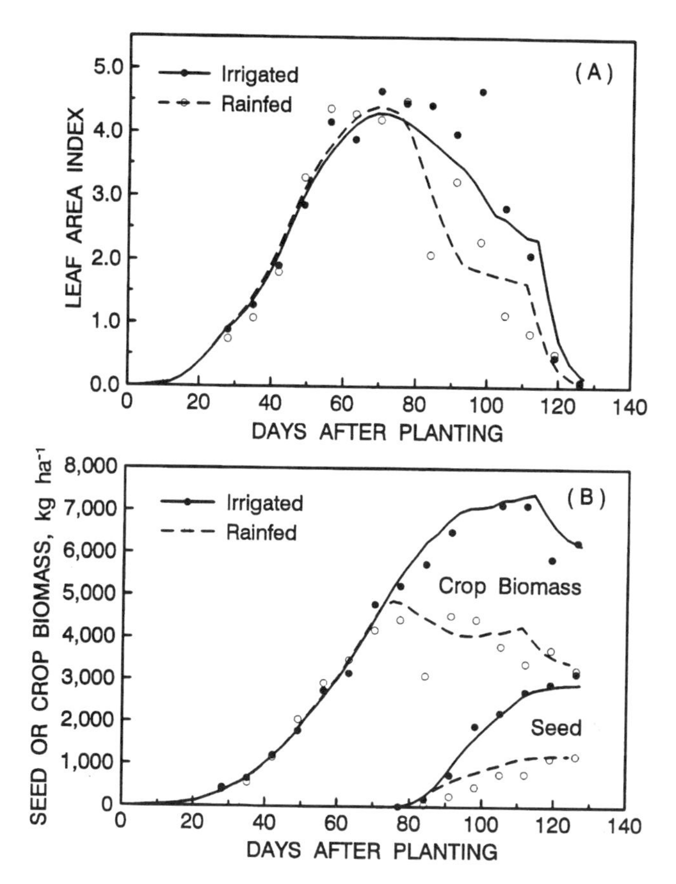

Figure 9. (A) Leaf area index and (B) Seed mass and total crop mass of Bragg soybean planted 15 June, 1978 at Gainesville, Florida, under irrigated versus rainfed conditions. Lines are simulations and symbols indicate field-observed values.

Crop responses to climatic factors

Growth responses to water deficit

Processes sensitive to water deficit include photosynthesis, transpiration, N_2-fixation, leaf area increase (via decreased specific leaf area, SLA), V-stage progress, internode elongation (height and width increase), and partitioning to roots. When root water uptake is unable to meet the transpirational demand of the foliage, then photosynthesis and transpiration are reduced in direct proportion for every decrease in water uptake (direct stomatal coupling). Decreases in V-stage development, SLA, internode elongation, and shift in partitioning to roots begins when the ratio of root water uptake to potential evapotranspiration first falls below 1.5. These processes are thus more sensitive to water supply than is photosynthesis. The N_2-fixation process is decreased as the fraction of available soil water in the topsoil (the top 30 cm) falls below 0.50. Senescence is enhanced (with a delay feature) following days when plant transpirational demand cannot be met.

The 1978 season is a good example of severe drought effects on LAI, biomass accumulation, and seed growth of soybean. A very severe drought (2 cm of rainfall over a 75-day period) increased leaf abscission (Figure 9A) and dramatically lowered dry matter accumulation during the seed filling period (Figure 9B). Despite severe water limitations, the plants managed to add a small pod load and produced about one-third of normal seed yield with seeds of almost normal size. The water stress was so severe that simulated N_2-fixation was reduced to almost zero; however, there was more than enough vegetative N to mobilize to produce normal seed protein composition (results not shown).

Yield response to rainfall and total evapotranspiration (Et)

Simulation runs were carried out using between 10 and 150% of the actual rainfall occurring over 10 weather years, 1978–1987, at Gainesville, for Bragg soybean planted on day 125. Simulations were initialized with the soil profile containing 50% available soil water to a depth of 180 cm. Soybean seed yield increased with increasing rainfall, gradually becoming asymptotic above 700 mm (Figure 10A). Seed yield plotted against total *Et* illustrates an initial curvilinear increase, with a transition toward a linear response (Figure 10B). In the curvilinear part of the relationship (at low rainfall), soil evaporation is initially a large fraction of total *Et*. As rainfall increases, soil evaporation becomes a smaller fraction of *Et* because a larger crop canopy is produced and shades the soil, and more of the water used goes to transpiration in the linear phase.

Vegetative growth responses to temperature

Vegetative processes that are sensitive to temperature include rate of germination and emergence, rate of vegetative node formation, duration of vegetative growth, specific leaf area, photosynthesis, and maintenance respiration. For example, soybean leaf appearance rate has a base temperature of 7°C and increases up to 30°C (Hesketh et al., 1973). Rate of germination-emergence has similar temperature sensitivity (Hesketh et al., 1973; Stucky, 1976). Leaf area expansion (area per unit leaf mass) increases up to 25 or 30°C (Hofstra and Hesketh, 1975; Thomas and Raper, 1978), leaf photosynthesis increases up to 35°C (Harley et al., 1985), and total biomass increases up to 28°C (Baker et al., 1989) or 32°C (Pan et al., 1994).

Temperature dependence of rate of node formation (V-stage progression) is illustrated in Figure 11 for soybean, peanut, and common bean. This temperature function is important during early vegetative growth because rate of node development limits the rate of leaf area growth and canopy coverage. This is especially important because the maximum leaf area increment per successive

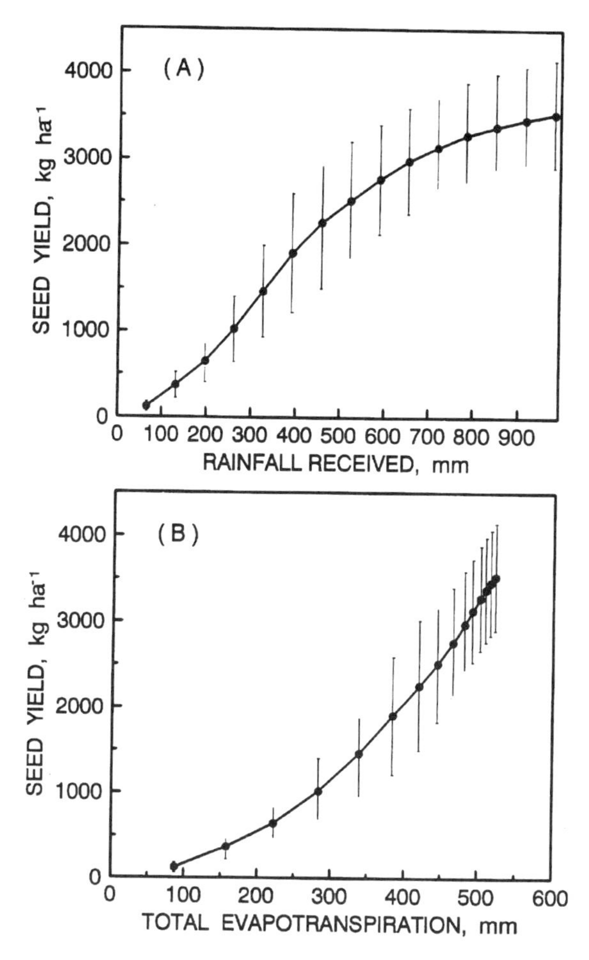

Figure 10. Predicted seed yield of Bragg soybean in response to (A) Rainfall received, and (B) Total evapotranspiration. Simulated with 10 weather years (1978–1987) at Gainesville by modifying rainfall in steps of 10%, from 10% to 150% of actual. Crop was planted on day 125, starting with 50% of available soil water in the soil profile to 180 cm.

V-stage is limited for the first 5 to 7 nodes produced. Temperature also influences relative internode elongation and the SLA (Figure 12A and 12B). In addition to temperature influence on elongation and SLA, temperature has primary effects on decreasing photosynthesis and rate of leaf node appearance. Together they act to reduce rate of early season growth and canopy coverage as illustrated in Figure 13 for soybean grown at mean temperatures of 15, 20, and 25°C. The decreased rate of leaf area expansion is particularly noticeable below 20°C mean temperature because all functions are acting together (decreased V-stage progress, shorter internodes, decreased SLA, decreased

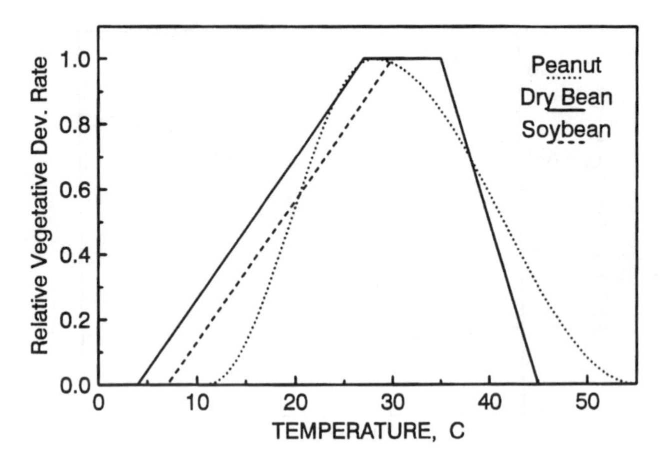

Figure 11. Relative rate of vegetative node appearance (V-stage progression) as a function of temperature for dry bean, peanut, and soybean.

photosynthesis, and decreased N-fixation). The resultant function can be tested against various controlled environment temperature studies, because CROPGRO's hedgerow approach now allows us to simulate single plants as well as hedgerows.

Reproductive growth responses to temperature

To predict seed yield response to future global climate change factors or to unusually cool or hot weather stresses, it is important to describe properly temperature effects on the duration of seed growth phase, seed growth rate, pod addition, and partitioning/pod abortion. Temperature effects on duration were described and shown in Figure 1. Figure 14A illustrates the relative shape of single seed growth rate versus temperature for the three legumes. The seed growth rate function for soybean has its optimum at 23.5°C based on the data of Egli and Wardlaw (1980), although we extended the lower optimum to 21°C. For soybean, final mass per seed and individual seed growth rate decline as temperature exceeds 23°C (Egli and Wardlaw, 1980; Baker et al., 1989; Pan et al., 1994). The same seed growth rate function works well for peanut (Cox, 1979). There are no data on single-seed growth rate versus temperature for dry bean, although this species is believed to be more sensitive to elevated temperature than soybean.

The relative shape of pod addition rate versus temperature (Figure 14B) for soybean is presumed similar to the single seed growth rate function, except that the upper optimum temperature is extended to 26.5°C, and pod addition is reduced at temperatures below 21.5°C to a cut-off of zero podset below 14°C. Several studies indicate that soybean does not form pods when night temperature is 14°C or lower (Thomas and Raper, 1978) or even 18°C and lower

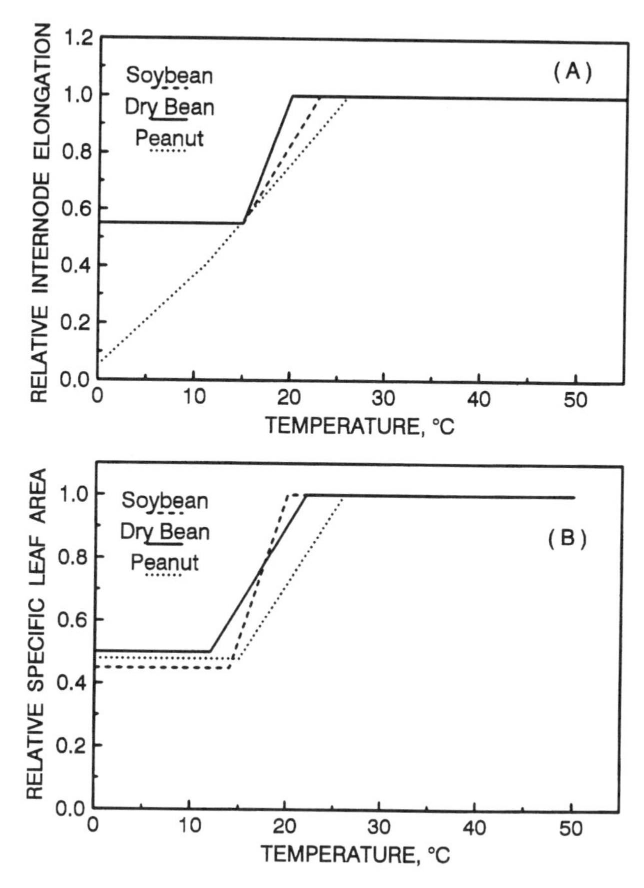

Figure 12. Relative effect of temperature on (A) Internode elongation and (B) Specific leaf area for soybean, dry bean, and peanut.

(Hesketh et al., 1973; Hume and Jackson, 1981), although the response is cultivar-dependent (Lawn and Hume, 1985). There is an offsetting effect of increasing day temperature, but the night temperature is more important (Lawn and Hume, 1985). According to Thomas and Raper (1977), a 18/14°C day/night regime caused short, fused, misshaped pods without seeds. The pod addition rate function for peanut is similar, except that the pod addition rate declines to zero as temperature drops from 21°C to 17°C, based on data of Campbell (1980). The pod addition function for dry bean is assumed to have its entire response curve shifted to cooler temperatures.

At higher than optimum temperature, these three grain legumes exhibit poor reproductive fruit formation, extended vegetative growth, and poor partitioning. Specific causes may be associated with increased production of vegetative primordia and expression of more vegetative sites, high temperature-induced delay in

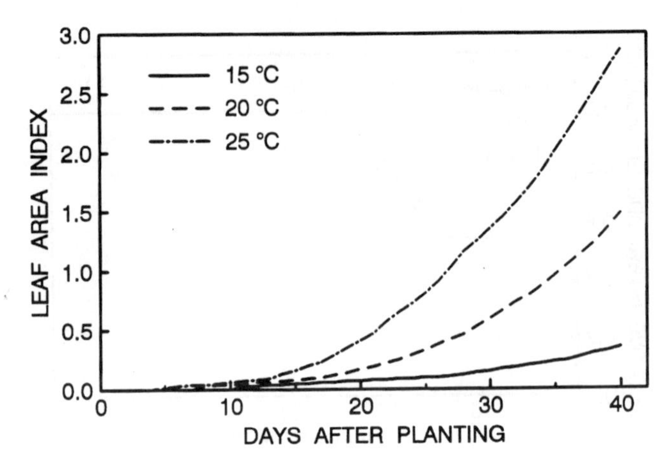

Figure 13. Predicted leaf area index over time for soybean grown at 15, 20, or 25°C mean temperature (20/10, 25/15, and 30/20°C max/min temperature, respectively).

onset of reproductive sites, and increased failure of successful fertilization of reproductive sites. We have thus included a 'partitioning' limit function (Figure 14C) that decreases the maximum fraction partitioning to pods as temperature exceeds 35, 33, and 28°C in soybean, peanut, and dry bean, respectively. We have since confirmed this general phenomenon with experimental data on soybean (Boote et al., 1994; Pan et al., 1994). Harvest index declines progressively as temperature increases above the optimum for soybean (Baker et al., 1989; Pan et al., 1994) and peanut (G. Hammer, 1995).

Seed yield response to temperature

Final predicted seed yield is the integrated result of all model processes and temperature influences on processes of vegetative development, reproductive development, photosynthesis, respiration, N_2-fixation, vegetative growth, and seed growth. The response in Figure 15 illustrates that predicted optimal seed yield for dry bean, soybean, and peanut occurs at 20, 22, and 24°C, respectively. These responses reflect optimal conditions for seed growth processes and nearly optimal conditions for photosynthesis. They do not reflect conditions optimal for either the shortest or longest crop life cycle duration. At temperatures optimal for yield, the life cycle durations are relatively short, and only 8 to 10 days longer than the shortest durations possible that occur at about 4°C higher mean temperatures. Long life cycle durations are predicted at quite cool (less than 20, 22, or 24°C for bean, soybean and peanut, respectively) or quite warm temperatures (above 30 or 32°C for soybean and peanut), yet yield is not enhanced because of non-optimal conditions for seed growth, pod addition, and photosynthesis. Soybean yield response to temperature is similar to that predicted by SOYGRO V5.42 (Boote et al., 1989a), in part because many of the temperature relationships are similar.

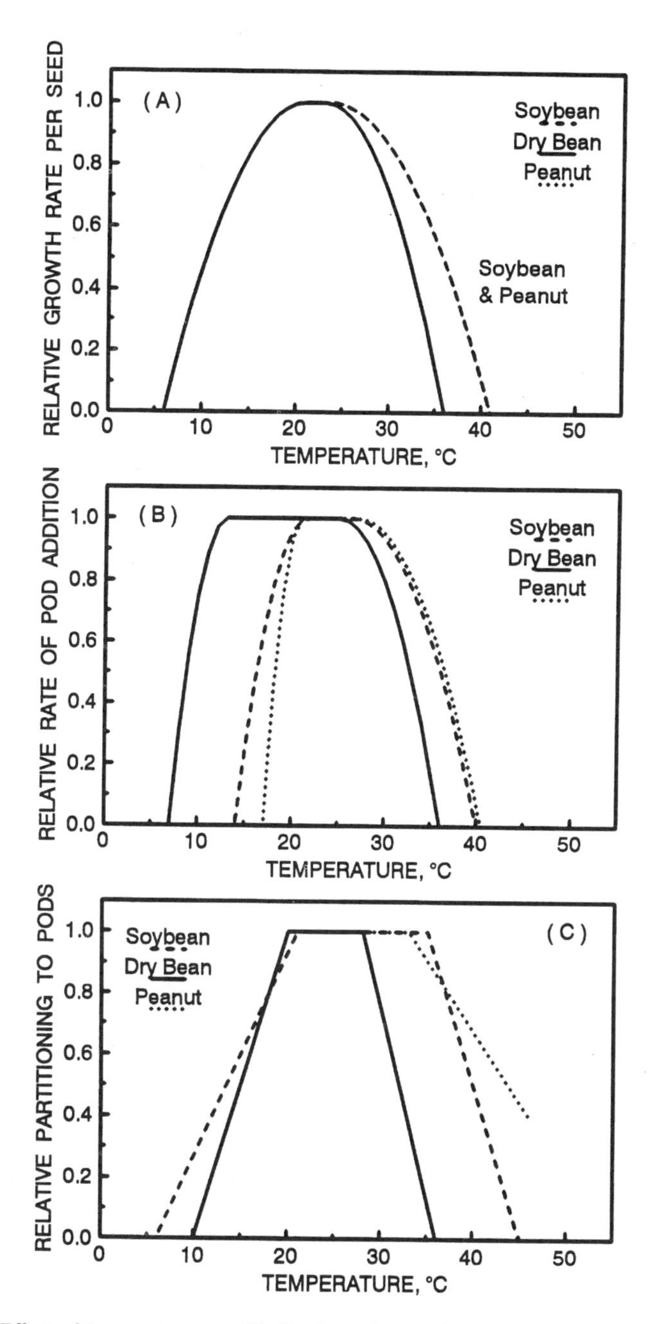

Figure 14. Effect of temperature on (A) Single seed growth rate, (B) Pod addition rate, and (c) Partitioning limit for soybean, dry bean, and peanut.

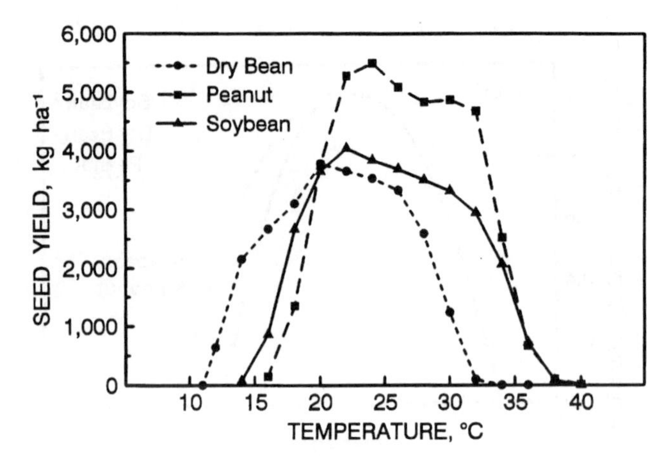

Figure 15. Predicted seed yield response to temperature for dry bean, peanut, and soybean using the daily photosynthesis option. Simulations assume automatic, optimal irrigation.

The decline in seed yield at higher than optimal temperature is consistent with recent experimental data (Pan et al., 1994).

Yield response to management factors

Response to planting date

One of the advantages of crop growth models is the ability to simulate yield response to management conditions over long-term historical weather records to determine the best management conditions such as planting date, row spacing, plant population, irrigation, and cultivar choice. For example, the optimal planting date response was simulated for soybean, peanut, and dry bean planted at Gainesville, Florida on a Millhopper fine sand soil, with and without irrigation (Figure 16). The optimal planting date for soybean, based on 20 years of weather data, is late April to early May (days 110–125), and the acceptable planting window was April, May, and early June (days 95–155). The optimal planting date for soybean was the same under either irrigated or rainfed conditions. It is encouraging to note that this is consistent with extension service recommendations and research reports for soybean in Florida. The optimum period is the result primarily of the optimal delaying effect of daylength (early or late planting produced early flowering and short plants) and optimal temperature conditions for growth and development processes.

Peanut, by contrast, has a much broader optimal planting date caused partly by its lack of photoperiod-sensitivity and greater drought tolerance. High peanut yields were predicted for planting from early March to mid-May (days 65–140), dates that take advantage of Florunner peanut's long life cycle and the higher radiation months. Planting dates after mid-July (day 200) resulted

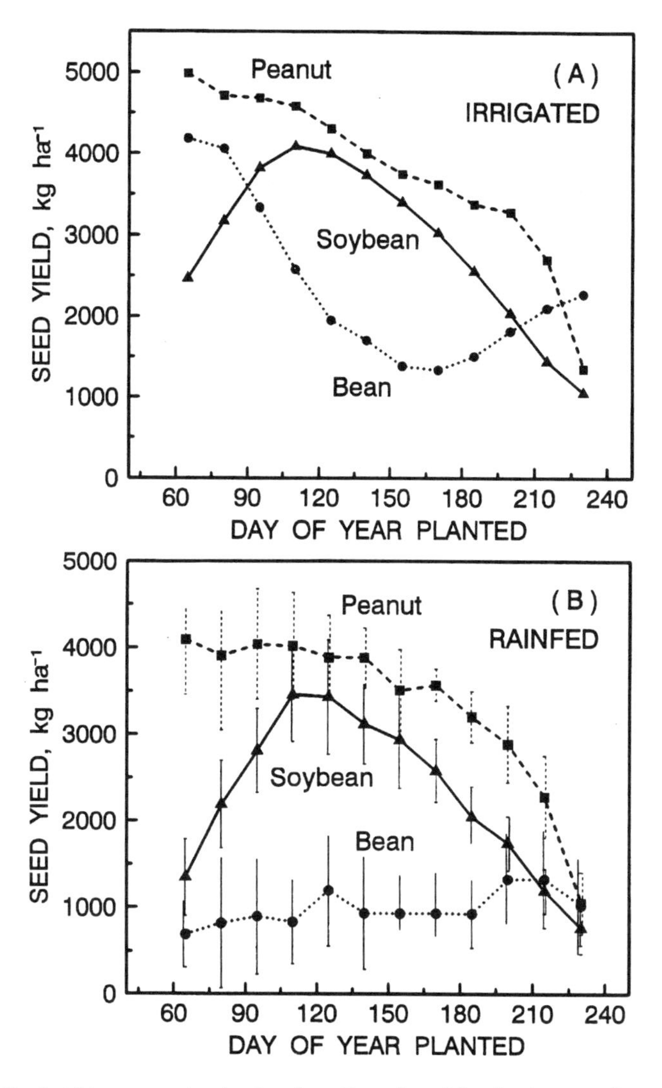

Figure 16. Seed yield response to planting date (day of year) for dry bean, soybean, and peanut planted at Gainesville, Florida, on Millhopper fine sand soil: (A) With optimal irrigation, and (B) Under rainfed conditions. Results are averaged over 20 years of simulated weather. Vertical bars reflect plus or minus one standard deviation under rainfed conditions.

in much lower yields and increased probability of freeze damage prior to peanut maturity.

Predicted dry bean yields showed a complex response to planting date that differed under rainfed or irrigated conditions, for several reasons: (1) temperatures of mid-summer in Florida are sufficiently high to decrease podset,

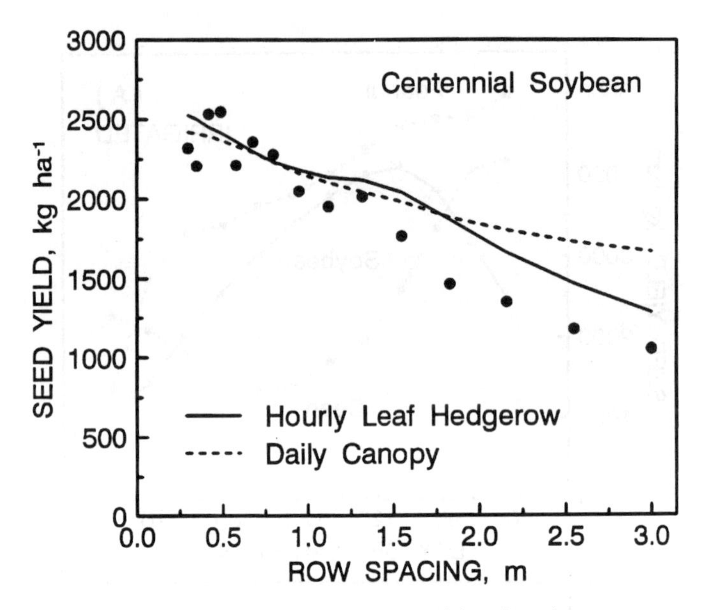

Figure 17. Simulated and observed seed yield of Centennial soybean in response to row spacing at Gainesville, Florida, in 1987, using the daily canopy and hourly leaf hedgerow photosynthesis options (original data from Hiebsch et al., 1990).

partitioning, and seed growth of bean; (2) dry bean is more severely affected by drought because of its smaller rooting depth and shorter life cycle, compared with soybean and peanut. Under full irrigation, early spring plantings of dry bean yielded best, but early summer plantings yielded poorly because of elevated temperature during pod-filling, with late summer plantings yielding somewhat better. Under rainfed conditions, plantings prior to day 125 yielded poorly because of low spring rainfall (compare irrigated and rainfed dry bean yields in Figure 16A and 16B); early summer plantings perform poorly because of elevated temperature. Predicted seed yields of 900 to 1300 kg ha^{-1} were achieved for rainfed plantings from day 125 to as late as day 230.

Response to row spacing

Soybean yield response to row spacing is shown in Figure 17 for an actual data set in which the cultivar Centennial was grown at row spacings from 0.3 to 3.0 m (Hiebsch et al., 1990). Both the daily and the hourly hedgerow photosynthesis options predict decreasing yield with wider row spacing; however, the hedgerow version gives a better prediction at quite wide row spacings. It appears that both versions do not predict sufficient drought stress at wide row spacings because roots have access to the total soil volume.

Table 1. Predicted soybean seed yield for six maturity groups simulated at Ames, Iowa, Urbana, Illinois, and Hoytville, Ohio, over three weather years (1985, 1986, and 1987) for planting on 2 May in 0.76 m rows at 30 plants per m^2. Rainfed in Iowa and Illinois and irrigated in Ohio.

Maturity group (MG)	R7 maturity date				Seed yield			
	Iowa	Illinois	Ohio	Mean	Iowa	Illinois	Ohio	Mean
	Day-month				kg ha^{-1}			
0	8 AUG	4 AUG	14 AUG	9 AUG	2214	2789	2464	2489
1	18 AUG	13 AUG	22 AUG	18 AUG	2542	3159	2914	2872
2	31 AUG	25 AUG	3 SEP	30 AUG	3128	3531	3157	3272
3	14 SEP	6 SEP	16 SEP	12 SEP	3465	3563	3017	3348
4	28 SEP	19 SEP	29 SEP	25 SEP	2871	3296	2741	2969
5	11 OCT	1 OCT	12 OCT	8 OCT	2149	2746	2087	2327

Hypothetical *MG 4* and *MG 5* suffered killing freeze prior to R7 in 1985 in Illinois; *MG 5* suffered killing freeze prior to R7 in 1987 in Illinois and Ohio.

Optimum cultivar selection for a given location

CROPGRO soybean has generic maturity group (*MG*) coefficients for *MG* 000 through 10. We evaluated the optimal soybean *MG* to plant at three sites in the Midwestern USA: Ames, Iowa, Urbana, Illinois, and Hoytville, Ohio, using actual weather years (1985, 1986, and 1987) from these sites, with appropriate soils data (Table 1). Six hypothetical *MG cultivars* (0, 1, 2, 3, 4, *and* 5) *were sown on* 2 *May at each site in* 0.76 *m row spacing at* 30 *plants* m^{-2}. The simulations indicated that optimal yield was predicted with *MG 2* and *MG 3*, a finding consistent with known practice at those three sites. Early MGs (0 and 1) yielded less because they did not utilize the whole growing season (predicted maturity occurred on 9 August and 18 August for *MG 0* and *MG 1*). The MGs 4 and 5 were later maturing and risked cool or freezing weather, particularly *MG 5* which yielded much less.

Similar analysis was conducted for MGs 1 to 10 planted 5 May at Gainesville, Florida, in 0.91 m rows at 30 plants m^{-2} without irrigation, using 20 years of simulated weather (results not shown). At this site, *MG 6* was the earliest to yield well (3030 kg ha^{-1}), with progressively lower yields for earlier MGs (as low as 1070 kg ha^{-1} for *MG 1*). MGs 7 to 10 gave the highest predicted yields in the range of 3280 to 3370 kg ha^{-1}. Predicted maturity for *MG 6* was 15 September, whereas MGs 7, 8, 9, and 10 were predicted to reach maturity on 28 September, 9 October, 21 October, and 27 October, respectively.

Conclusions

The CROPGRO model is a considerable improvement over the previous SOYGRO, PNUTGRO, and BEANGRO crop models in several ways. The code includes leaf-level photosynthesis, hedgerow canopy light interception, soil N balance, N uptake, N_2-fixation, added evapotranspiration methods, enhanced crop development routines, and pest-coupling approaches. The code is more modular and will simulate all three legumes, having species characteristics in a read-in 'species' file. The models have also benefitted from experience with additional data sets obtained over the past six years. The models are released as part of the DSSAT V3.0 decision support software which has a number of new features that facilitate a greater degree of sensitivity analysis and allows crop rotations with the legume and cereal models to be simulated (Hoogenboom et al., 1994a, 1994b; Tsuji et al., 1994). Improvements and updated releases of the CROPGRO model are planned every two to three years, including adaptation for other crops such as tomato, pasture, and sugarcane.

References

Baker J T, Allen L H Jr, Boote K J, Jones P H, Jones J W (1989) Response of soybean to air temperature and CO_2 concentration. Crop Science 29:98–105.

Boote K J, Jones J W, Mishoe J W, Wilkerson G G (1986) Modeling growth and yield of groundnut. Pages 243–254 in Agrometeorology of Groundnut: Proceedings of an International Symposium, ICRISAT Sahelian Center, Niamey, Niger. ICRISAT, Patancheru, AP 502 324, India.

Boote K J, Jones J W, Hoogenboom G, Wilkerson G G, Jagtap S S (1987) PNUTGRO V1.0, Peanut crop growth simulation model, user's guide. Florida Agricultural Experiment Station Journal No. 8420. University of Florida, Gainesville, Florida, USA.

Boote K J, Jones J W, Hoogenboom G (1989a) Simulating growth and yield response of soybean to temperature and photoperiod. Pages 273–278 in Proceedings of the World Soybean Research Conference IV, 5–9 March 1989, Buenos Aires, Argentina.

Boote K J, Jones J W, Hoogenboom G (1989b) Simulating crop growth and photosynthesis response to row spacing. Agronomy Abstracts:11.

Boote K J, Jones J W, Hoogenboom G, Wilkerson G G, Jagtap S S (1989c) PNUTGRO V1.02, Peanut crop growth simulation model, user's guide. Florida Agricultural Experiment Station Journal No. 8420. University of Florida, Gainesville, Florida, USA.

Boote K J, Jones J W, Singh P (1991) Modeling growth and yield of groundnut–state of the art. Pages 331–343 in Groundnut: A global perspective: Proceedings of an International Workshop, 25–29 Nov. 1991, ICRISAT Center, India. ICRISAT, Patancheru, AP 502 324, India.

Boote K J, Pickering N B (1994) Modeling photosynthesis of row crop canopies. HortScience 29:1423–1434.

Boote K J, Tollenaar M (1994) Modeling genetic yield potential. Pages 533–565 in Boote K J, Bennett J M, Sinclair T R, Paulsen G M (Eds.) Physiology and determination of crop yield. ASA-CSSA-SSSA, Madison, Wisconsin, USA.

Boote K J, Kirkham M B, Allen L H Jr, Baker J T (1994) Effects of temperature, light, and elevated CO_2 on assimilate allocation. Agronomy Abstracts:150.

Campbell G S, Norman J M (1989) The description and measurement of plant canopy structure. Pages 1–19 in Russell G, Marshall B, Jarvis P G (Eds.) Plant canopies: their growth, form and function. Cambridge University Press, Cambridge, UK.

Campbell I S (1980) Growth and development of Florunner peanuts as affected by temperature. Ph.D. dissertation, University of Florida, Gainesville, Florida, USA.

Cox F R (1979) Effect of temperature treatment on peanut vegetative and fruit growth. Peanut Science 6:14–17.

Egli D B, and Wardlaw I F (1980) Temperature response of seed growth characteristics of soybeans. Agronomy Journal 72:560–564.

Erbs D G, Klein S A, Duffie J A (1982) Estimation of the diffuse radiation fraction for hourly, daily and monthly-average global radiation. Solar Energy 28:293–302.

Farquhar G D, von Caemmerer S (1982) Modelling of photosynthetic response to environment. Pages 549–587 in Lange O L, Nobel P S, Osmond C B, Ziegler H (Eds.) Encyclopedia of plant physiology, NS vol. 12B: Physiological plant ecology II. Springer-Verlag, Berlin, Germany.

Godwin D C, Jones C A (1991) Nitrogen dynamics in soil-plant systems. Pages 287–321 in Hanks J, Ritchie J T (Eds.) Modeling soil and plant systems. ASA Monograph 31, American Society of Agronomy, Madison, Wisconsin, USA.
Godwin D C, Ritchie J T, Singh U, Hunt L A (1989) A user's guide to CERES Wheat: V2.10. International Fertilizer Development Center, Muscle Shoals, Alabama, USA.
Goudriaan J (1977) Crop micrometeorology: a simulation study. Centre for Agricultural Publishing and Documentation, Wageningen, The Netherlands.
Grimm S S, Jones J W, Boote K J, Hesketh J D (1993) Parameter estimation for predicting flowering date of soybean cultivars. Crop Science 33:137–144.
Grimm S S, Jones J W, Boote K J, Herzog D C (1994) Modeling the occurrence of reproductive stages after flowering for four soybean cultivars. Agronomy Journal 86:31–38.
Hammer G L, Sinclair T R, Boote K J, Wright G C, Meinke H, Bell M J. 1995. A peanut simulation model: I. Model development and testing. Agronomy Journal 87: 1085–1093.
Harley P C, Weber J A, Gates D M (1985) Interactive effects of light, leaf temperature, CO_2 and O_2 on photosynthesis in soybean. Planta 165:249–263.
Hesketh J D, Myhre D L, Willey C R (1973) Temperature control of time intervals between vegetative and reproductive events in soybeans. Crop Science 13:250–254.
Hiebsch C, Kembolo Salumu K U, Gardner F P, Boote K J (1990) Soybean canopy structure, light interception, and yield as influenced by plant height, row spacing, and row orientation. Soil and Crop Science Society of Florida Proceedings 49:117–124.
Hofstra G, Hesketh J D (1975) The effects of temperature and CO_2 enrichment on photosynthesis in soybean. Pages 71–80 in Marcelle R (Ed.) Environmental and biological control of photosynthesis. Dr. W Junk, b.v. Publishers, The Hague, The Netherlands.
Hoogenboom G, White J W, Jones J W, Boote K J (1990). BEANGRO V1.0: Dry bean crop growth simulation model, user's guide. Florida Agricultural Experiment Station Journal No. N–00379. University of Florida, Gainesville, Florida, USA.
Hoogenboom G, Jones J W, Boote K J (1991) Predicting growth and development of grain legumes with a generic model. Paper No. 91–4501. American Society of Agricultural Engineers, St. Joseph, Michigan, USA.
Hoogenboom G, Jones J W, Boote K J (1992) Modeling growth, development, and yield of grain legumes using SOYGRO, PNUTGRO, and BEANGRO: A review. Transactions of the American Society of Agricultural Engineers 35:2043–2056.
Hoogenboom G, Jones J W, Boote K J, Bowen W T, Pickering N B, Batchelor W D (1993) Advancement in modeling grain legume crops. Paper No. 93–4511. American Society of Agricultural Engineers, St. Joseph, Michigan, USA.
Hoogenboom G, Jones J W, Hunt L A, Thornton P K, Tsuji G Y (1994a) An integrated decision support system for crop model applications. Paper No. 94–3025. American Society of Agricultural Engineers, St. Joseph, Michigan, USA.
Hoogenboom G, Jones J W, Wilkens P W, Batchelor W D, Bowen W T, Hunt L A, Pickering N B, Singh U, Godwin D C, Baer B, Boote K J, Ritchie J T, White J W (1994b) Crop models. Pages 95–244 in Tsuji G Y, Uehara G, Balas S (Eds.) DSSAT Version 3, Volume 2. University of Hawaii, Honolulu, Hawaii, USA.
Hoogenboom G, White J W, Jones J W, Boote K J (1994c) BEANGRO, a process-oriented dry bean model with a versatile user interface. Agronomy Journal 86:182–190.
Hume D J, Jackson A K H (1981) Pod formation in soybean at low temperatures. Crop Science 21:933–937.
International Benchmark Sites Network for Agrotechnology Transfer Project (1989) Decision support system for agrotechnology transfer Version 2.1 (DSSAT V2.1). Department of Agronomy and Soil Science, College of Tropical Agriculture and Human Resources, University of Hawaii, Honolulu, Hawaii, USA.
Jensen M E, Burman R D, Allen R G (Eds.) (1990) Evapo-transpiration and irrigation water requirements: a manual. American Society of Civil Engineers, New York, USA.
Jones J W, Boote K J, Hoogenboom G, Jagtap S S, Wilkerson G G (1989) SOYGRO V5.42, Soybean crop growth simulation model, user's guide. Florida Agricultural Experiment Station Journal No. 8304. University of Florida, Gainesville, Florida, USA.
Jones J W, Boote K J, Jagtap S S, Hoogenboom G, Wilkerson G G (1987) SOYGRO V5.4, Soybean crop growth model, user's guide. Florida Agricultural Experiment Station Journal No. 8304. University of Florida, Gainesville, Florida, USA.

Kimball B A, Bellamy L A (1986) Generation of diurnal solar radiation, temperature, and humidity patterns. Energy in Agriculture 5:185–197.

Lawn R J, Hume D J (1985) Response of tropical and temperate soybean genotypes to temperature during early reproductive growth. Crop Science 25:137–142.

McCree K J (1974) Equations for the rate of dark respiration of white clover and grain sorghum as functions of dry weight, photosynthetic rate, and temperature. Crop Science 14:509–514.

Pan D, Boote K J, Baker J T, Allen L H Jr, Pickering N B (1994) Effects of elevated temperature and CO_2 on soybean growth, yield, and photosynthesis. Agronomy Abstracts:150.

Parton W J, Logan J A (1981) A model for diurnal variation in soil and air temperature. Agricultural and Forest Meteorology 23:205–216.

Penman H L (1948) Natural evaporation from open water, bare soil and grass. Proceedings of the Royal Society, London A 193:120–145.

Penning de Vries F W T, van Laar H H (1982) Simulation of growth processes and the model BACROS. Pages 114–136 in Simulation of plant growth and crop production. PUDOC, Wageningen, The Netherlands.

Penning de Vries F W T, Brunsting A H M, van Laar H H (1974) Products, requirements and efficiency of biosynthesis: a quantitative approach. Journal of Theoretical Biology 45:339–377.

Pickering N B, Jones J W, Boote K J (1990) A moisture- and CO_2–sensitive model of evapotranspiration and photosynthesis. Paper No. 90–2519. American Society of Agricultural Engineers, St. Joseph, Michigan, USA.

Pickering N B, Jones J W, Boote K J, Hoogenboom G, Allen L H Jr, Baker J T (1993) Modeling soybean growth under climate change conditions. Paper No. 93–4510. American Society of Agricultural Engineers, St. Joseph, Michigan, USA.

Pickering N B, Jones J W, Boote K J (1995) Adapting SOYGRO V5.42 for prediction under climate change conditions. In Rosenzweig C E, Allen L H Jr, Harper L A, Hollinger S E, Jones J W (Eds.) Proceedings of a symposium on climate change and international impacts. American Society of Agronomy Special Publication. American Society of Agronomy, Madison, Wisconsin, USA.

Priestley C H B, Taylor R J (1972) On the assessment of surface heat flux and evaporation using large-scale parameters. Monthly Weather Review 100:81–92.

Ritchie J T (1985) A user-oriented model of the soil water balance in wheat. Pages 293–305 in Fry E, Atkin T K (Eds.) Wheat growth and modeling. Plenum Publishing Corporation, NATO-ASI Series.

Spitters C J T (1986) Separating the diffuse and direct component of global radiation and its implication for modeling canopy photosynthesis. Part II. Calculation of canopy photosynthesis. Agricultural and Forest Meteorology 38:231–242.

Spitters C J T, Toussaint H A J M, Goudriaan J (1986) Separating the diffuse and direct components of global radiation and its implications for model canopy photosynthesis. I. Components of incoming radiation. Agricultural and Forest Meteorology 38:217–229.

Stucky D J (1976) Effect of planting depth, temperature, and cultivars on emergence and yield of double cropped soybeans. Agronomy Journal 68:291–294.

Thomas J F, Raper C D Jr (1977) Morphological response of soybeans as governed by photoperiod, temperature, and age at treatment. Botanical Gazette 138:321–328.

Thomas J F, Raper C D Jr (1978) Effect of day and night temperature during floral induction on morphology of soybean. Agronomy Journal 70:893–898.

Tsuji G Y, Uehara G, Balas S (Eds.) (1994) DSSAT Version 3. University of Hawaii, Honolulu, Hawaii, USA.

Wilkerson G G, Jones J W, Boote K J, Ingram K T, Mishoe J W (1983) Modeling soybean growth for crop management. Transactions of the American Society of Agricultural Engineers 26:63–73.

Wilkerson G G, Jones J W, Boote K J, Mishoe J W (1985) SOYGRO V5.0: Soybean crop growth and yield model. Technical documentation. Agricultural Engineering Department, University of Florida, Gainesville, Florida, USA.

Modeling growth and development of root and tuber crops

U. SINGH[1], R.B. MATTHEWS[2], T.S. GRIFFIN[3], J.T. RITCHIE, L.A. HUNT[5] and R. GOENAGA[6]

[1]*International Fertilizer Development Center, Muscle Shoals, Alabama 35662, USA*
[2]*Department of Water Management, Silsoe College, Cranfield University, Silsoe, Bedford, MK45 4DT, UK*
[3]*Department of Applied and Environmental Sciences, University of Maine, Orono Maine 04469, USA*
[4]*Homer Nowlin Chair, Department of Soil and Crop Science, Michigan State University, East Lansing, Michigan 48824, USA*
[5]*Department of Crop Science, University of Guelph, Guelph, Ontario, Canada NIG 2W1*
[6]*Tropical Agriculture Research Station, USDA-ARS, Mayaguez, Puerto Rico 00681, USA*

Key words: taro, tanier, cassava, potato, aroids, phenology, genetic coefficients, calibration, evaluation

Abstract

Root and tuber crops are physiologically and botanically a diverse group of plants with a common underground storage organ for carbohydrates. Among all the IBSNAT crop simulation models, the root and tuber family of models have the least amount in common. Crop growth simulation models exist for edible aroids, taro (*Colocasia esculenta* L. Schott) and tanier (*Xanthosoma* spp.), potato (*Solanum tuberosum* L.) and cassava (*Manihot esculenta* L. Crantz). Technically, the economic important products harvested are cassava roots, potato tubers, and aroid corms and cormels. The aroids, potato and cassava models simulate growth and development as affected by environmental factors and cultural practices. All models calculate growth using a capacity model for carbon fixation constrained by solar radiation, temperature, soil water deficit, and nitrogen deficit. They each simulate the effect of soil, water, irrigation, N fertilization, planting date, planting density, row spacing, and the method of planting on plant growth, development and yield. The models assume that during early growth the leaf and stem (petiole in aroids) are the dominant sinks for assimilate. As plants mature most of the assimilate is translocated to storage organs. In the evaluation presented the models show great potential for simulating growth to aid in the interpretation of experimental data, and subsequently, following refinement, help in the evaluation of potential changes in management in diverse environments.

Introduction

Crops such as potato, taro, tanier and cassava are basic foods for many people in Latin America, Africa, and the Caribbean and Pacific Islands. Additionally, these crops comprise an important fraction of the energy source in the diet of people of developed nations. About 6 percent of the world's dietary calories are provided by root and tuber crops. They make their greatest contribution in Africa and Oceania (islands of the North and South Pacific), where they

G.Y. Tsuji et al. *(eds): Understanding Options for Agricultural Production, 129–156*

Table 1. Characteristics of taro, tanier, cassava and potato.

Characteristics	Taro	Tanier	Cassava	Potato
Growth period (months)	6–18	9–18	9–24	3–7
Life cycle	perennial	perennial	perennial	annual
Optimal rainfall (cm)	250	140–200	100–150	50–75
Optimal temperature (°C)	21–27	23–29	25–29	15–18
Drought resistant	no	no	yes	no
Optimal pH	5.5–6.5	5.5–6.5	5–6	5.5–6.0
Fertility requirement	high	high	low	high
Organic matter requirement	high	high	low	high
Waterlogging tolerant	yes	no	no	no
Planting material	corms cormels	corms cormels	stem cutting	tuber suckers
Storage organ	corms cormels	corms cormels	roots	tubers
Storage period in ground	moderate	long	long	short
Post-harvest storage life	variable	2–3 months	short	long

Source: Kay (1973).

provide 15 percent and 7 percent of the dietary calories, respectively. Roots and tubers are often considered emergency foods since their production may determine whether subsistence is possible for inhabitants in some regions of the world.

The edible aroids, taro and tanier, are important staples in many Pacific and Caribbean islands, Africa, and Southeast Asia. Over six million tonnes of taro and tanier are produced annually in the world (FAO, 1991). They are grown as commercial crops only in the United States (Hawaii and Puerto Rico), Nigeria, Indonesia, the Philippines, Egypt, and a host of islands in the Pacific and Caribbean.

Under relatively dry conditions, cassava produces more starch per unit area per year than any known crop. One of cassava's principal characteristics is its ability to produce economic yields under relatively marginal rainfall and soil conditions and with a minimum amount of purchased inputs such as fertilizers. Although cassava has been widely characterized as a subsistence crop, about 70 percent of its production is marketed in the Americas and Thailand (Horton, 1988). Cassava consumption is heavily concentrated in Africa and South America, where it contributes to 9.6 percent and 4 percent of dietary calories, respectively.

After rice, wheat and maize, potato is the next most valuable food crop grown in the developing world (FAO, 1991). Potato production has spread from traditional mountainous and temperate environments into warmer areas. Typical of these nontraditional areas are the plains of India, Bangladesh and Pakistan, Peru's coastal valleys, Philippines, Thailand and Vietnam. Potato generally has a shorter vegetative period than cassava or aroids, and it performs best under cool temperatures (Table 1). Taro can be cultivated under widely varying hydrological regimes, ranging from paddies with continually flowing water to upland cropping with supplemental irrigation. Taro, tanier and cassava have a long flexible growing season without a definite maturation time (Table 1). The growing season length is influenced by environmental, management, and socio-economic factors and there is considerable flexibility in the timing of harvest operations. Predicting a definite maturity date is therefore difficult. When harvesting is delayed yield reductions normally occur because of rotting or sprouting of cormels in tanier. The edible and marketable organs in tanier are only the unsprouted cormels.

Traditionally, agricultural research and development efforts have focused on export crops and the major cereals. Relatively little attention has been paid to root and tuber crops. Contrary to conventional wisdom, these crops are not always inexpensive or easily cultivated. Although both crops have a fundamental role in the diets of the people of Africa, the Caribbean and the Pacific, policy makers and researchers have disregarded them or minimized their importance. Today, however, their considerable potential as abundant sources of energy is commanding more attention in the vital fight against hunger.

Average yield of root and tuber crops vary widely. Cassava yields range from 1.5 t ha^{-1} in Sudan to more than 32 t ha^{-1} in Fiji (FAO, 1991). Under farm conditions, yield of taro and tanier are also very low, ranging from 1.4 to 6 t ha^{-1} (Lyonga and Nzietcheung, 1986). Under research station conditions, yields of 50–75 t ha^{-1} for both upland and lowland taro and 40 t ha^{-1} for tanier have been recorded (Goenaga and Chardon, 1993; Silva, et al., 1992). Bridging the yield gap requires a thorough understanding of the physiology and the development of the root crops and the impact of various abiotic, biotic, and management factors on crop growth and development. The objective of this chapter is to present the status of the Simulation of Underground Bulking Storage Organs (SUBSTOR) models for aroids (taro and tanier) and potato, and the CROPSIM Cassava model. Growth and development are described in detail in this chapter, while the soil water and nitrogen dynamics are being presented in other chapters.

Root crop models

The accumulation and partitioning of biomass and the phenological development of root and tuber crops are influenced by environmental variables, particularly by temperature (Irikura et al., 1979; Paradales et al., 1982; Prange et al., 1990; Prasad and Singh, 1992; Snyder and Ewing, 1989), photoperiod (Bolhuis, 1966; De Bruijn, 1977; Keating et al., 1985; Tsukamoto and Inaba, 1961) and intercepted solar radiation (Fukai et al., 1984; MacKerron and Waister, 1985). We are not aware of any crop growth models that simulate the growth of edible aroids, both tanier and taro. On the other hand, numerous efforts have been made to develop cassava and potato simulation models.

Potato growth models range from very simple regression models based on temperature (Iritani, 1963; Manrique and Bartholomew, 1991) to more complex temperature and dry matter partitioning models (Hartz and Moore, 1978; Ingram and McCloud, 1984; Sands et al., 1979), to the mechanistic, organ-level model of Ng and Loomis (1984). Cassava growth models (Boerboom, 1978; Gijzen et al., 1990; Gutierrez et al., 1988) have generally been based on fixed allocation of dry matter with very little environmental control. A leaf area-driven model by Cock et al. (1979) simulated crop growth rate and overcame the fixed partitioning of assimilate to various organs. However, it did not take into account the effects of temperature, solar radiation, or water stress. Fukai and Hammer (1987) included these effects in their model.

The effect of photoperiod has not been included in many models. Also the effect of vapor pressure deficit on stomatal closure and growth has not been considered by any of the current cassava models. A common feature of the above potato and cassava models is that they are site- and cultivar-specific. Hence, when either of these are changed, the models have to be recalibrated. The basic prerequisite of a model as envisioned under the International Benchmark Sites Network for Agrotechnology Transfer (IBSNAT) Project is that it performs well for a wide range of environmental and management conditions as well as different cultivars with minimum recalibration of the model.

Modeling philosophy

The models developed under the auspices of the IBSNAT Project are characterized by a balance in the amount of detail included in their different components. This is done because predictions can generally be no better than the level of detail of the least understood part of the model and it cannot be made more accurate by including information from disciplines in which there is a wealth of information. For example, the SUBSTOR-Aroids model treats development and growth processes in aroids with equal importance, although there is dearth of information pertaining to development. To produce realistic predictions, growth and development must be treated in balance as these processes are affected by environment and stresses in different ways.

A model's usefulness, particularly in the developing countries, could be limited by the availability of data for running and evaluating it. The SUBSTOR-models for potato and aroids and the CROPSIM model for cassava use the minimum data set (MDS) concept as identified by the IBSNAT Project (IBSNAT, 1986). The MDS include data for site, weather, soil, and crop management. The models also follow the standard input and output file structures (Hunt et al.,1994; IBSNAT, 1988). To accommodate varietal differences, the models require input data for quantifying varietal characteristics. These varietal characteristics are summarized through the genetic coefficients.

Accurate input information is a prerequisite for a reliable simulation. Spatial variability in plant size (and below ground organs) is more widespread in root and tuber crops than cereals. The factors that lead to such variability must be understood before developing complex and very detailed models for root crops.

Model background

The root crop models, SUBSTOR-Aroids, SUBSTOR-Potato, and CROPSIM-Cassava, developed under the auspices of the IBSNAT Project are dynamic, process-oriented computer models that simulate growth, development, and yield as a function of varietal characteristics, weather, soil water and nutrient status, and crop management. The models have been written in FORTRAN 77 computer language and run on IBM-compatible microcomputers. They have a similar programming structure to the CERES models (Godwin et al., 1990; Singh et al., 1993). SUBSTOR also shares water and nitrogen balance routines with the CERES models.

The SUBSTOR-Aroids and SUBSTOR-Potato Models are designed to simulate growth and development under: (1) nonlimiting conditions, (2) water limited conditions, or (3) water and nitrogen limited conditions. The water balance sub-model used in the SUBSTOR and CROPSIM models, has been described in detail by Ritchie (1985). The water balance sub-model used in the SUBSTOR-Aroids also has a lowland component with the ability to simulate a flooded field with a bund around it, as in CERES-Rice (Singh, 1994). The nitrogen submodel for the upland conditions in the SUBSTOR models is based on Godwin and Jones (1991). The N submodel used in the SUBSTOR-Aroids, with the capacity to simulate N dynamics under both aerobic and anaerobic conditions, has been described by Godwin and Singh (1991) and Singh (1994).

Crop development

Although there is very little in common among potatoes and aroids, the crop development and growth submodels for these crops have similar structure and variable names.

Phenological development in aroids

The developmental phases are organized around times in the plant's life cycle when changes occur in the partitioning of assimilates among different plant organs. These specific times are not as well defined nor as clearly distinct as in cereals and grain legumes. The SUBSTOR-Aroids model simulates seven growth stages. With the exception of the fallow period, which is management dependent, all other stages are driven by environment and genotype (Table 2).

In SUBSTOR-Aroids, temperature is the primary environmental variable that influences development. Development rates are assumed to be directly proportional to temperature when the daily minimum temperature (*TEMPMN*) is greater than the base temperature (*TBASE*) of 10°C and the daily maximum temperature (*TEMPMX*) is less than 33°C. Under such conditions, the daily

Table 2. Development phase for taro and tanier as defined in SUBSTOR-Aroids.

Phase Event	Duration	Genetic/Crop coefficient	Environmental/Management control
7	Fallow	Planting date	
8	Root formation	P8*	Thermal time, moisture and extreme temperatures
9	First leaf emergence	P9*	Thermal time
1	Establishment	P1	Thermal time
2	Rapid vegetative growth	P3, P4	Thermal time
5	Rapid corm and cormel growth	P5	Thermal time, assimilate supply
6	Post maturity		

*Crop-specific coefficient for establishment with 'hulis": corm-petiole cutting in taro and mini-sets (eye-sprouts) in tanier.

temperature accumulation or daily thermal time (DTT) for a day is defined as:

$$DTT = ((TEMPMX + TEMPMN)/2) - TBASE$$

When the minimum temperature falls below the base or the maximum temperature exceeds 33°C then a weighted 3-hourly temperatures ($TTMP$) is used to calculate thermal time. For the periods when $TTMP$ exceeds 42°C or falls below $TBASE$, thermal time is not incremented. Otherwise, the thermal time is calculated as follows:

$$DTT = \sum (TTMP - TBASE)/8$$

$$\text{for } TBASE < TTMP < 33^{\circ}C$$

$$DTT = \sum (33 - TBASE) \times ((1 - (TTMP - 33))/TBASE)/8$$

$$\text{for } 33^{\circ}C < TTMP < 42^{\circ}C$$

The above equations provide reasonable accuracy for modeling both phenological and leaf development. Four cultivar and two crop specific inputs, expressed in terms of accumulated thermal time (°C), influence phasic development in aroids (Table 3). Maturity is not only influenced by genotype (P5) and temperature, but also by assimilate availability. Root formation (germination) is affected by soil moisture as well. For version 2.1, all other growth durations are completely temperature driven.

The SUBSTOR-Aroids model calculates the number of leaves, suckers, and cormels produced. In general, photoperiod does not affect the leaf appearance rate, leaf opening rate, leaf number and growth duration. However, the number of cormels increases with increasing photoperiod (Prasad and Singh, 1992). The principal environmental factor affecting morphogenesis is temperature. The model simulates the formation of roots followed by leaves, petioles, and corms (Figure 1) Leaves and corm develop synchronously in aroids with organ development closely tied to leaf appearance. In taro, sucker formation begins in the rapid vegetative growth phase and only after the appearance of the sixth leaf. The establishment of leaf area is important because it is the site for biomass

Table 3. Cultivar and crop specific coefficients for some selected taro and tanier cultivars.

Phase/ plant component	Coefficient symbol	Taro Cultivars			Tanier Cultivars			
		Lehua	Bunlong	Samoa	Tausala	Kelly	Blanca	Morada
				Hybrid	–ni-			
Root formation	P8 (°C)	35	35	35	35	250	250	250
First leaf emergence	P9 (°C)	50	50	50	50	500	500	500
Establishment	P1 (°C)	800	500	800	1200	1500	1200	1440
Rapid vegetative								
Early stage	P3 (°C)	1300	700	1000	1000	1200	1500	3000
Late stage	P4 (°C)	500	1200	900	900	400	600	1000
Rapid corm and cormel growth to maturity	P5 (°C)	1050	1200	1200	700	600	900	1000
Leaf appearance rate:								
Phyllochron interval	PHINT (°C)	150	150	150	150	–	–	–
Intercept (Tanier)	PCINT	–	–	–	–	150	122	150
Gradient (Tanier)	PCGRD	–	–	–	–	0.0	1.75	2.03
Sucker and cormel number (scaler value)	G3	0.85	1.30	0.90	1.00	1.50	0.92	1.00
Max. leaf size	G4	1.00	1.50	1.00	1.00	1.00	1.00	1.00

production through the conversion of CO_2 and light energy. It is therefore imperative that the model predicts leaf appearance reliably.

The leaf appearance rate in the SUBSTOR-Aroids model is a function of thermal time and is dependent on genotype. The phyllochron interval (*PHINT*) or the thermal time required for a single leaftip to appear is 150°C for Colocasia. This was found to be the case with at least four taro cultivars (Lehua, Bunlong, Samoa Hybrid, and Tausala-ni-Samoa (Niue)). In *Xanthosoma*, however, the phyllochron interval is cultivar-specific, ranging from 122°C to 180°C (Table 3);

$$PHINT = PCINT + \text{Leaf No.} \times PCGRD$$

The SUBSTOR-Aroids model simulates a faster rate of leaf appearance when the plants are young. The phyllochron interval approach works well except under conditions where extreme water and nutrient stress occur. Light quality under shaded conditions may also influence leaf appearance rate. However, there is a need for additional research before the effects of stress and light level or quality on leaf appearance rate can be incorporated into the SUBSTOR-Aroids model.

Phenological development in potato

The brief description of SUBSTOR-Potato presented here is based on Griffin et al. (1993). Five growth phases were defined and are listed as follows: (i) pre-planting, (ii) planting to sprout germination, (iii) sprout germination to emergence, (iv) emergence to tuber initiation, and (v) tuber initiation to maturity. All growth stages are affected by temperature: the temperature factor affects either vine growth or root and tuber growth (*RTFSOIL*, Figure 1). When the observed emergence date is not an input, the model simulates bud sprouting and elongation as a function of *RTFSOIL*. The emergence date is also a function of planting depth. Tuber initiation (*TI*) is a function of daylength and temperature and is also modified by plant N status and soil water deficit. Based on the findings of Ewing (1981), and Wheeler and Tibbetts (1986), the model assumes that tuber initiation by 'early' cultivars is less sensitive to high temperature and/or long daylengths than tuber initiation by 'late' cultivars. This effect is simulated in the model by a cultivar specific coefficient for critical temperature, *TC* (Table 4), above which tuber initiation is inhibited to some extent. A higher *TC* (lower sensitivity to high temperature) corresponds to 'early' cultivars. The *TC* together with daily mean temperature determine the relative temperature factor for tuber initiation, *RTFTI* (0–1) as:

$$RTFTI = 0.0 \qquad \text{for } TEMPM > TC + 8 \quad \text{or } TEMPM < TC - 8$$

$$RTFTI = 1.0 - 0.0156 \times (TEMPM - TC)^2$$

$$\text{for } TC < TEMPM < TC + 8$$

For photoperiods greater than 12 h, early cultivars with lower sensitivity to

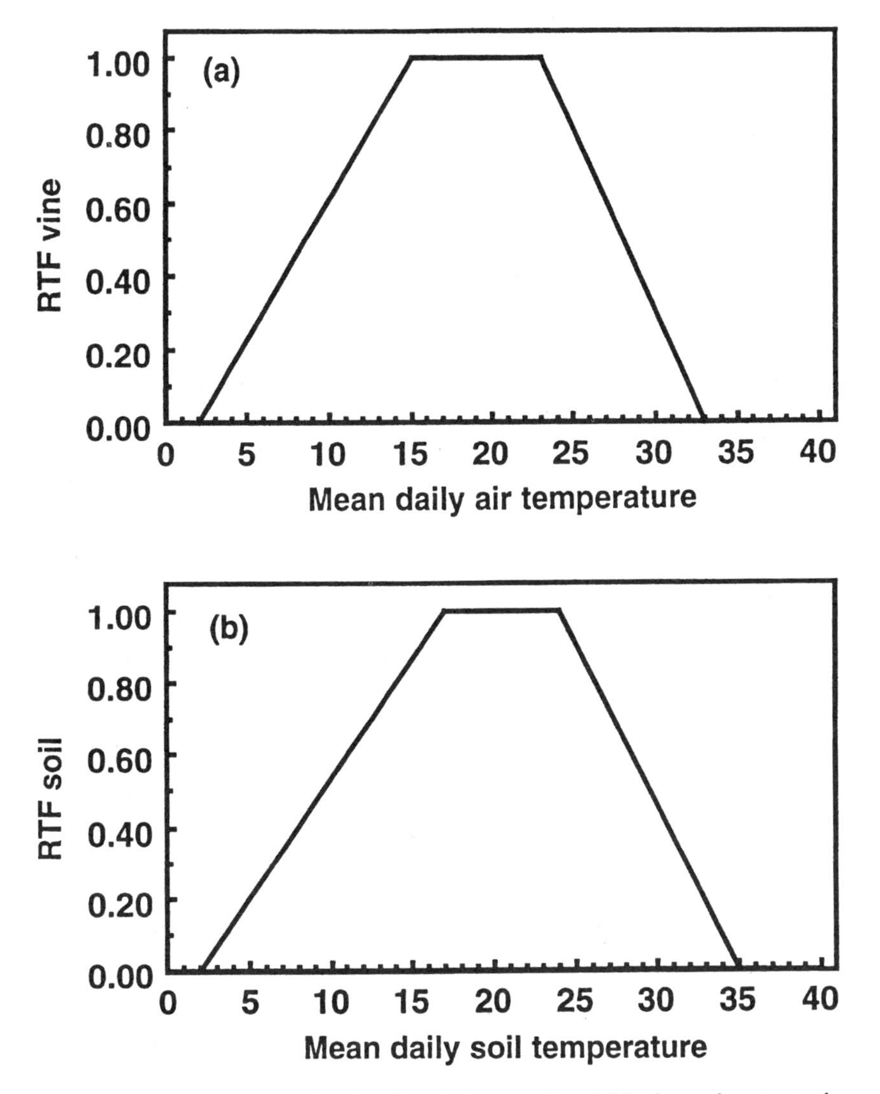

Figure 1. Relative thermal time functions for (a) vine growth and (b) tuber and root growth.

tuber initiation are assigned low values for coefficient P2 (Table 4). The model uses this coefficient and photoperiod (*PHPER*) to determine the relative daylength factor for tuber initiation, *RDLFTI* (0–1) as:

$$RDLFTI = (1.0 - P2) + 0.00694 \times P2 \times (24.0 - PHPER)^2$$

On a daily basis, beginning at emergence, the model calculates a tuber induction index (*TII*) as:

$$TII = (RTFTI \times RDLFTI) + 0.5 \times (1.0 - AMIN1(SWDF2, NDEF2))$$

Table 4. Cultivar specific coefficients for SUBSTOR-Potato.

Description	Coefficient symbol	Cultivar					
		Segago Burbank	Russet	Katahdin Piper	Maris	Desiree	Norchip
Leaf expansion rate ($cm^2\ m^{-2}\ d^{-1}$)	G2	2000	2000	2000	2000	2000	2000
Tuber growth rate ($g\ m^{-2}\ d^{-1}$)	G3	22.5	22.5	25.0	25.0	25.0	25.0
Determinacy	PD	0.7	0.6	0.7	0.8	0.9	1.0
Sensitivity of TI to:							
Photoperiod	P2	0.8	0.6	0.6	0.4	0.6	0.4
Temperature (°C)	TC	15.0	17.0	19.0	17.0	17.0	17.0

Table 5. Cultivar specific coefficients for CROPSIM-Cassava (Source: Tsuji et al., 1994).

Variable	Definition
DUB1	Duration of branch 1 phase (Bd from germination to first branch)
DUBR	Duration of branch 2 and greater phases (Bd between branches)
DESP	Development, sensitivity to photoperiod (h^{-1}) (0 = insensitive)
PHCX	Canopy photosynthesis maximum rate ($g\ dm\ m^{-1}\ d^{-1}$)
SNOPE	Stem number per plant at emergence
SNOFX	Shoot number per fork, maximum
SNOPX	Shoot number per plant, maximum
SWNX	Stem weight to node weight ratio
LNOIS	Leaf number increase rate, standard (leaves Bd^{-1}) where leaves = $shoot^{-1}$.
LALX	Leaf area, maximum ($cm^2\ leaf^{-1}$)
LAXA	Age at which maximum leaf area is reached (Bd after emergence)
LAL3	Leaf area at 300 days after emergence ($cm^2\ leaf^{-1}$)
LAWS	Leaf area to weight ratio, standard ($cm^2\ g^{-1}$)
LFL1	Leaf life (day)

A FORTRAN function, *AMIN*1, uses the most limiting (minimum) of the water stress (*SWDF*2) or nitrogen deficiency (*NDEF*2) factors to hasten tuber initiation under stress conditions. At a cumulative tuber induction value of 20 actual tuber initiation occurs.

Phenological development in cassava

The crop phenology submodel as used in CROPSIM-Cassava has been fully described by Matthews and Hunt (1994). Three phases are defined in the model: (i) planting to emergence, (ii) emergence to first branching (reproductive phase), and (iii) first branching to maturity or final harvest. The duration of the last two phases are cultivar-specific characteristics (Table 5). The model independently tracks both the vegetative and reproductive development. The main factors affecting both vegetative and reproductive development are temperature and drought stress. A photoperiod effect on reproductive development is also simulated. Suboptimal conditions may affect the biological requirement (*Bd*). Under optimum temperature and photoperiod, with nonlimiting water and nutrient conditions, *Bd* is equivalent to a chronological day (Matthews

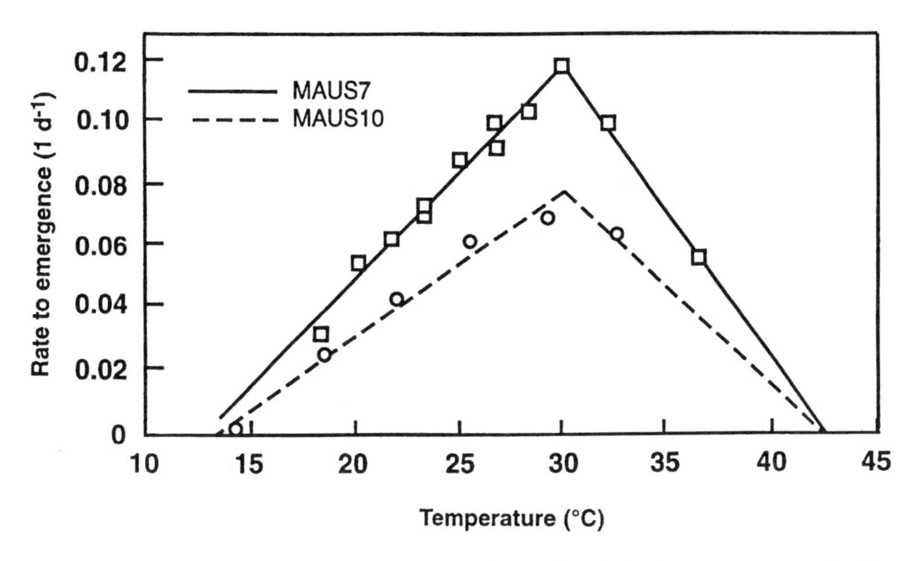

Figure 2. Temperature function for developmental rate from planting to emergence for two cassava cultivars (Source: Keating and Evenson, 1979).

and Hunt, 1994). The temperature function in CROPSIM-Cassava is similar to the SUBSTOR models, requiring base, optimum, and high temperatures. For example, the temperature effect on developmental rate from planting to emergence for all cassava cultivars is based on a *TBASE* of 13°C, optimum temperature of 30°C and a maximum temperature of 42.5°C (Figure 2).

An important feature of the *CROPSIM* model is its ability to simulate both lateral and sympodial branching. The former is important in determining the initial number of stems per unit area (*SNOPE*), while the latter influences canopy leaf area (Matthews and Hunt, 1994). The model uses both temperature and photoperiod to simulate the time to first branching, *DUB1*, a cultivar-specific characteristic (Table 5). Subsequent branching proceeds at a linear rate with the duration between branches, DUBR, being a genotypic characteristic (Table 5). Two other genotypic coefficients describe branching: the maximum number of axils per fork and the maximum number of axils per plant. Branching is reduced by drought stress.

Crop growth

The growth submodels in SUBSTOR-Aroids, SUBSTOR-Potato and CROPSIM-Cassava describe four processes: expansion growth, mass accumulation, assimilate partitioning, and senescence. Their development has been a major challenge because the partitioning of assimilates is a dynamic process, requiring several feedback mechanisms.

Growth in SUBSTOR-Aroids

Expansion growth. Plant leaf area has a crucial role in determining light interception and biomass accumulation. Leaf expansion growth is highly sensitive to environmental stresses, with leaf expansion being reduced more than photosynthesis. This reduction in expansion growth without a concomitant decrease in photosynthesis could increase the specific leaf weight or increase the proportion of assimilate partitioned to the roots or corms. The plant responses are simulated by using separate water and nitrogen deficit functions for reducing leaf expansion growth ($SWDF2$, $NDEF2$) and mass accumulation ($SWDF1$, $NDEF1$). The rate of leaf expansion already decreases when the maximum possible root water absorption on a day ($TRWU$) is less than 1.5 times the potential transpiration (EP), that is $SWDF2 = 0.67$, while the biomass accumulation is unaffected ($SWDF1 = 1.0$).

$$SWDF1 = TRWU/EP \times 1.5 \qquad \text{for } SWDF1 \leq 1.0$$

$$SWDF1 = 1.0 \qquad \text{for } SWDF1 > 1.0$$

$$SWDF2 = TRWU/EP$$

The leaf area of the main plant ($PLAG$) is calculated using the Gompertz function with the rate of leaf appearance and the rate of leaf expansion of the growing leaves as principal components. The model determines the number of fully expanded leaves or cumulative phyllochrons ($CUMPH$) from the daily thermal time and phyllochron interval ($PHINT$) as:

$$CUMPH = CUMPH + TI \qquad \text{where} TI = DTT/PHINT$$

$$PLAG = A \times \exp(-K \times CUMPH) \times \exp(b \times \exp(-K \times CUMPH)) \times AMIN1(SWDF2, NDEF2) \times TI$$

where A is maximum leaf area at infinite time. It is a function of the maximum leaf size coefficient, G4 (Table 3). K and b are crop-specific constants with taro and tanier having different values. The model uses the specific leaf weight ratio (SLW) ranging from 38 to 42 g m^{-2} for taro and 50 to 71 g m^{-2} for tanier to determine leaf growth (g $plant^{-1}$). Tanier with thicker leaves has higher values for SLW.

Mass accumulation. Potential biomass accumulation (PCARB in g $plant^{-1}$) is a non-linear function of intercepted photosynthetically active radiation ($IPAR$). The radiation conversion function was derived from data of Prasad and Singh (1992) and Goenaga and Singh (1992). The percentage of incoming radiation (PAR) intercepted by the canopy, $IPAR$, is an exponential function of leaf area index (LAI) with an extinction coefficient (k) of 0.603,

$$IPAR/PAR = 1.0 - \exp(-k \times LAI).$$

Division of $PCARB$ by planting density ($PLANTS$, plants m^{-2}) converts biomass accumulation from an area to a plant basis, i.e.,

$$PCARB = (6.200 \times IPAR \times 0.65)/PLANTS$$

The actual rate of biomass production (*CARBO*) in the model can be less than the potential rate due to the effects of non-optimal temperature (*PRFT*), water stress (*SWDF1*), or N stress (*NDEF1*). The optimum daytime temperature for biomass accumulation is 28°C. Water stress also reduces *CARBO* whenever crop extraction of soil water falls below the potential transpiration rate for that day. The reduction factor *SWDF1* is based on the water balance subroutine of the CERES models (Ritchie, 1985). The estimation of N deficiency is based on the N concentration in the tops (leaves and petioles), with the N deficiency factors (*NDEF1* and *NDEF2*) being determined using the function described by Godwin and Jones (1991). The actual biomass production is thus determined as:

$$CARBO = PCARB \times AMIN1(PRFT, SWDF1, NDEF1)$$

Assimilate partitioning. Once the first leaf has appeared, assimilate is partitioned to roots, leaves, petiole, and corm. Assimilate partitioning is dynamic with 90–100% of assimilates in the early phases of main plant growth going to roots, leaves, and petioles. In sharp contrast, this value drops to less than 20% late in the season (Figure 3). Under nonlimiting conditions leaf growth is source-driven and affected by factors influencing leaf expansion and appearance rates. However, under stress conditions the model simulates leaf growth as limited by sink or assimilate supply. The proportion of assimilates partitioned to roots affects root density and the ability of the aroids root system to supply the shoot with water and nutrients. When water is limiting, the proportion of assimilates partitioned to roots increases.

Partitioning of assimilates in the SUBSTOR-Aroids models is thus dependent on growth stage (Figure 3), on cultivar (Singh et al., 1992), and on water and N stress (Figure 4). The genotypic variation in partitioning may occur due to differences in growth duration or suckering habit (sucker formation). A greater proportion of the assimilates in tanier than in taro is channeled to the cormels. This trait is desirable because cormels form the marketable yield. Sucker growth and development is a function of leaf appearance rate, assimilate availability, temperature, and water and N stress. As illustrated in Figure 4, assimilate partitioning for sucker formation in taro is not only reduced but also delayed under N stress.

Senescence. Senescence of leaves and petioles is dependent on stage of growth, shading, non-optimal temperatures, and on water and N stress. As expected, the model predicts high amounts of leaf and petiole senescence in the growth stage leading to maturity. Reliable prediction of senescence, particularly late in the season, is important as declining LAI has a significant impact on biomass production and corm yield.

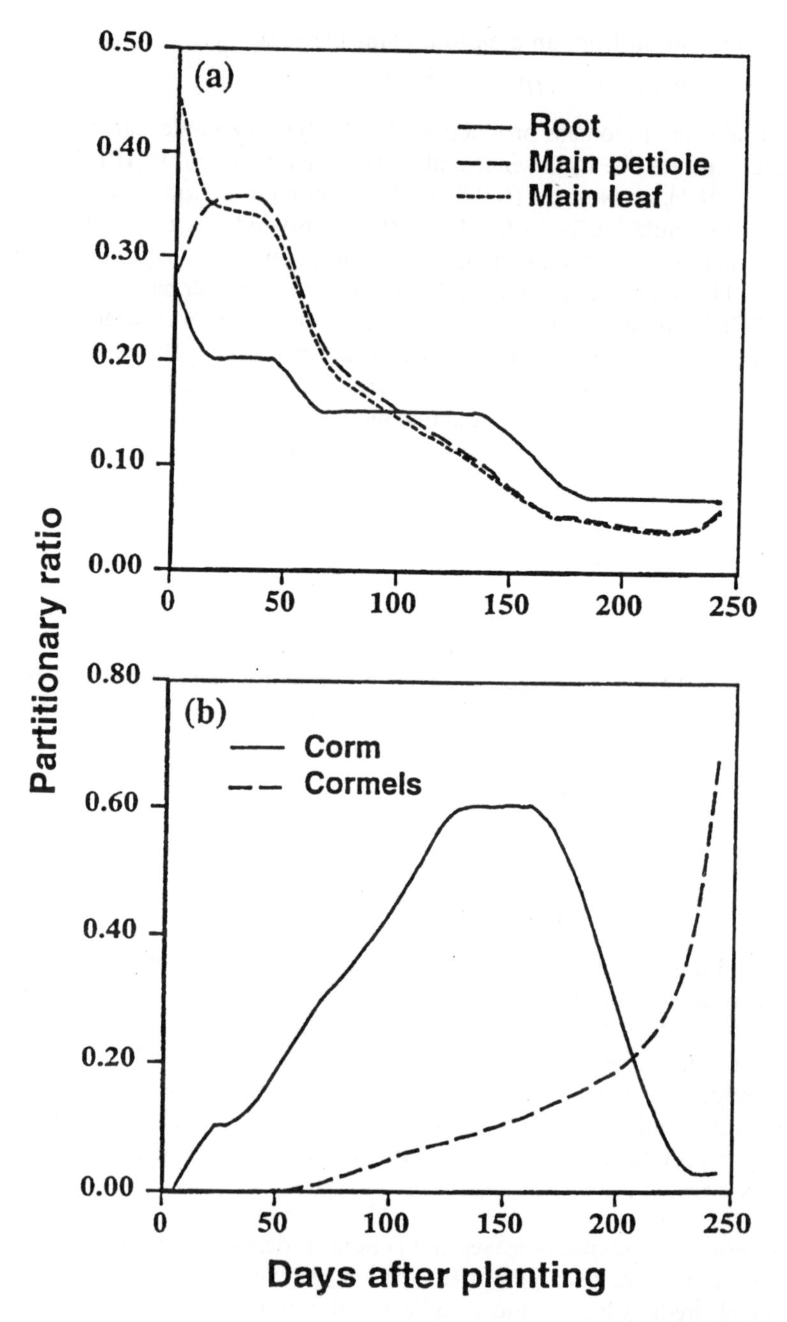

Figure 3. Assimilate partitioning to (a) root and main plant petiole and leaf, and (b) corm and cormels under non-limiting water and N conditions.

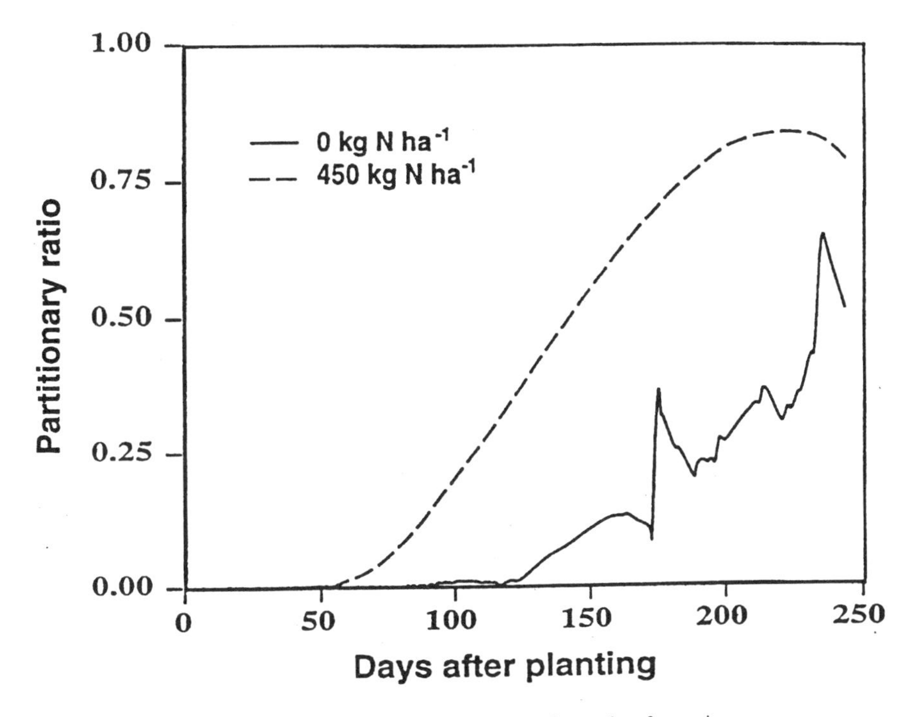

Figure 4. Effect of N stress on partitioning of assimilates for sucker formation.

Growth in SUBSTOR-Potato

During the pre-emergence phase, growth in SUBSTOR-Potato is entirely supported by available carbohydrate from the seed piece (*SEEDAV*). Daily root and pre-emergence sprout growth are a linear function of *RTFSOIL*. If the sum of sprout and root growth exceeds *SEEDAV*, the growth of both components is reduced proportionally to equal *SEEDAV*. *SEEDAV* continues to support post-emergence growth up to a plant leaf area (PLA) of 400 cm^2 $plant^{-1}$.

During the post-emergence vegetative growth stage, potential photosynthetic carbohydrate assimilation is calculated as:

$$PCARB = RUE \times PAR/PLANTS \times (1.0 - \exp(-k \times LAI))$$

where the variables are as described previously for SUBSTOR-Aroids. The radiation use efficiency (RUE) of 3.5 g MJ^{-1} PAR is based on Allen and Scott (1980), Jeffries and MacKerron (1989), Manrique et al. (1991) and Sale (1973). The extinction coefficient (k) of 0.55 is also based on Allen and Scott (1980). The actual daily carbohydrate accumulation is given as:

$$CARBO = PCARB \times AMIN1(PRFT, SWDF1, NDEF1) + 0.5 \times DDEADLF$$

where $0.5 \times DDEADLF$ represents translocation of 50% of the carbohydrate in senesced leaves prior to abscission (Johnson et al., 1986). If *CARBO* exceeds the daily growth demand, excess is stored in a soluble carbohydrate pool (*RVCHO*), which could be up to 10% of haulm dry weight (Ng and Loomis, 1984). This pool is allowed to exist only when PLA > 400 cm^2 $plant^{-1}$ and when seed reserve is no longer available.

Leaf growth is a function of a relative temperature factor (*RTFVINE*) and based on a leaf area to leaf weight ratio of 270 $cm^2\ g^{-1}$. Stem growth (*GROSTM*) is assumed to be the same as *GROLF*. Growth of leaves, stem and roots occur with equal priority, thus growth of all organs are equally affected during carbohydrate shortage.

Simulating tuber bulking is a key process in the growth subroutine. Several fundamental changes occur once bulking starts: *RUE* is increased to 4.0 g MJ^{-1} PAR and carbohydrate is partitioned to competitive organs which no longer have equal priority. Tubers are given the first priority for available assimilate. This assumption is not critical under non-limiting conditions or when tuber sink is small because all growth demand could be met. However, under stress conditions or when tuber sink demand is very high, growth of vines and roots can be drastically reduced.

Potential tuber growth (*PTUBGR*) is a function of maximum tuber growth rate, G3 (Table 4) and temperature,

$$PTUBGR = G3 \times RTFSOIL/PLANTS$$

The actual tuber growth (*GROTUB*), is modified by water and N stress and tuber sink strength (*TIND*) as:

$$GROTUB = PTUBGR \times AMIN1(SWDF2, NDEF2, 1.0) \times TIND$$

The sink strength is a function of temperature, water and N status. It also takes into account the effect of determinate and indeterminate cultivars, cultivar coefficient *PD*, on tuber sink strength (Griffin et al., 1993).

The genetic coefficient for maximum leaf expansion (G2), a temperature factor, and water and N stress indices are used to determine potential leaf expansion (*PLAG*),

$$PLAG = G2 \times RTFVINE/PLANTS \times AMIN1(SWDF2, NDEF2, 1.0)$$

Water stress and N deficiency. The model simulates reduced expansion growth and photosynthesis, increased allocation of assimilate to tubers, and increased rate of phenological development under stress conditions. Nitrogen deficiency or excess in the model is based on vine N concentration. Under conditions of excess N, SUSTOR-Potato actually simulates a delay in development.

Growth in CROPSIM-Cassava

Expansion growth. Leaf expansion growth in CROPSIM-Cassava is, as in the other models, a function of leaf appearance and expansion rates. Temperature is the key environmental factor affecting the leaf appearance rate (LAR, leaves day^{-1}). Other factors affecting LAR are: cultivar-specific leaf number increase rate (*LNOIS*, leaves Bd^{-1}) at emergence and the leaf number increase period (LNOIP, Bd after emergence) (Table 5). LAR is also a function of crop development age since emergence (*DAGE*, *Bd*),

$$LAR = f(TEMPMN, TEMPMAX) \times LNOIS \times (1 - DAGE/LNOIP)$$

Added feature of the *CROPSIM* model are capabilities to simulate the reduction in LAR due to drought stress, and the compensatory increase in LAR to values above those obtained had there been no stress. For leaf size, the model uses the empirical relationship of Cock et al. (1979) with three cultivar genetic coefficients: maximum leaf area (*LALX*, cm^{-2} leaf), the age at which *LALX* is reached (*LALXA*, Bd after emergence), and the leaf area at 300 days after emergence (*LAL3*) to determine leaf size (Table 5). The model also accounts for the compensatory increase in leaf size on release of drought stress in a similar position. Cultivar-specific leaf area to weight ratio (*LAWS*, cm^2 g^{-1}) (Table 5), temperature and drought stress affect the specific leaf area in the CROPSIM-Cassava model. Specific leaf area is used to estimate leaf growth.

Biomass production. Potential growth (*PCARB*) for a given LAI is determined as a function of the maximum canopy photosynthesis rate (*PHCX*), a cultivar coefficient-input, which ranges from 25–27 g m^{-2} day^{-1} as:

$$PCARB = PHCX \times (1 - \exp(-0.27 \times LAI))$$

The actual growth is then estimated using temperature, light, water stress and vapor pressure deficit (VPD) factors.

Assimilate partitioning. The model assumes that the growth of leaves and stem occur at a rate below that of assimilate supply, and the remaining assimilate goes to the storage roots (Matthews and Hunt, 1994). Assimilate partitioning to the fibrous roots decreases as the crop ages. Root distribution in the soil profile is modeled similar to the CERES models (Ritchie, 1985).

Senescence. Leaf senescence is modeled as a function of temperature and shading. As evident from Table 5, a cultivar-specific coefficient, potential leaf life (*LFLI*, day) also controls senescence. The model simulates the effect of both the duration and intensity of exposure of the crop to temperature and shading stress. Under moderately low temperatures the model simulates leaf shedding in older plants but not in younger plants.

Model evaluation

The SUBSTOR and CROPSIM models have been tested under nonlimiting conditions and with water and N stresses for growth, development and yield.

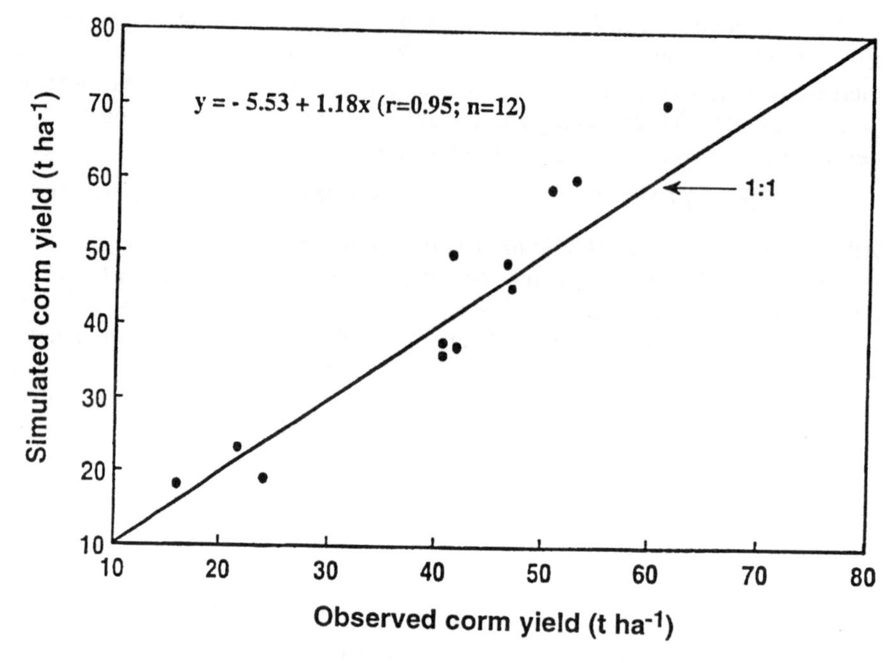

Figure 5. Simulated versus observed corm yield for taro and tanier.

No independent evaluation has been conducted for the water balance and nitrogen dynamic submodels. On-going field trials in Hawaii and Puerto Rico, USA, for aroids, and cassava trials in Thailand will provide data for testing of the water and N balance. In general, the SUBSTOR-Aroids model has not been rigorously tested against independent field data from diverse environments. Results presented here for the root crop models, are thus preliminary and await more rigorous testing.

Performance of SUBSTOR-Aroids

In the following tests, the model's performance was compared with taro experiments conducted in: (i) 1989–1990 at the Haleakala site, Maui, Hawaii, USA (Prasad and Singh, 1992), (ii) 1991–1992 at Waimanalo Research Station, Oahu, Hawaii, USA (Singh et al., 1995), and (iii) 1987–1988 Koronivia Research Station, Fiji (Prasad and Singh, unpublished data) and for tanier experiments conducted during 1989–1990 at Isabela, Puerto Rico, USA (Goenaga and Singh., 1992). Figure 5 shows observed and simulated corm yield for the above experiments. The 'corm yield' is comprised of total corm and cormel fresh weight for taro and only the cormel fresh weight for tanier. The agreement for yield is good ($r = 0.95$, $n = 12$). Model development/calibration and evaluation data were from independent treatments and experiments from the same experimental site. Hence, for a more thorough evaluation of the model, additional independent and more diverse sites are needed.

The model simulated the changes in LAI during the growing cycle reliably for taro cultivar Bunlong grown at the Haleakala site in Maui, Hawaii (Figure 6). It simulated a pronounced increase in LAI late in the season due to rapid sucker growth in Bunlong. High tillering and sucker formation is a cultivar specific characteristic (Table 3). The total corm (corm + cormel) growth, in general, was over-predicted by the model (Figure 6). Such a lack of correspondence could indicate that growth was affected by factors not simulated by the model. Simulations assumed ample nutrients (other than N) and micronutrients, absence of pests and diseases, and adequate irrigation. In the field experiment the latter assumption was not always met. When the above simulation was carried out under rainfed conditions, the LAI remained low and as a consequence corm growth was drastically reduced (Figure 6).

The model's sensitivity to nitrogen applications is shown in Table 6. The percentage changes are relative to 150 kg N ha^{-1} applied in five equal applications for cultivar Bunlong planted on 8 April 1991 using weather and soil data from the Waimanalo Research Station in Oahu, Hawaii, USA. The soil NH_4–N and NO_3–N was 40 kg N ha^{-1}. The simulation was run using non-limiting water conditions. Under high N rates the prolific sucker formation characteristics of Bunlong is evident. There was a two-fold increase in number of cormels when the N rate was increased from 150 kg N ha^{-1} to 300 kg N ha^{-1}. The N response even up to 300 kg N ha^{-1} as simulated by the model may explain the large gap between yields observed in farmers' fields and yields recorded at research stations.

Performance of SUBSTOR-Potato

SUBSTOR-Potato has undergone more rigorous evaluations under diverse conditions than either the SUBSTOR-Aroids or the CROPSIM-Cassava. Table 7 provides a description of some of the data sets used for evaluation purposes. The overall performance of the crop development component of the model is indicated in a comparison of the observed versus simulated time to tuber initiation (*TI*, Figure 7) for some of the experiments described in Table 7. The model's performance was very good ($r = 0.92$, $n = 41$), given the unknown differences in the physiological age of the seed piece. The model was equally good in simulating tuber yield ($r = 0.90$, $n = 54$) from a very diverse data set, where the observed tuber yield ranged from less than 2 t dry matter ha^{-1} to more than 20 t dry matter ha^{-1} (Figure 8). A detailed presentation of the evaluation of the SUBSTOR-Potato model is given in Griffin et al. (1993).

Performance of CROPSIM-Cassava

Evaluation of the cassava model has not been conducted for a diverse array of data sets. Matthews and Hunt (1994) tested the performance of the model

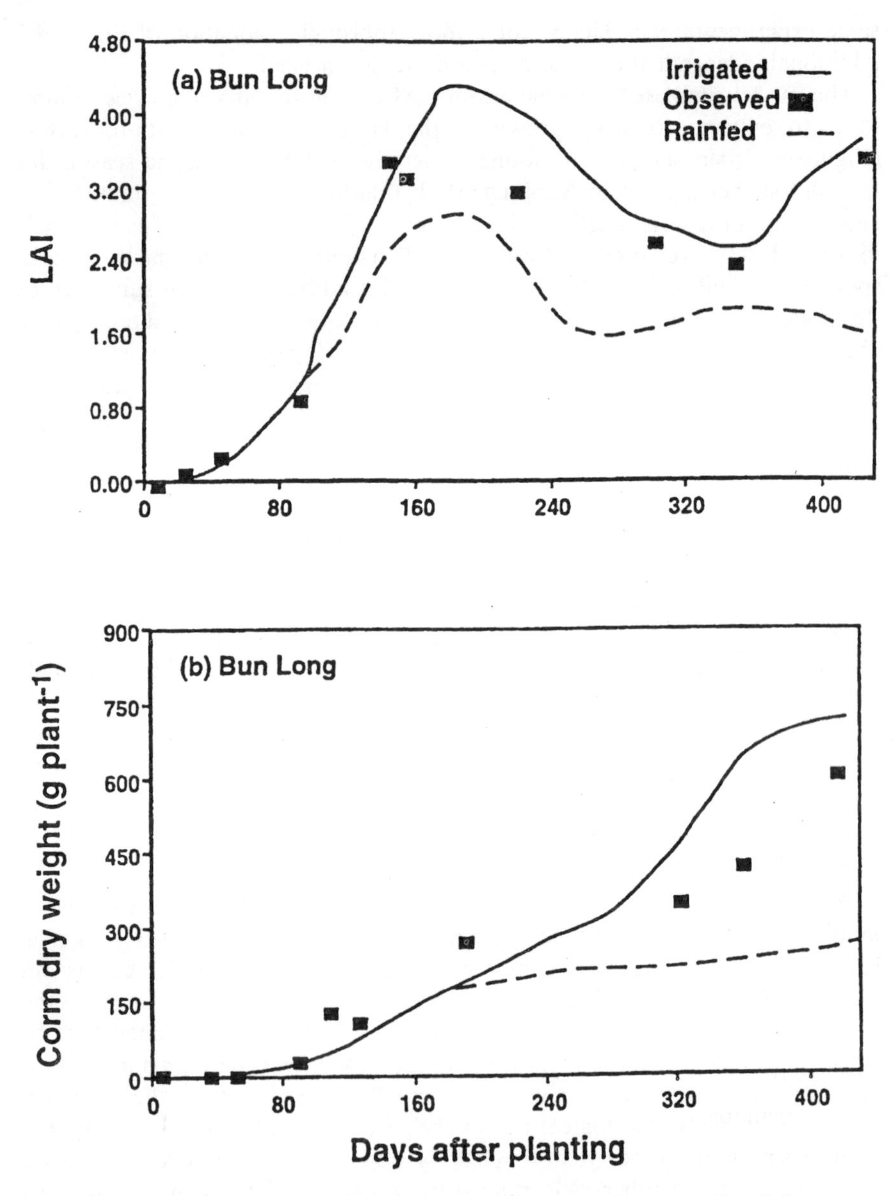

Figure 6. Simulated response of LAI and total corm growth to irrigation and rainfed treatments. Observed data is for irrigated treatments (Source: Singh et al., 1992).

with experiments conducted in Queensland, Australia, by Keating et al. (1982) and at the Centro Internacional de Agricultura Tropical (CIAT), Colombia, by Connor et al. (1981) and Veltkamp (1986). As shown in Figure 9, the model performed well for a range of planting dates (Keating et al., 1982) and hence

Table 6. Sensitivity analysis for final harvest components, relative to 150 kg N ha^{-1} rate at Waimanalo, Oahu, Hawaii.

N rate (kg ha^{-1})	Corm yield	Total biomass	LAI	Cormel number	Mean corm weight	N uptake (kg ha^{-1}
0	48	42	34	61	48	24
50	74	74	60	69	75	62
150	100	100	100	100	100	100
300	125	109	116	200	125	144
450	127	110	118	216	127	147

the effect of temperature and photoperiod on total crop dry matter and storage root dry weight.

The model's capacity to simulate the effect of drought stress was evaluated using the data reported by Connor et al. (1981) and Veltkamp (1986). The model performed well under both non-stress and drought conditions (Figure 10). The sensitivity of many of the parameters used in the cassava model was analyzed by Matthews and Hunt (1994). One of the key findings was a change in the ranking of parameter sensitivities with drought stress. Sensitivity of the models to soil input data was generally greater for water and nitrogen limiting conditions.

From the sensitivity tests it is evident that the performance of all root crop models is highly dependent on the availability of good quality input data for weather, soil, and crop management.

Knowledge gap

From the evaluation of the models it is evident that additional data sets must be collected, particularly for the aroids and cassava models, so that their applicability can be extended to a greater diversity of environments and genotypes. Only a few MDS experiments have been conducted for aroids and cassava. Many vital data for understanding crop growth and development have not been accurately determined and reported in the literature. This is not surprising, especially for taro and tanier, as these crop are predominantly grown in the developing world and in general are not a high value crop. Some data are available for taro, tanier, and cassava with respect to plant competition for resources such as light, water, and nutrients. However, in none of these studies has a complete data set been collected to help comprehend physiological processes, the interactions within the soil-plant-atmosphere continuum, and the influence of cultural practices on plant growth. While the SUBSTOR-Potato model has been extensively tested, it is in general confined to the traditional temperate region and with a limited testing in the highlands of Haleakala in Maui, Hawaii. The model has not been evaluated for the coastal areas, the plains of Indian subcontinent and the South-East Asian region. In order to raise the status of the tropical root crops from underexploited or

Table 7. Description of validation data for SUBSTOR-Potato (Griffin et al., 1993).

Source of data	Location	Latitude	Cultivars	Years
Sale (1973)	Murrumbridge, Australia	35.0 S	Sebago	1970, 1971
Manrique and Bartholomew (1991)	Hamakuapoko, Hawaii (91 m)	20.6 N	Katahdin, Desiree	1986
	Haleakala, Hawaii (640 m)	23.0 N	Katahdin, Desiree	1986
	Olinda, Hawaii (1097 m)	20.5 N	Katahdin, Desiree	1986
Ewing et al. (1990)	Ithaca, New York		Katahdin	1980, 1981, 1982, 1985, 1986
Ritchie (unpublished)	Kimberly, Idaho	42.3 N	Russet Burbank	1978
	Aberdeen, Idaho	43.0 N	Russet Burbank	1978
	Entrican, Michigan	43.2 N	Russet Burbank	1985, 1986, 1987, 1988
	Hermiston, Oregon	45.8 N	Russet Burbank	1988
	Grand Forks, North Dakota	47.9 N	Russet Burbank Norchip	1985, 1986 1987
Jeffries and MacKerron (1987, 1989)	Invergowrie, Scotland	56.5 N	Maris Piper	1984, 1985 1986, 1987

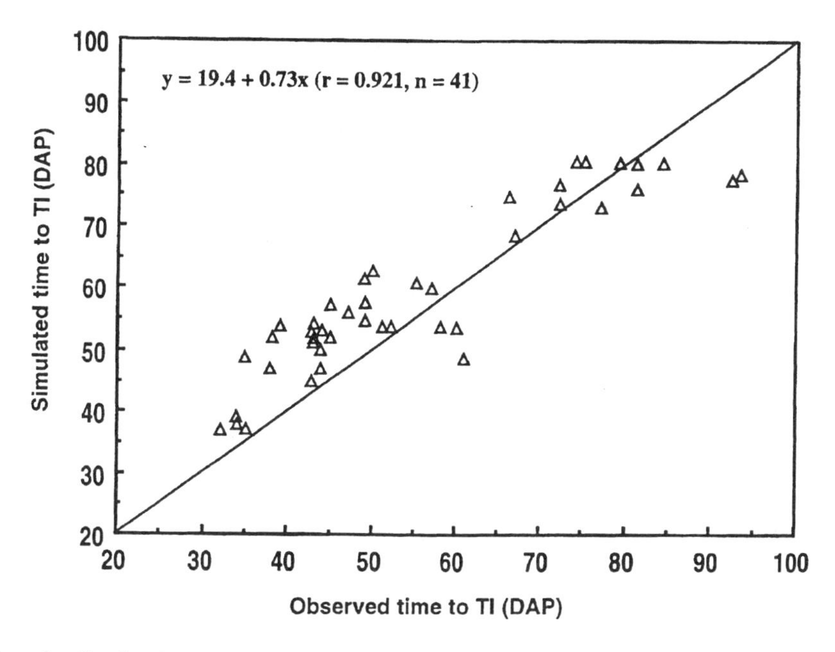

Figure 7. Simulated versus observed time to tuber initiation (TI) in potato.

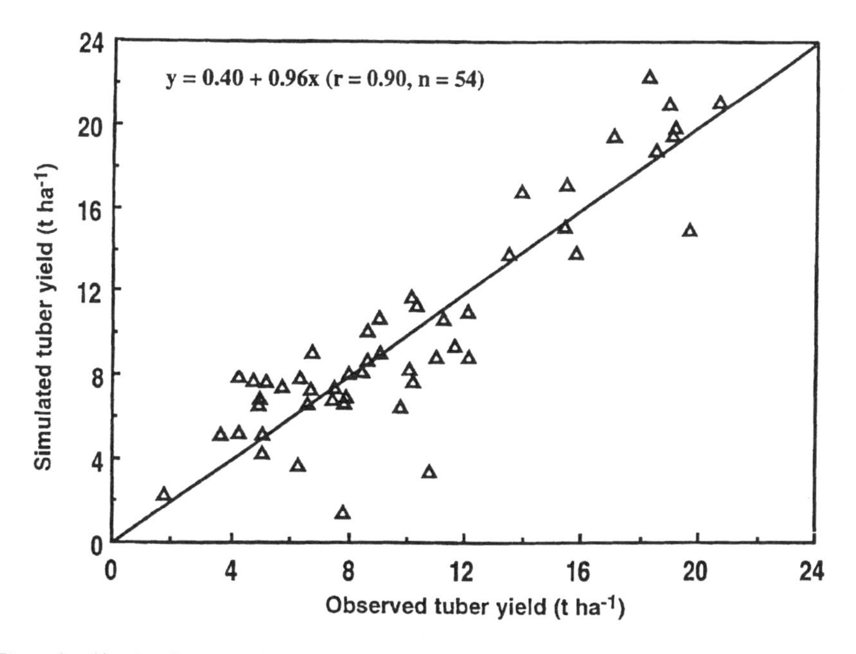

Figure 8. Simulated versus observed potato tuber yield.

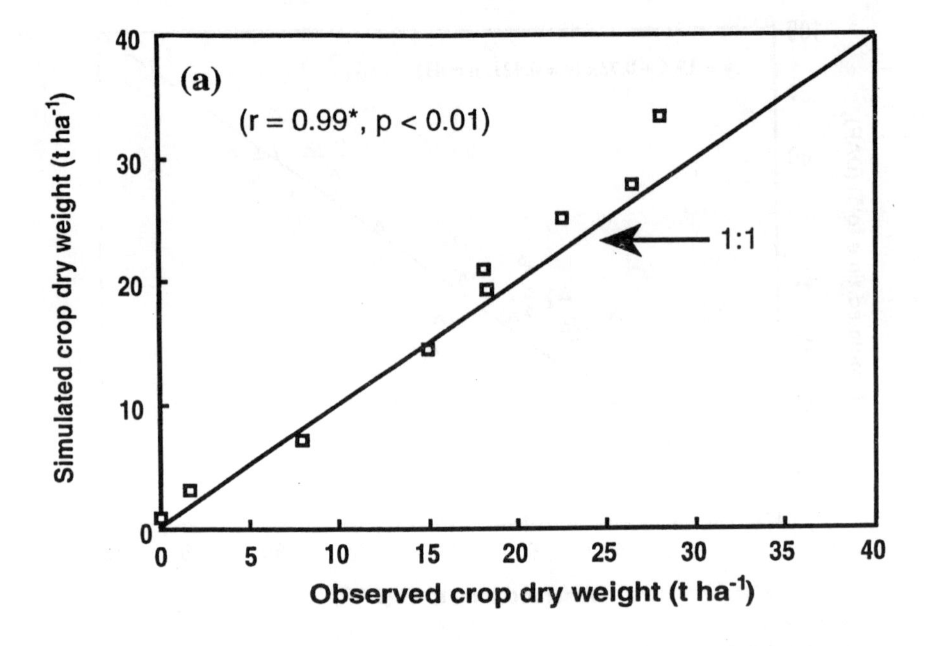

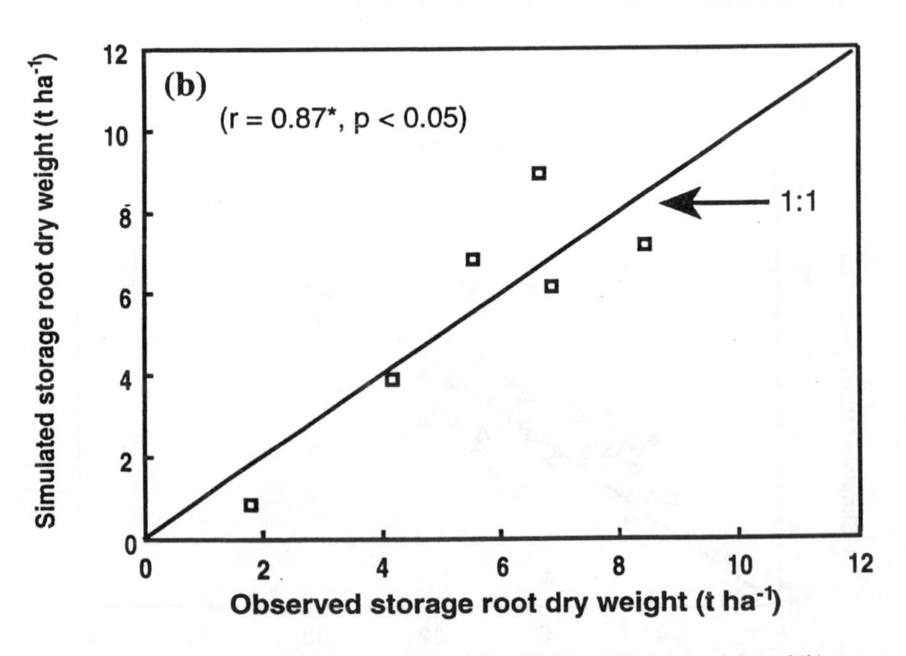

Figure 9. Comparison of observed and simulated values of (a) total crop dry weight and (b) storage root dry weight for cassava (Source of experimental data: Keating et al., 1982).

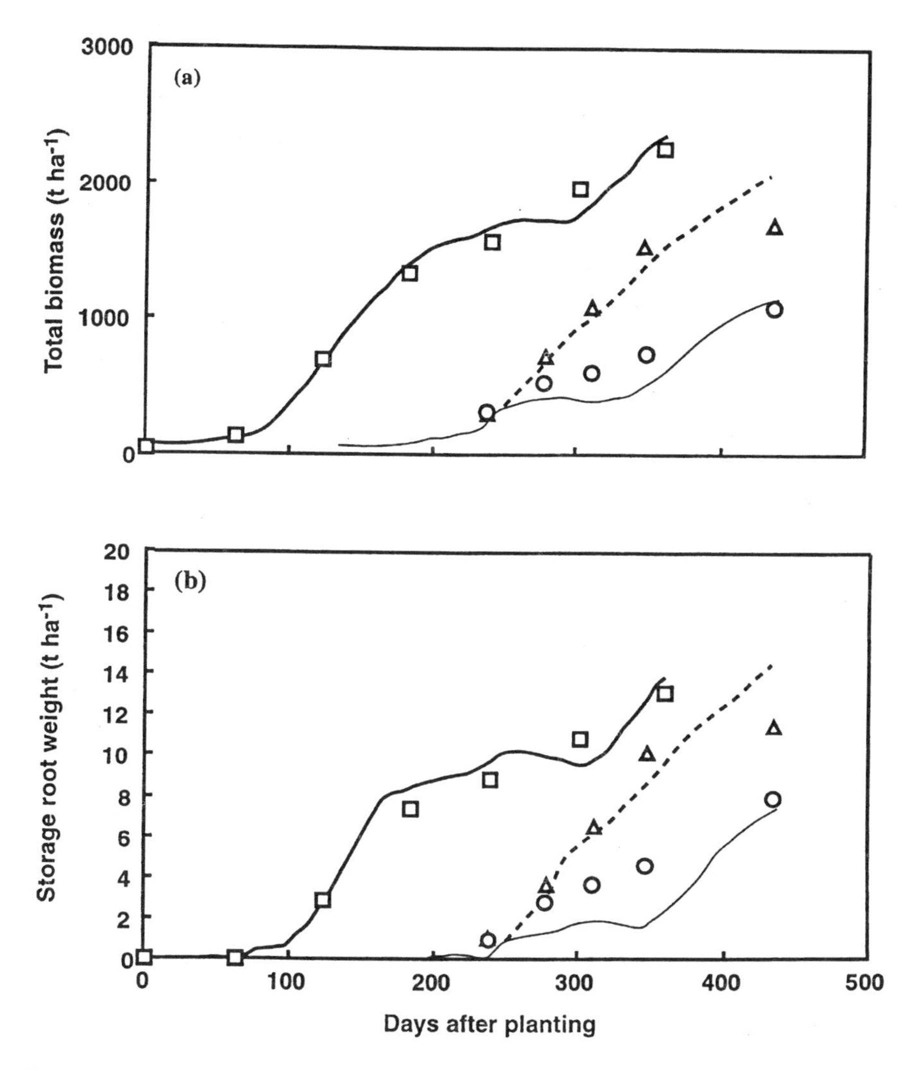

Figure 10. Comparison of simulated (lines) and observed values (symbols) for cassava cultivar MCol22 grown at CIAT, Colombia; (a) total crop dryweight and (b) storage root dry weight.

famine crops to one that is a major energy and food source, coordinated research among scientists in the developing world is essential. Future experiments must be designed to collect experimental data based on the MDS concept (IBSNAT, 1988; Singh, 1992), enhance basic research, and facilitate model development and improvement.

A database on varietal traits, cultural practices, and socioeconomic aspect is nonexistent. Yet farmers in developing countries have identified appropriate cultivars for upland versus lowland conditions, in terms of starch quality, and have developed unique cultural practices. The need for a scientific knowledge base for many of the underground storage organ crops (cassava, yams, sweet-potato, taro and tanier) is critical. Further development, improvement, and evaluation of SUBSTOR and CROPSIM models are thus dependent on establishing a network of coordinated experimental trials and improving and/or creating an information data base for the root crops.

Conclusions

The IBSNAT-type models have proven to be useful research and management tools. The root crop models SUBSTOR-Potato, SUBSTOR-Aroids and CROPSIM-Cassava are no exception. These models have simulated the differences among cultivars, effect of weather, soil, water stress, and N supply on crop growth and development. However, none of the models have been evaluated at a process-level. It is only with additional evaluation and in a more diverse environment and with a wide array of genotypes that the usefulness of the models as vital research and management tools can be confirmed. For the moment, however, the models can serve as tools to aid in the interpretation of field experiments. Used in this way, the models will help accelerate the acquisition of knowledge concerning the control of growth and development of these vitally important crops.

References

Allen E J, Scott R K (1980) An analysis of growth of the potato crop. Journal of Agricultural Science 94:583–606.

Boerboom B W J (1978) A model of dry matter distribution in cassava (*Manihot esculenta* Crantz). Netherlands Journal of Agricultural Science 26:267–277.

Bolhuis G G (1966) Influence of length of the illumination period on root formation in cassava, *Manihot utillissima* Pohl. Netherlands Journal of Agricultural Sciences 14:251–254.

Connor D J, Cock J H, Parra G E (1981) Response of cassava to water shortage. I. Growth and yield. Field Crops Research 4:181–200.

Cock J H, Franklin D, Sandoval G, Juri P (1979) The ideal cassava plant for maximum yield. Crop Science 19:271–279.

De Bruijn G H (1977) Influence of daylength on the flowering of cassava. Tropical Root Tuber Crops Newsletter 10:1–3.

Ewing E E (1981) Heat stress and the tuberization stimulus. American Potato Journal 58:31–49.

Ewing E E, Heym W D, Batutis E J, Snyder R G, Ben Khedher M, Sandlan K P, Turner A D (1990) Modifications to the simulation model POTATO for use in New York. Agricultural Systems 33:173–192.

FAO (1991) FAO production yearbook 1990. Vol. 44. FAO Statistics Series No.99. Food and Agriculture Organization of the United Nations (FAO), Rome, Italy.

Fukai S, Alcoy A B, Llamelo A B, Patterson R D (1984) Effect of solar radiation on growth of cassava (*Manihot esculenta* Crantz). I. Canopy development and dry matter growth. Field Crops Research 9:347–360.

Fukai S, Hammer G L (1987) A simulation model of the growth of the cassava crop and its use to estimate cassava productivity in Northern Australia. Agricultural Systems 23:253–257.

Gijzen H, Veltkamp H J, Goudriaan J, De Bruijn G H (1990) Simulation of dry matter production and distribution in cassava (*Manihot esculenta* Crantz). Netherlands Journal Agricultural Science 38:159–173.

Godwin D C, Jones C A (1991) Nitrogen dynamics in soil-plant systems. Pages 287–321 in Hanks R J and Ritchie J T (Eds.) Modeling plant and soil systems. Agronomy 31, American Society of Agronomy, Madison, Wisconsin, USA.

Godwin, D.C. and U. Singh (1991) Modeling of nitrogen dynamics in rice cropping systems. Pages 287–294 in Deturk P, Ponnamperuma F N (Eds.) Rice production on acid soils of the tropics. Institute of Fundamental Studies, Kandy, Sri Lanka.

Godwin D C, Ritchie J T, Singh U, Hunt L A (1990) A user's guide to CERES-Wheat V2.10. Simulation Manual IFDC-SM-2, International Fertilizer Development Center, Muscle Shoals, Alabama, USA.

Goenaga R, Chardon U (1993) Nutrient uptake, growth and yield performance of three tanier (*Xanthosoma* spp.) cultivars grown under intensive management. Journal of Agricultural Science University of Puerto Rico 77:1–10.

Goenaga R, Singh U (1992) Accumulation and partition of dry matter in tanier (*Xanthosoma* spp.). Pages 37–43 in Singh U (Ed.) Proceedings of the workshop on taro and tanier modeling University of Hawaii, Honolulu, Hawaii, USA.

Griffin T S, Johnson B S, Ritchie J T (1993) A simulation model for potato growth and development: SUBSTOR-Potato Version 2.0. Department of Agronomy and Soil Science, College of Tropical Agriculture and Human Resources, University of Hawaii, Honolulu, Hawaii, USA.

Gutierrez A P, Wermelinger B, Schulthess F, Baumgaertner J U, Herren H R, Ellis C K, Yaninek J S (1988) Analysis of biological control of cassava pests in Africa. I. Simulation of carbon, nitrogen and water dynamics in cassava. Journal of Applied Ecology 25:901–920.

Hartz T K, Moore F D III (1978) Prediction of potato yield using temperature and insolation data. American Potato Journal 55:431–436.

Horton D (1988) Underground crops: Long-term trends in production of roots and tubers. Winrock International, Morrilton, Arkansas, USA.

Hunt L A, Jones J W, Hoogenboom G, Godwin D C, Singh U, Pickering N B, Thornton P K, Boote K J, Ritchie J T (1994) Input and output file structures for crop simulation models. Pages 35–72 in Uhlir P F, Carter G C (Eds.) Crop modeling and related environmental data. A focus on applications for arid and semiarid regions in developing countries. CODATA, International Council of Scientific Unions, Paris, France.

International Benchmark Sites Network for Agrotechnology Transfer (IBSNAT) Project (1986) Decision support system for agrotechnology transfer. Agrotechnology Transfer 2:1–5.

International Benchmark Sites Network for Agrotechnology Transfer (IBSNAT) (1988) Technical Report 1: Experimental Design and Data Collection Procedures for IBSNAT. Third Edition, Department of Agronomy and Soil Science, College of Tropical Agriculture and Human Resources, University of Hawaii, Honolulu, Hawaii, USA.

Ingram K T, McCloud D E (1984) Simulation of potato growth and development. Crop Science 24:21–27.

Irikura Y, Cock J H, Kawano K (1979) The physiological basis of genotype-temperature interactions in cassava. Fields Crop Research 2:227–239.

Iritani W M (1963) The effect of summer temperature in Idaho on yield of Russet Burbank potatoes. American Potato Journal 40:47–52.

Jeffries J A, MacKerron D K L (1987) Aspects of physiological basis of cultivar differences in yield of potato under droughted and irrigated conditions. Potato Research 30:201–207.

Jeffries J A, MacKerron D K L (1989) Radiation interception and growth of irrigated and droughted potato (*Solanum tuberosum*). Field Crops Research 22:101–112.

Johnson K B, Johnson S B, Teng P S (1986) Development of a simple potato growth model for use in crop-pest management. Agricultural Systems 19:189–209.

Kay D E (1973) Root crops. Tropical Products Institute, London.

Keating B A, Evenson J P (1979) Effect of soil temperature on sprouting and sprout elongation of stem cuttings of cassava (*Manihot esculenta* Crantz). Field Crops Research 2:241–251.

Keating B A, Evenson J P, Fukai S (1982) Environmental effects of growth and development of cassava (*Manihot esculenta* Crantz). I. Crop development. Field Crops Research 5:271–281.

Keating B A, Wilson G L, Evenson J P (1985) Effect of photoperiod on growth and development of cassava (*Manihot esculenta* Crantz). Australian Journal of Physiology 12:631–640.

Lyonga S N, Nzietchueng S (1986) Cocoyam and the African food crisis. In Terry E R, Akoroda M O, Arena O B (Eds.) Tropical root crops: Root crops and the African crisis. Proceedings Third Triennial Symposium of the International Society for Tropical Root Crops, Owerri, Nigeria.

MacKerron D K L, Waister P D (1985) A simple model of potato growth and yield. Part I. Model development and sensitivity analysis. Agricultural and Forest Meteorology 34:241–252.

Manrique L A, Bartholomew D P (1991) Growth and yield performance of potato grown at three elevations in Hawaii. II. Dry matter production and efficiency of partitioning. Crop Science 31:367–372.

Manrique L A, Kiniry J R, Hodges T, Axness D S (1991) Dry matter production and radiation interception of potato. Crop Science 31:1044–1049.

Matthews R B, Hunt L A (1994) GUMCAS: model describing the growth of cassava (*Manihot esculenta* L. Crantz). Field Crops Research 36:69–84.

Ng E, Loomis R S (1984) Simulation of growth and yield of the potato crop. Simulation Monographs Pudoc, Wageningen, the Netherlands.

Paradales J R, Melchor F M, de la Pena R S (1982) Effect of water temperature on the early growth and development of taro. Annals Tropical Research 4(4);231–238.

Prange R K, McRae K B, Midmore D J, Deng R (1990) Reduction in potato growth at high temperature: role of photosynthesis and dark respiration. American Potato Journal 67:357–369.

Prasad H K, Singh U (1992) Effect of photoperiod and temperature on growth and development of taro (*Colocasia esculenta* (L.) Schott). Pages 29–35 in Singh U (Ed.) Proceedings of the workshop on taro and tanier modeling, University of Hawaii, Honolulu, Hawaii, USA.

Ritchie J T (1985) A user-oriented model of the soil water balance in wheat. Pages 293–305. In Day W, Atkin R K (Eds) Wheat Growth and Modeling. Plenum Press, New York, New York, USA.

Sale P J M (1973) Productivity of vegetable crops in a region of high solar input. II. Yields and efficiencies of water use and energy. Australian Journal of Agricultural Research 24:751–762.

Sands P J, Hackett C, Nix H A (1979) A model of the development and bulking of potatoes (*Solanum tuberosum* L.). I. Derivation from well-managed field crops. Field Crops Research 2:309–331.

Silva J A, Coltman R, Paul R, Arakaki A (1992) Response of chinese taro to nitrogen fertilization and plant population. Pages 13–16 in Singh U (Ed.) Proceedings of the workshop on taro and tanier modeling, University of Hawaii, Honolulu, Hawaii, USA.

Singh U (Ed) (1992) Proceedings of the workshop on taro and tanier modeling, 8–14 August 1991, University of Hawaii, Honolulu, Hawaii, USA.

Singh U (1994) Nitrogen management strategies for lowland rice cropping system. Pages 110–130 in Proceedings International Conference on Fertilizer Usage in the Tropics, Kuala Lumpur, Malaysia.

Singh U, Tsuji G Y, Goenaga R, Prasad H K (1992) Modeling growth and development of taro and tanier. p. 45–56. In U. Singh (ed.) Proceedings of the workshop on taro and tanier modeling, 8–14 August 1991, University of Hawaii, Honolulu, Hawaii, USA.

Singh U, Ritchie J T, Godwin D C (1993) A user's guide to CERES-RICE–V2.10. Simulation Manual IFDC-SM-4, International Fertilizer Development Center, Muscle Shoals, Alabama, USA.

Singh U, Prasad H K, Goenaga R, Ritchie J T (1995) A user's guide to SUBSTOR-Aroids V2.10. Department of Agronomy and Soil Science, College of Tropical Agriculture and Human Resources, University of Hawaii, Honolulu, Hawaii, USA.

Snyder R G, Ewing E E (1989) Interactive effects of temperature, photoperiod, and cultivar tuberization of potato cuttings. Horticultural Science 24:336–338.

Tsuji G Y, Uehara G, Balas S (1994) Decision Support System for Agrotechnology Transfer DSSAT v3 International Benchmark Sites Network for Agrotechnology Transfer, University of Hawaii, Honolulu, Hawaii, USA.

Tsukamoto Y, Inaba, K (1961) The effect of daylength upon the cormel formation in taro. Memoirs of the Research Institute of Food Science Kyoto University 23:15–22.

Veltkamp H J (1986) Physiological causes of yield variation in cassava (*Manihot esculenta* L. Crantz). Agricultural University Wageningen Papers 85–6 (1985).

Wheeler R M, Tibbetts T W (1986) Utilization of potatoes for life support systems in space: I. Cultivar photoperiod interactions. American Potato Journal 63:315–323.

Decision support system for agrotechnology transfer: DSSAT v3

J. W. JONES[1], G.Y. TSUJI[2], G. HOOGENBOOM[3], L.A. HUNT[4], P.K. THORNTON[5], P.W. WILKENS[6], D.T. IMAMURA[2], W.T. BOWEN[7] and U. SINGH[6]

[1]*Department of Agricultural and Biological Engineering, P.O. Box 110570, University of Florida, Gainesville, FL32611–0570, USA*
[2]*Department of Agronomy and Soil Science, 2500 Dole Street, Kr 22, University of Hawaii, Honolulu HI 96822, USA*
[3]*Biological and Agricultural Engineering Department, University of Georgia, Griffin, Georgia, USA*
[4]*Crop Science Department, University of Guelph, Guelph, Ontario, Canada*
[5]*International Livestock Research Institute, P.O. Box 30709, Nairobi, Kenya*
[6]*International Fertilizer Development Center, P.O. Box 2040, Muscle Shoals, AL35662, USA;*
[7]*Centro International de la Papa, Apartado 1558, Lima 100, Peru*

Key words: models, decision support system, DSSAT, sustainability, technology transfer, risk management

Abstract

Agricultural decision makers at all levels need an increasing amount of information to better understand the possible outcomes of their decisions to help them develop plans and policies that meet their goals. An international team of scientists developed a decision support system for agrotechnology transfer (DSSAT) to estimate production, resource use, and risks associated with different crop production practices. The DSSAT is a microcomputer software package that contains crop-soil simulation models, data bases for weather, soil, and crops, and strategy evaluation programs integrated with a 'shell' program which is the main user interface. In this paper, an overview of the DSSAT is given along with rationale for its design and its main limitations. Concepts for using the DSSAT in spatial decision support systems (for site-specific farming, farm planning, and regional policy) are presented. DSSAT provides a framework for scientific cooperation through research to enhance its capabilities and apply it to research questions. It also has considerable potential to help decision makers by reducing the time and human resources required for analyzing complex alternative decisions.

Introduction

The International Benchmark Sites Network for Agrotechnology Transfer (IBSNAT) project successfully completed ten years of activities with two major accomplishments: (i) a functional computerized product called the Decision Support System for Agrotechnology Transfer (DSSAT), and (ii) a global network of cooperators active in developing, testing, and applying agricultural system models to solve agricultural problems worldwide (IBSNAT 1993). The rationale for the approach taken by this global project was founded on the premise that the Earth's land resources are finite, whereas the number of people

G.Y. Tsuji et al. *(eds): Understanding Options for Agricultural Production, 157–177*

that the land must support continues to grow rapidly. This creates a major problem for agriculture; production must be increased to meet rapidly growing demands while natural resources must be protected. Now, more than ever, decision makers at all levels need an increasing amount of information to help them understand the possible outcomes of their decisions and develop plans and policies for achieving their goals. New agricultural research is needed to supply information to farmers, policy makers, and other decision makers on how to accomplish sustainable agriculture over the wide variations in climate, soils, social, political, and economic environments around the world. Traditional research can not meet the needs for such information. There are several reasons for this. First, traditional experiments are conducted at particular points in time and space, and results are thus site- and season-specific. Weather and soil variabilities are very important factors in experimental results. Second, traditional research is time consuming and expensive, and thus it is unlikely that sufficient amounts of traditional research could supply needed information in the time frame needed, even if it could be financed. Third, agricultural research is often lacking where it is needed most, for example in the developing countries of the tropics.

The IBSNAT project proposed to use a systems approach to transfer knowledge because of the unrealistic amounts of trial and error, traditional research that would be needed without it (Uehara, 1989). The systems approach provides a framework in which research is conducted to *understand* how the system and its components function. This understanding is then integrated into system models that allow one to *predict* the behavior of the system for a given set of conditions. After one is confident that the models can simulate the real world adequately, computer experiments can be performed hundreds or even thousands of times for given environments to determine how to best manage or *control* the system. The IBSNAT project applied these concepts to crop production systems; DSSAT was developed to operationalize this approach and make it available for application on a global basis.

Before proceeding, some explanation of the use of the term decision support system is needed. Decision support systems (DSS) are 'interactive computer-based systems that help decision makers utilize data and models to solve unstructured problems' (Sprague and Carlson, 1982). The goal of such systems is to improve the performance of decision makers while reducing the time and human resources required for analyzing complex decisions. In concept, DSS should support all phases of a decision making process, characterized by Simon (1960) as: (1) searching for conditions calling for a decision by identifying possible problems or opportunities, (2) creating and analyzing possible courses of action, and (3) suggesting a course or courses of action from those analyzed.

Most of the relatively few agricultural decision aids that have been developed are aimed at the farmer as a decision maker, and attempt to improve operational decisions such as pest control (Hearn, 1987; Michalski et al., 1983) fertilizer management (Yost et al., 1988), or a wider set of crop management decisions

(Plant 1989; McKinion et al., 1989). The need for agricultural decision support systems extends beyond those for field level operational decisions with the farmer as a decision maker. A different approach was taken by the cooperating scientists in the IBSNAT project. The DSSAT was designed for users to easily create 'experiments' to simulate, on computers, outcomes of the complex interactions between various agricultural practices, soil, and weather conditions and to suggest appropriate solutions to site specific problems (Jones, 1986; Uehara, 1989). This system relies heavily on simulation models to predict the performance of crops for making a wide range of decisions. The original DSSAT (version 2.1) was released in 1989 (IBSNAT 1989). A second release (version 3.0) was made late in 1994 (Tsuji et al., 1994). The purposes of this chapter are to describe the latest version of DSSAT and to assess its future potential and impact.

Description of DSSAT v3

The DSSAT was designed to allow users to (1) input, organize, and store data on crops, soils, and weather, (2) retrieve, analyze and display data, (3) calibrate and evaluate crop growth models, and (4) evaluate different management practices at a site. In adapting and applying the DSSAT to a location, users typically use the following procedures (Jones, 1993):

(1) Conduct field experiments on one or more crops, and collect a minimum data set (MDS) required for running and evaluating a crop model. Field experiments may be needed to calibrate local cultivars. Run the model using the new data to evaluate the ability of the model to predict performance of crops in the region of interest. In many cases, data from previous experiments are used. Modify model if evaluation shows that it does not reach the level of precision required.
(2) Enter other soil data for the region and historical weather data for sites in the region. Quality check of soil and weather data. Conduct sensitivity analysis on the crop model(s) to gain an overview of model responses to alternative management practices and weather conditions.
(3) Select a set of new management practices and simulate each of these together with existing practices, over a number of years to predict the performance and uncertainty associated with each practice. Compare the alternative practices using means, variances, and cumulative probability distributions of simulated yield, water use, season length, nitrogen uptake, net profit and other responses. Provide results and recommendations for decision making.

The functional capabilities of DSSAT v3 can be summarized by Figure 1. Crop simulation models are at the center of the system. Data bases describe weather, soil, experiments, and genotype information for applying the models

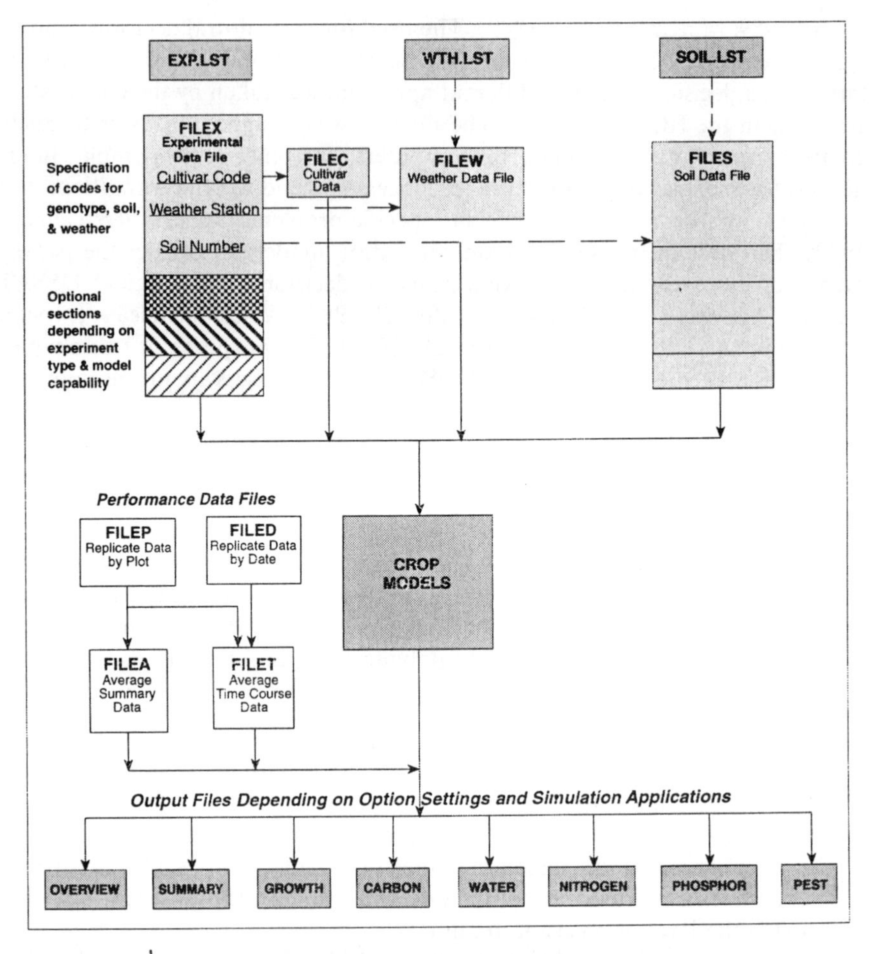

Figure 1. Schematic of the main components of DSSAT v3.

to different situations. Software in the system helps users prepare these data bases for their own fields or farms so that DSSAT can simulate performance of real or proposed experiments at their sites. Outcomes from the application software allow users to compare simulated results with their own measured results to give them confidence that the models work adequately, or to determine if modifications are needed to improve their capabilities or accuracy. Outputs can be printed or graphically displayed for conducting sensitivity analyses In addition, programs are contained in DSSAT to allow users to simulate options for crop management over a number of years to assess the risks associated with each option (Thornton et al., 1995). Table 1 lists the options that users can choose in the DSSAT to create different management strategies, and the simulated performance indicators that can be analyzed.

Table 1. Listing of the crop management options to create different strategies, and the crop performance variables that can be studied in DSSAT v3.

Management options	Variables available for analysis
Crop cultivar	Grain yield
Planting	Pod yield
Plant population	Biomass
Row Spacing	Season length
Soil type	Reproductive season length
Irrigation	Seasonal rainfall
Fertilization (nitrogen)	Seasonal evapotranspiration
Initial conditions	Water stress, vegetative
Crop residue management	Water stress, reproductive
Crop rotations	Number of irrigations
Harvesting	Total amount of irrigation
Pest damage	Number of nitrogen applications
	Nitrogen applied
	Nitrogen uptake
	Nitrogen leached
Weather factors	Nitrogen stress, vegetative
	Nitrogen stress, reproductive
Temperature	Net returns
Solar radiation	Phosphorus applied
Precipitation	Seed used
Carbon dioxide	Runoff
Wind[1]	Soil organic C
Relative humidity[1]	Soil organic N
	Residue applied
	Nitrogen fixation

[1]Optional daily weather data requirements. These variables are used in the Penman method for computing daily potential evapotranspiration, if they are available.

The programs to perform these functions in DSSAT v3 are written in various computer languages. A shell program, the user interface, provides access to these programs using pop-up menus (Figure 2). Thus, users do not have to worry about which language is being used or how to execute that particular program. Arrow keys are used to select specific tasks to perform. A program is included to install DSSAT. This system is available through the IBSNAT office at the University of Hawaii (Tsuji et al., 1994). Source codes for each of the crop models are not included in the DSSAT package, but can be obtained by request from IBSNAT or any of the model authors.

Data base component

Data are essential for all aspects of a systems approach; for quantifying system responses, for developing and testing models, and for analyzing options to assist in decision making. The IBSNAT project emphasized this facet of the systems approach from the very start, which led to two major contributions essential to the project. First, the concept of a minimum data set was developed (IBSNAT 1988). This development provided needed guidance to researchers as to the type of data needed to validate predictive capabilities of the models

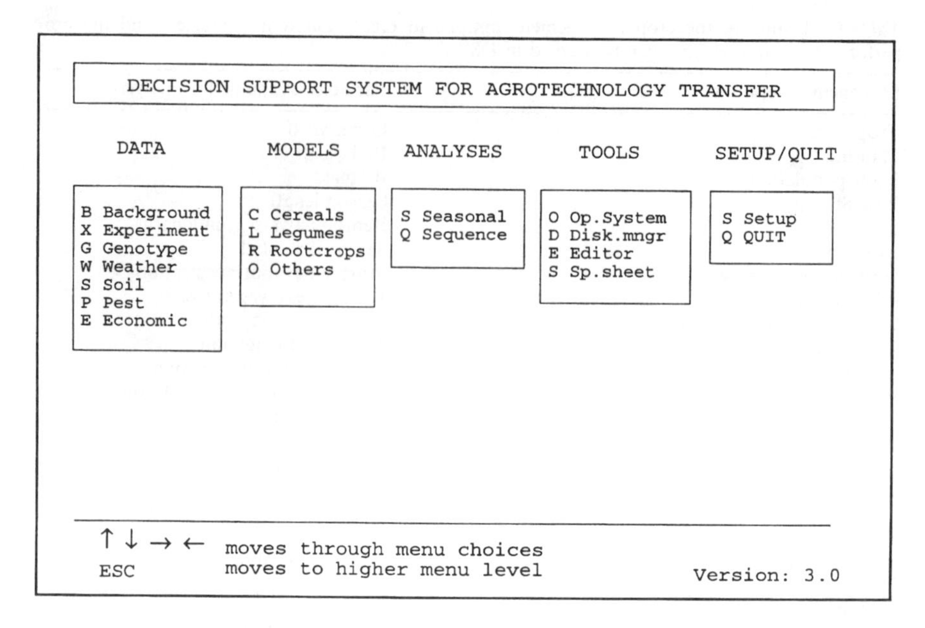

Figure 2. Main screen of DSSAT v3 showing menu options that guide users to different functions.

and to apply these data to their own conditions. Secondly, data file organization and data formats for crop systems were defined and documented (Jones et al., 1994; Hunt et al., 1994a). The DSSAT 'standard' data files and formats provide a very useful framework for documenting experiments in a uniform way; data can be read by researchers as well as by computers for use as model inputs. Data definitions include crop management, soil, and weather information to characterize how, when, and where the crop was grown and to characterize measured crop and soil performance. These DSSAT standards provide all definitions needed as inputs to operate the crop models as well as model outputs for use in various applications. DSSAT v3 fully utilizes these data structures. This feature facilitates easy addition of new data sets by experimenters and easy addition of additional crop models and application programs if they adhere to the standard inputs and outputs. For example, DSSAT has recently been linked with GIS (Engel and Jones, 1995; Beinroth et al., 1997) creating capabilities for use of any crop model in DSSAT for analysis of crop production over different spatial scales. These standard data files and structures are fully documented (Jones et al., 1994; Hunt et al., 1994a).

The data files and data formats used in DSSAT v3 are considerably different than those in the original DSSAT v2.1. Originally, commercial data base management software was used to enter and store the data in DSSAT. Utilities were included with v2.1 to extract data into ASCII files specifically for input to the crop models. However, we found that most researchers were not using this feature of DSSAT v2.1 because of the difficulty of accessing data for other purposes. If users needed to access data, either they had to have the data base software and know how to work with it, or they had to use only the data extracted for use by crop models. This was too restrictive for general use. The new DSSAT v3 data standards are in ASCII files which can easily be accessed by any text editor or spread sheet program. Thus, all data can now be entered, manipulated, exchanged, and analyzed using familiar software.

Figure 3 is a schematic of the DSSAT v3 data base components, showing how the data are used by crop models. The experiment details file (referred to as FILEX) contains information to document a field experiment or management of a farmer's field. This file is also read by a module (the input module) of the crop models to simulate the crop growing under the specified management conditions. Soil profile data are stored in a file that is accessed by the input module as well; the profile data are selected using a soil number from FILEX. Cultivar information are read from crop-specific files that contain information about a number of different cultivars; FILEX specifies which cultivar to select from this file. Weather data are stored in yearly files using a naming convention defined by Jones et al. (1994). Output files from the crop models in DSSAT v3 depend on the options selected in a simulation control section in FILEX. Measured field data are stored in column format in ASCII files. Data from these files are read by various DSSAT v3 programs to compare with simulated results for validation purposes.

The DSSAT v3 has various programs to help users create new experiments (real or hypothetical). These are described in detail by Hunt et al. (1994a). A few of the major utilities are presented below.

(1) XCREATE. The purpose of this utility is to enable users to create an experiment description file, i.e., to create a FILEX, which is used to document an experiment and also to provide inputs to the crop models (Imamura et al., 1994). The program has a menu that allows users to easily enter data for defining treatments as well as crop management for each treatment. Users select menus for entry of general descriptive information for the experiment, field information, initial soil conditions, irrigation, fertilizer management, residue management, cultivar, tillage, and other information. Lists in DSSAT v3 contain information that helps users create new experiments, such as lists of available soils, cultivars, etc., and the software shows these lists when needed for user convenience. Another feature of the FILEX creation software is that users can select an existing experiment and make modifications in it to create a new one so that they do not have to start with an empty file each time. When all data have been entered, a new FILEX is saved in the exact format needed, relieving the user from having to worry about details of the format.

(2) WEATHERMAN. The purpose of this program is to enable users to reformat weather data files to make them compatible with the DSSAT v3

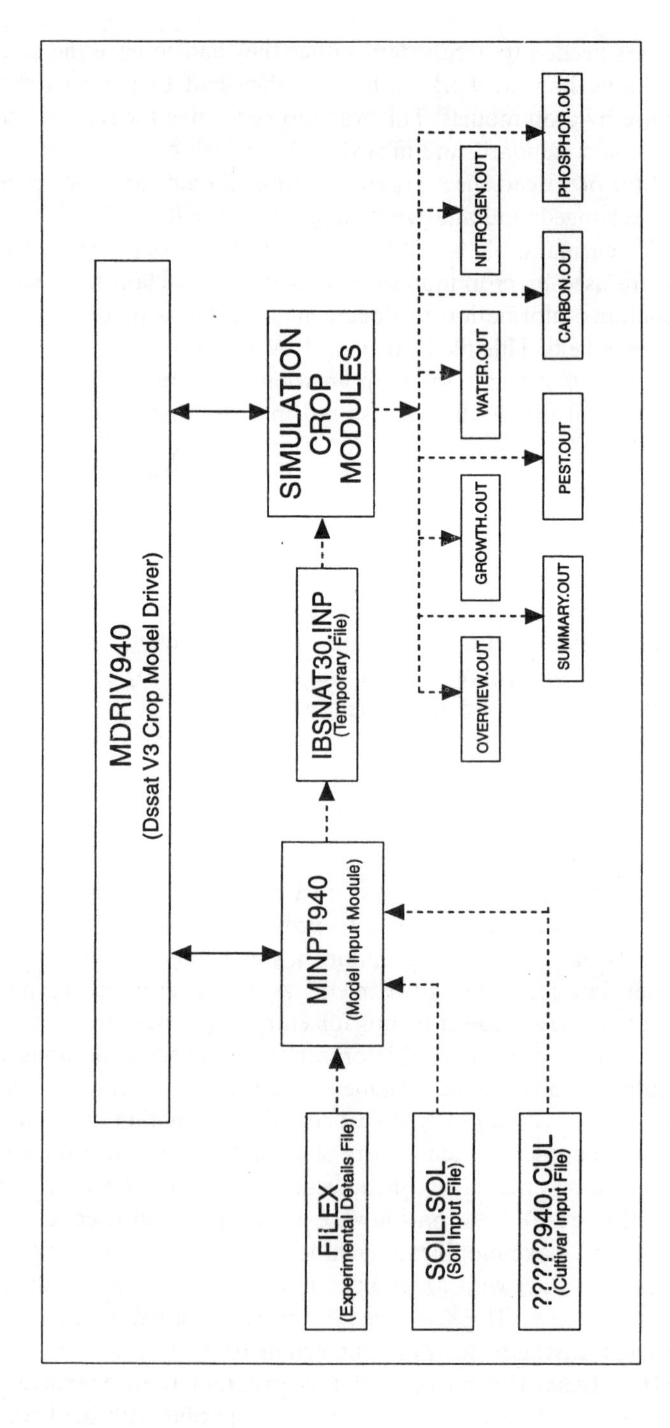

Figure 3. Diagram of DSSAT v3 data base components and how they provide information for crop model applications.

models, fill in missing data, generate weather data, compute statistics, and graph weather data (Pickering et al., 1994; Hansen et al., 1994). The program allows one to define weather station information in climate files, which include monthly averages of daily weather data. These climate data files are used to generate daily data using the SIMMETEO model of Geng et al. (1988), adapted to run within the DSSAT WeatherMan program. When daily weather data are available for a location, they are 'imported' into an internal file where data are checked for out-of-range values, missing values, or missing time periods of data. Formats of the raw data files can be defined and saved for future use when additional data in the same format are obtained. Statistics are computed for daily weather generation using the WGEN (Richardson and Wright, 1984) model. When users wish to create complete files for use by the models, various options are available for filling in missing values or out-of-range data. Real weather data and generated weather data can be analyzed and compared by computing a number of statistics.

(3) WINGRAF. This utility is used to graph simulated and observed data. After any crop simulation, users can select 'Graph' from the DSSAT menu to graph simulated and observed (if they are available) time series graphs, or graph observed vs. simulated yield or other variables for validation purposes.

(4) GENCALC. The purpose of this program is to allow users to calculate cultivar coefficients for different crops (Hunt et al., 1994b). This program runs the crop models for existing experiments a number of times, modifying cultivar coefficients each time until simulated and observed crop traits are equal or within a defined tolerance. Rules are specified for each crop as to the possible range of each coefficient, the measured crop trait that is used as a target value, and how much to change the coefficient on each successive run. The program allows coefficients to be computed for a number of treatments or experiments, then averages and coefficients of variability are computed to allow users to obtain estimates best suited for their conditions. This procedure terminates by entering new cultivar values into the crop's cultivar file.

(5) LIST/EDIT. The List/Edit feature of DSSAT allows users to obtain various lists of data of different kinds (Hunt et al., 1994a). For example, the list/edit function under the 'Experiment' menu of DSSAT lists all available experiments in a particular directory. For each experiment, it lists the crop, the experiment name, and a brief description of the experiment. The List/Edit program obtains this information from FILEX's, in this case, and creates an internal relational data base for user access. Users can highlight a particular experiment using the arrow keys, then by pressing the <F8> key, they can edit FILEX, or any of the data files containing observed data. Users can search to locate experiments with certain characteristics. A large number of experiments may be stored on

the computer, so users can 'toggle' on or off a number of experiments with which he/she is working. The List/Edit program then creates a file with a list of those experiments selected (the EXP.LST file). This file is accessed by crop models to provide a list of real or hypothetical experiments that can be simulated.

(6) SOIL PROFILE CREATE. The purpose of this program is to allow users to create new soil profiles for their experiments or analyses. Users are asked for specific information, such as number of layers, percent sand, silt, and clay, bulk density, etc., that are usually available from soil characterization data. Soil water storage and other variables are computed by a program to provide a full set of soil profile data required by the crop models (Jones et al., 1994). Users can edit this information before it is written to the soil file (SOIL.SOL) for use by the crop models. Although it is recommended that soil data be collected from the fields in which experiments are conducted, this is sometimes not possible; thus this soil profile creation utility provides a useful approximation for users.

Crop simulation models

The functions of DSSAT were selected primarily to support the use of crop simulation models in decision making applications. The utility of this system depends on the ability of the crop models to provide realistic estimates of crop performance for a wide range of environment and management conditions and on the availability of data required to operate the models. The first release of the DSSAT (V2.1, IBSNAT 1989) contained models of the following four crops: maize (CERES-Maize V2.10; Ritchie et al., 1989), wheat (CERES-Wheat V2.10; Godwin et al., 1989), soybean (SOYGRO V5.42; Jones et al., 1989) and groundnut (PNUTGRO V1.02; Boote et al., 1989). Eight additional crop models have since been added: rice (CERES-Rice; Godwin et al., 1990, Singh et al., 1993) drybean (BEANGRO V1.01; Hoogenboom et al., 1991), sorghum (CERES-Sorghum; Alargarswamy and Ritchie, 1991), millet (CERES-Millet; Singh et al., 1991), barley (CERES-Barley; Otter-Näcke et al., 1991), potato (SUBSTOR-Potato; Griffin et al., 1993), aroid (SUBSTOR-Aroid; Prasad et al., 1991), and cassava (CROPSIM-Cassava; described by Matthews and Hunt, 1994). These models are process oriented, and are designed to have global applications; i.e., to be independent of location, season, and management system. The models simulate effects of weather, soil water, cultivar, and nitrogen dynamics in the soil and crop, on crop growth and yield for well drained soils. On modern personal computers they each require only a few seconds to simulate one growing season.

One of the lessons learned from the implementation of a number of crop models in DSSAT v2.1 was that each model had to contain similar computer code to read input data, allow users to conduct sensitivity analyses, and provide other capabilities needed for routine tasks. This situation created possibilities for errors in programming, required duplication of effort, and resulted in the need to maintain a number of different programs that performed basically the same functions. Thus, Hoogenboom et al. (1994) developed a modular system for implementing crop models. It contains a model driver and a model input module for use by all crop models in DSSAT v3 (Figure 4). The model driver (MDRIV940.EXE) controls access to the input module and to the correct crop model. It is executed every time any crop is to be simulated. The input module (MINPT940.EXE) reads crop management conditions from FILEX. It checks for errors and allows users to modify any input for sensitivity analysis. This module then writes to disk a simple, temporary file with all information needed to simulate a crop at a particular site. After writing this temporary file, the input module is finished; the driver then executes the appropriate crop model. The crop model reads the temporary file, simulates growth and yield, writes outputs (Figure 4), then returns control to the driver. This design separates the task of model execution from the task of reading the standard data files, thus allowing crop model developers to focus on the models as opposed to programming to link the model with the data structures.

Cereals: The CERES family of crop models

The original, individual CERES crop models were combined into a single module to simulate wheat, maize, barley, sorghum, and millet (Singh et al., 1991). Only the CERES-Rice model was kept separate because of its major differences in soil water and nitrogen balance routines and the need to simulate transplanting effects. The CERES models increment growth on a daily time step and require daily weather data (maximum and minimum temperature, solar radiation, and precipitation). They compute crop phasic and morphological development using temperature, day length, and cultivar characteristics. Daily dry matter growth is based on light intercepted by the leaf area index multiplied by a conversion factor. Biomass partitioning into various plant components is based on potential growth of organs and daily amount of growth produced. Soil water and nitrogen balance submodels provide daily values of supply to demand ratios of water and nitrogen, respectively, which are used to influence growth and development rates.

The CERES models simulate growth by taking into account the following processes:

(1) Phenological development, especially as it is affected by genotype, temperature, and daylength. The models simulate the timing of panicle initiation and the duration of each major growth stage.
(2) Extension growth of leaves, stems, and roots (morphological development)
(3) Biomass accumulation and partitioning
(4) Soil water balance that simulates daily soil evaporation, plant transpiration, runoff, percolation, and infiltration under rainfed and irrigated condi-

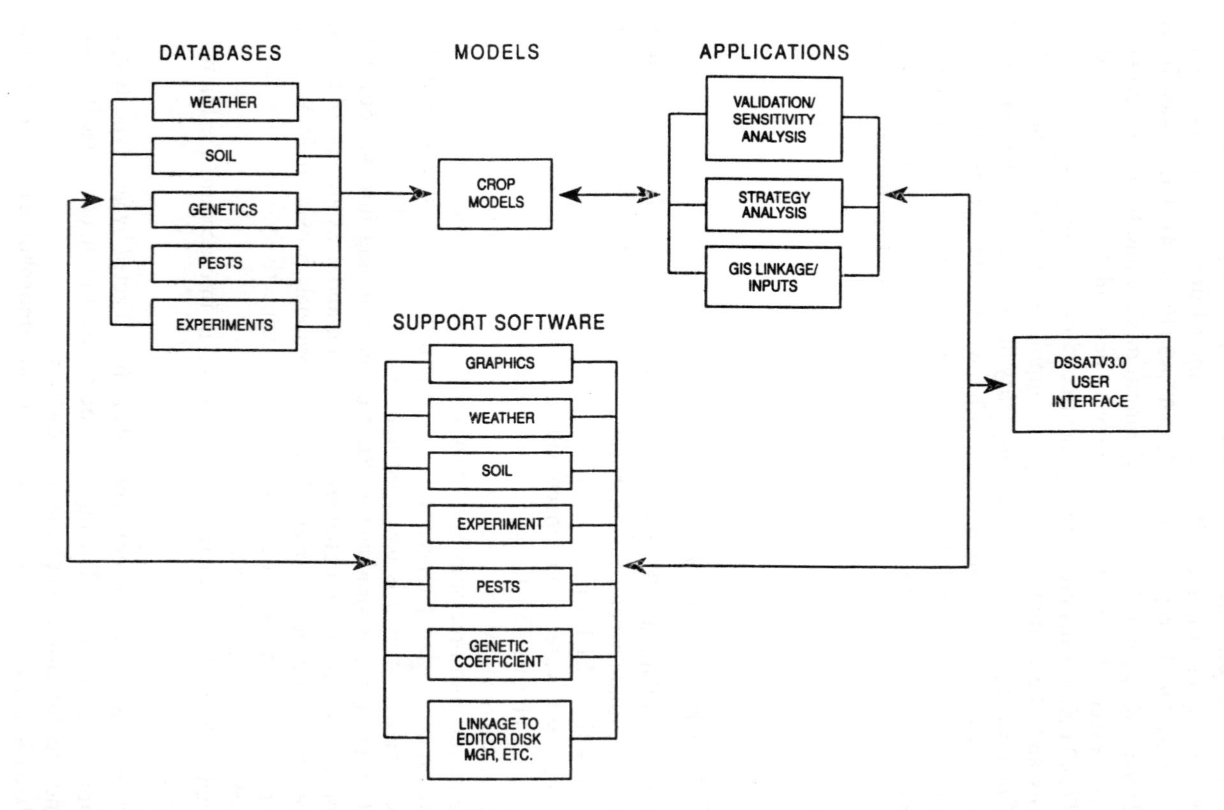

Figure 4. Schematic of relationships between the crop model driver program (MDRIV940.EXE), the crop model input program (MINPT940.EXE), the temporary file read by the crop model modules (IBSNAT30.INP), and the crop simulation model modules.

tions. Water deficiency affects leaf expansion and, if sufficiently severe, dry matter production.

(5) Soil nitrogen transformations associated with mineralization/immobilization, urea hydrolysis, nitrification, denitrification, ammonia volatilization, N uptake and use by the crop, and losses of N associated with runoff and percolation. N limitations affect leaf area development, tillering, photosynthesis, and senescence of leaves during grain filling.

The CERES-Rice model is similar to other CERES models, with several additional features. In CERES-Rice, the effect of transplanting on growth and development has been added. Another feature allows the water balance to simulate crop water use under flooded rice systems, under upland conditions, or under intermittent flooding and dry soil surface conditions. Finally, the nitrogen submodel required rather major modifications to simulate N transformations under flooded and intermittant flooded and upland conditions.

Grain legumes: The CROPGRO family of models. Originally, the models for soybean, peanut, and dry bean were programmed as separate computer codes, and were referred to respectively as SOYGRO, PNUTGRO, and BEANGRO (Hoogenboom et al., 1992). Because of their similarity and the difficulty of maintaining separate codes for each, the authors amalgamated the models into a single set of computer codes (Hoogenboom et al., 1994, Boote et al., 1997) for use in DSSAT v3. These models are now referred to as CROPGRO-Soybean, CROPGRO-Peanut, and CROPGRO-Dry Bean, for soybean, peanut, and dry bean, respectively (Hoogenboom et al., 1994). The models simulate the timing of phenological stages as affected by temperature and daylength. Dry matter production is computed by daily or hourly canopy photosynthesis models. Dry matter partitioning is based on source-sink relationships during early vegetative growth and during reproductive growth. Partitioning during the later vegetative phase is based on empirical functions that vary by crop. Crop-specific data files provide coefficients that characterize functional responses of each crop to its environment. Cultivar-specific data files provide coefficients to simulate the response of different cultivars to environment. For example, cultivar-specific coefficients quantify the photoperiod and temperature responsiveness of a cultivar as well as characteristics of vegetative and reproductive growth.

The CROPGRO family of models includes the same soil water and soil nitrogen modules as the CERES family. Growth of each crop is based on carbon, water, and nitrogen balances in the plant. In the process of creating a single set of codes, other changes were made to facilitate more applications, such as climate change impacts on crops (Curry et al., 1990), pest effects on crops (Batchelor et al., 1993). Currently, only the grain legume models in DSSAT can simulate effects of pest damage.

Root crop models: CROPSIM and SUBSTOR. Models for cassava, aroid, and potato were developed by Matthews and Hunt (1994), Prasad et al. (1991), and Griffin et al. (1993), respectively. The models for aroid and potato are named SUBSTOR models to reflect the SUBterranean STORage product, and are referred to as SUBSTOR-Aroid and SUBSTOR-Potato, respectively. The cassava model is named CROPSIM-Cassava. These models all operate similarly to the CERES and CROPGRO families of models. The same driver and input modules (Hoogenboom et al., 1994) link each model with the DSSAT v3 data base to enable users to select experiments, treatments, and perform sensitivity analyses. The models compute development and growth as the other models, and use the same soil water and nitrogen models. They were originally based on the CERES family of crop models.

Other crop models. In addition to the twelve crops currently in DSSAT v3, additional research has produced new models that are being integrated into a future version of DSSAT (sunflower by Villalobos et al., 1995; sugarcane by Inman-Bamber, 1991; and tomato by Scholberg et al., 1997). After these models are integrated as planned, users will be able to use existing soil and weather data in the DSSAT as well as existing utility programs to perform a range of computer studies. This clearly demonstrates the benefits of having a well-defined system, open to users who want to add models of other crops or. to add other models of the same crops.

Applications for decision making

The real power of the DSSAT for decision making lies in its ability to analyze many different management strategies. When a user is convinced that the model can accurately simulate local results, a more comprehensive analysis of crop performance can be conducted for different soil types, cultivars, planting dates, planting densities, and irrigation and fertilizer strategies to determine those practices that are most promising and least risky. Two programs in DSSAT v3 allow users to analyze options for management. The first is the Seasonal Analysis program described by Thornton et al. (1994a) and Thornton and Hoogenboom (1994). This program enables users to run multiple year simulations with one or more crop models using actual or generated weather data. With this program, the soil initial conditions are re-initialized before each season to simulate the uncertainty in crop performance due to uncertainty in weather, given the existing starting conditions of the soil. Users can create new options for analysis using the same XCREATE program described above and analyze both biophysical variability and economic risks of each practice. They can also modify economic information to conduct sensitivity analyses on economic variables, and include uncertainty in economic data. The second program under the ANALYSES menu in DSSAT v3 allows users to create crop rotations and analyze the changes over time in production of the crops in the sequence

(Thornton et al., 1994b, 1995). Multiple year simulations are run using generated weather sequences to simulate the effects of uncertain future weather patterns. Under this option, soil initial conditions are set at the start of the first crop, then they are carried over from one crop to the next to study the effects of changes in soil carbon, nutrient, and water states on crop production over time. Users can analyze biophysical and economic variable time trends using graphical and statistical analysis programs developed for DSSAT (Thornton et al., 1994b, 1995). Regression analyses are performed if desired to determine how simulated crop performance indicators change over time. This program provides time series graphs, box plots, and cumulative probability plots for comparisons of results.

There are many examples in which scientists have used DSSAT and its components to provide information for decisions ranging from those that deal with field-scale crop management options to improve crop production and minimize risk, to those that identify sites in a region where production of a particular crop could be expanded, and to those that estimate changes in crop performance due to potential changes in climate. Most of these applications have involved scientists at national and international research organizations and universities and have focused on possible solutions to problems of local or regional interest. In the climate change studies, scientists from nineteen countries used the DSSAT to estimate the effects of possible changes in climate on crop production for use by national and international policy makers (Rosenzweig et al., 1995). Thus, the direct users of DSSAT have mostly been agriculturists to whom policy makers, farmers, extension workers and private businesses and others would typically go for recommendations, assessments, advice, and other information for making decisions. Its success is due to the impact that it is having as a tool for researchers and others who provide information for decisions. The crop models have limitations, and interpretation of results is made by these users. The DSSAT facilitates their analysis and interpretation of results from specific experiments and extension of those results for decision making under a wider set of conditions.

Spatial scale applications of DSSAT v3

The DSSAT provides users with an analysis of the uncertainty in crop performance due to year-to-year weather variability, for each defined management strategy. For spatially varying fields as well as farm and regional scales, there are additional uncertainties, such as spatial variabilities of soil and weather, and uncertainties in the selection of crops and practices over space. New technology is providing farmers information on spatial variability of yields across their fields as well as technology to vary management within a field. Global Positioning Systems (GPS) are now available on farm combines and other farming equipment. This provides many opportunities for farmers to optimize management using so-called site-specific farming techniques. Decisions that farmers make about which crop to plant in a particular field or portion of a field and the practices to use depend on the availability of land, known spatial variability in the field, previous history, preferences and other information on economics and resource availability (Thornton, 1991). In addition, many problems faced by agricultural decision makers extend beyond the boundaries of individual fields. Policy makers and agribusiness decision makers may need to know the aggregate agricultural productivity and resource demands over a region under various plans or policies. For the homogenous field scale, DSSAT can be used to organize weather and soil data for many sites in a field, farm, or region, and it can simulate various alternatives for each of the sites, which could then be analyzed to estimate field, farm, or regional performance. However, information not contained in the DSSAT is also required, such as the spatial variability of current land use, weather, and soils, and the proposed arrangements over space of crops and their management practices.

Applications of DSSAT have been extended from a uniform field concept to analyze options for spatially-variable areas by combining it with geographic information systems (GIS). Booltink and Verhagen (1997) integrated spatial analysis software they developed with DSSAT and a GIS component to analyze effects of within-field spatial variability and potential for site-specific farming. Engel and Jones (1995) combined widely-used GIS software with DSSAT to produce a desktop computer tool for analyzing and viewing simulated results over regional spatial scales. Thornton et al. (1997) developed a regional yield forecasting application for early warning of famines by combining components of DSSAT with GIS. Whereas much more work needs to be done to create robust, easy-to-use spatial decision support systems, these early efforts clearly demonstrate their potential value for decision makers and the practicality of using DSSAT in such systems.

Limitations

The most important limitations in DSSAT v3 are related to limitations in the crop models. Models for only a few crops are contained in the system, and those models do not respond to all environment and management factors. Missing are components to predict the effects of tillage, pests, intercropping, excess soil water and other factors on crop performance. They are most useful in the major regions of the world where weather, water, and nitrogen are the major factors that affect crop performance. Their value to date has also been in demonstrating potential applications and in teaching. Performance of the models may not be good under conditions of severe environmental stresses.

One of the questions faced by crop production scientists is whether to create more comprehensive crop simulation models, and if so, how to efficiently and effectively to do this. The models currently simulate potential production, and water and nitrogen limited production, but do not consider many factors that determine yield limitations in many agricultural fields. For example, DSSAT models do not currently simulate soil phosphorus (P) availability nor P stress effects on crop growth and yield, although P limits production in many soils. Adding a P component to the current models would increase input requirements. Ultimately, it seems logical that scientists should develop a P component as their understanding of P processes in soils and plants increases. This is true of other factors as well. However, how many factors should be added?

Most existing crop models are programmed in FORTRAN, and the programming style does not lend themselves to simple integration of new components. New programming concepts, such as object-oriented programming, may provide a basis for considerable improvement in modularity and new component integration in the future. Progress in the future requires modules that are easily documented, exchanged, and integrated into existing models. One approach that shows considerable promise is described by van Kraalingen (1995). Process modules can be created using these concepts and incorporated into existing models with minimal changes to other program components. This creates the potential for the creation of a 'library' of modules which can be 'plugged into' crop models as needed to account for factors that are most limiting or that are of most interest to users. This concept has been suggested in the past (Hesketh and Jones, 1976; McCown et al., 1994), but little progress has occurred to date that allows broad exchange and modular use of crop model components. This is a limitation that should be addressed by the international scientific community through development of more widely accepted protocols for programming, data standards, documentation, and model quality.

Other limitations exist in the design and implementation of the data base system and other components of the current DSSAT. Although most functions are easily accessed by research users with minimal computer expertise, some functions have been criticized as being too cumbersome, confusing, or limiting. Other user interface designs are clearly needed for other types of users. A final limitation is that the DSSAT is restricted to homogenous field scale analyses. Although it does not include spatial landscape features or capabilities to simulate performance over a spatial scale, it can be linked with GIS to create spatial scale decision support systems

Conclusions

The need for information for agricultural decision making at all levels is increasing rapidly. The generation of new data and its publication is not sufficient to meet these increasing needs. Unless it is put into a format that is easily and quickly accessible, new data and research findings may not be used effectively. Model-based decision support systems are needed to increase the performance of agricultural decision makers while reducing the time and human resources for analyzing complex decisions. The possibility and value of such systems has been demonstrated. A field scale, model-based decision support system has been developed through the close cooperation of a group of interdisciplinary scientists at different institutions. None of the institutions had the resources nor the expertise to develop this system alone, and it is a notable achievement that such cooperation led to this integrated system.

In view of these needs and what has been accomplished to date, it is clear that agricultural decision support systems are still in an early stage of development. Problems faced by decision makers are far more comprehensive than those that can be addressed by the current DSSAT. Further developments are needed in three basic directions. First, the crop models should be improved to account for factors not yet included. This is already being done by the developers of the models in the DSSAT and by other crop modeling groups. These efforts need to be focused to produce fewer crop models with broader capabilities and increased accuracy (not to create new crop models). Secondly, models of other crops and of other agricultural enterprises should be developed with capabilities to link into the same data bases. Modeling efforts exist for animal, aquatic, and forest systems, and efforts are needed to integrate them for describing the biophysical performances of each of those production systems. Finally, farm and regional decision support systems need to be designed that integrate a broader range of models and analysis capabilities to provide decision makers with predictions of farm and regional performance and resource requirements, and to assess the variability and sustainability of agricultural systems (Hansen and Jones, 1996). This will require an exceptional effort by scientists from various disciplines on an international scale.

References

Alargarswamy G, Ritchie J T (1991) Phasic development in CERES-Sorghum model. Pages 143–152 in Hodges T (ed.) Predicting Crop Phenology. CRC Press, Boca Raton, FL.

Batchelor W D, Jones J W, Boote K J, and Pinnschmidt H O (1993) Extending the use of crop models to study pest damage. Transaction of the American Society of Agricultural Engineering 36(2):551–558.

Beinroth F H, Jones J W, Knapp E B, Papajorjgi P, Luyten J (1997) Application of DSSAT to the evaluation of land resources. Chapter 14 in Tsuji G Y, Hoogenboom G, Thornton P K (eds.) International Benchmark Sites Network for Agrotechnology Transfer: A Systems Approach to Research and Decision Making. Kluwer Academic Publishers (in press).

Booltink H W G, Verhagen J (1997) Using decision support systems to optimize barley management on spatial variable soil. Pages 219–233 in Kropff M J, Teng P S, Aggarwal P K, Bouma J, Bouman B A M, Jones J W, van Laar H H (Eds.) Applications of Systems Approaches at the Field Level, Vol 2. Proceedings of the second international symposium on Systems Approaches for Agricultural Development, IRRI, Los Banos, Philippines, 6–8 December 1995, Kluwer Academic Publishers, London.

Boote K J, Jones J W, Hoogenboom G, Wilkerson G G, Jagtap S S (1989) PNUTGRO V1.02 Peanut Crop Growth Simulation Model: User's Guide. Florida Agricultural Experiment Station Journal No. 8420, Agricultural Engineering Department and Agronomy Department, University of Florida, Gainesville, FL.

Boote K J, Jones J W, Hoogenboom G, Wilkerson G G (1997) Evaluation of the CROPGRO-Soybean model over a wide range of experiments. Pages 113–133 in Teng P S, Kropff M J, ten Berge H F M, Dent J B, Lansigan F P, van Laar H H (eds.) Applications of Systems Approaches at the Field Level, volume 2. Kluwer Academic Publishers, London, UK.

Curry R B, Peart R M, Jones J W, Boots K J, Allen L H (1990) Simulation as a tool for analyzing crop response to climate change. Trans ASAE 33:981–990.

Engel T, Jones J W (1995) AEGIS/WIN version 3.0: User's manual. Agricultural and Biological Engineering Department, Gainesville, FL. 63 pp.

Geng S, Auburn J, Brandsletter E, Li B (1988) A program to simulate meteorological variables: documentation for SAMBAED. Agronomy Report No. 204, Univ of California, Crop Extension, Davis, California.

Godwin D, Ritchie J T, Singh U and Hunt L A (1989) A user's guide to CERES-Wheat V2.10. International Fertilizer Development Center. Muscle Shoals, AL.

Godwin D C, Singh U, Buresh R J and De Datta S K (1990) Modeling of nitrogen dynamics in relation to rice growth and yield. Transactions 14th Int. Congress Soil Sci. Kyoto, Japan. Vol. IV:320–325.

Griffin T S, Johnson B S, Ritchie J T (1993) A simulation model for potato growth and development: SUBSTOR-Potato V2.0. Research Report Series 02. IBSNAT Project. Dept of Agron and Soil Sci, College of Trop Agr and Human Resources, University of Hawaii, Honolulu, HI.

Hansen J W, Pickering N B, Jones J W, Wells C, Chan H V K, and Godwin D C (1994) Managing and generating daily weather data. Pages 132–197 in Tsuji G Y, Uehara G, Balas S (eds.) DSSAT v3 Vol 3-3. University of Hawaii, Honolulu, HI.

Hansen J W, Jones J W (1996) A systems framework for characterizing farm sustainability. Agricultural Systems 51:185–201.

Hesketh J D, Jones J W (1976) Some comments on computer simulators for plant growth – 1975, Ecological Modeling 2:235–247.

Hearn A B (1987) SIRATAC: A decision support system for cotton management. Rev. of Marketing and Agr. Econ. 55:170–173.

Hoogenboom G, White J W, Jones J W and Boote K J (1991) BEANGRO V1.01: Dry bean crop growth simulation model. Florida Agr. Exp. Sta. Journal No. N-00379. University of Florida, Gainesville, FL.

Hoogenboom G, Jones J W, Boote K J (1992) Modeling growth, development and yield of grain legumes using SOYGRO, PNUTGRO, and BEANGRO: A review. Trans ASAE 35(6)2043–2056.

Hoogenboom G, Jones J W, Wilkens P W, Batchelor W D, Bowen W T, Hunt L A, Pickering N B, Singh U, Godwin D C, Baer B, Boote K J, Ritchie J T, White J W (1994) Crop models. Pages 94–244 in Tsuji G Y, Uehara G, Balas S (Eds.) DSSAT v3 Vol 2-2. University of Hawaii, Honolulu, HI.

Hunt L A, Jones J W, Thornton P K, Hoogenboom G, Imamura D T, Tsuji G Y, Singh U (1994a) Accessing data, models, and application programs. Pages 21–110 in Tsuji G Y, Uehara G, Balas S (eds.) DSSAT v3 Vol 1-3. University of Hawaii, Honolulu, HI.

Hunt L A, Pararajasingham S, Jones J W, Hoogenboom G, Imamura D T, Ogoshi R M (1994b) GenCalc – Software to facilitate the use of crop models for analyzing field experiments. Agronomy Journal 85:1090–1094.

IBSNAT (1988) Technical Report 1. Experimental Design and Data Collection Procedures for IBSNAT, 3rd ed., Revised. Department of Agronomy and Soil Science, College of Tropical Agriculture and Human Resources. University of Hawaii, Honolulu, HI.

IBSNAT (1989) Decision Support System for Agricultural Transfer V2.10 (DSSAT V2.10). Department of Agronomy and Soil Science, University of Hawaii, Honolulu, HI.

IBSNAT (1993) The IBSNAT Decade. Department of Agronomy and Soil Science, College of Tropical Agriculture and Human Resources, University of Hawaii, Honolulu, HI.

Imamura D T (1994) Creating management files to run crop models and document experiments. Pages 111–144 in Tsuji G Y, Uehara G, Balas S (eds.) DSSAT v3 Vol 1-4. University of Hawaii, Honolulu, HI.

Inman-Bamber N G (1991) A growth model for sugar-cane based on a simple carbon balance and the CERES-Maize water balance. South African Journal Plant and Soil 8:93–99.

Jones J W (1986) Decision support system for agrotechnology transfer. Agrotechnology Transfer. 2:1–5. Department of Agronomy and Soil Science, University of Hawaii, Honolulu, HI.

Jones J W, Boote K J, Hoogenboom G, Jagtap S S, Wilkerson G G (1989) SOYGRO V5.42 Soybean Crop Growth Simulation Model: Users Guide. Florida Agricultural Experiment Station

Journal No. 8304. Agricultural Engineering Department and Agronomy Department, University of Florida, Gainesville, FL.

Jones J W (1993) Decision support system for agricultural development. Pages 459–472 in Penning de Vries F W T, Teng P S, Metselaar K (eds.) Systems Approaches to Agricultural Development. Kluwer Academic Publishers, Boston.

Jones J W, Hunt L A, Hoogenboom G, Godwin D C, Singh U, Tsuji G Y, Pickering N P, Thornton P K, Bowen W T, Boote K J, Ritchie J T (1994) Input and output files. Pages 1–94 in Tsuji G Y, Uehara G, Balas S (eds.) DSSAT v3 Vol 2–1. University of Hawaii, Honolulu, HI.

Matthews R B, Hunt L A (1994) GUMCAS: A model describing the growth of cassava (Manihot esculenta L. Crantz). Field Crops Research 36:69–84.

McCown R. L P, Cox G, Keating B A, Hammer G L, Carberry P S, Probert M E, Freebairn D M (1994) The development of strategies for improving agricultural systems and land-use management. Pages 81–96 in Goldsworthy P, Penning de Vries F W T (eds.), Opportunities, Use, and Transfer of Systems Research Methods in Agriculture to Developing Countries. Kluwer Academic Publishers. Boston.

McKinion, J M, Baker D N, Whisler F D, Lambert J R, Landivar J A and Mullendor G P (1989) Application of the GOSSYM/COMAX system to cotton crop management. Agricultural Systems 31:55–65.

Michalski R S, Davis J H, Bisht V S and Sinclair J B (1983) A computer-based advisory system for diagnosing soybean diseases in Illinois. Plant Disease 4:459–463.

Otter-Näcke S, Ritchie J T, Godwin D C, Singh U (1991) A user's guide to CERES-Barley – V2.10. International Fertilizer Development Center, Muscle Shoals, AL.

Pickering N B, Hansen J W, Jones J W, Wells C, Chan H V K, Godwin D C (1994) WeatherMan: A utility for managing and generating daily weather data. Agronomy Journal 86(2):332–337.

Plant R E (1989) An integrated expert system decision support system for agricultural management. Agricultural Systems 29:49–66.

Prasad H K, Singh U, Goenaga R (1991) A simulation model for aroid growth and development. Agronomy Abstracts (1991):77.

Richardson C W and Wright D A (1984) WGEN: A model for generating daily weather variables. U.S. Department of Agriculture, Agricultural Research Service, ARS – 8, 80 pp.

Ritchie J T, Singh U, Godwin D and Hunt L (1989) A user's guide to CERES-Maize V2.10. International Fertilizer Development Center. Muscle Shoals, AL.

Rosenzweig C, Ritchie J T, Jones J W, Tsuji G Y, Hildebrand P (eds.) (1995) Climate Change and Agriculture: Analysis of Potential International Impacts. ASA Special Publication Number 59, American Society of Agronomy, Madison, WI.

Scholberg J, Boote K J, Jones J W, McNeal B L (1997) Adaptation of the CROPGRO model to simulate field-grown tomato. Pages 135–151 in Kropff M J, Teng P S, Aggarwal P K, Bouma J, Bouman B A M, Jones J W, van Laar H H (eds.) Applications of Systems Approaches at the Field Level, Volume 2, Kluwer Academic Publishers, Boston.

Simon, H (1960) The New Science of Management Decisions. Harper & Row, Publishers. New York.

Singh U, Ritchie J T, Godwin D C (1993) A user's guide to CERES-Rice – V2.10. International Fertilizer Development Center, Muscle Shoals, AL.

Singh U, Ritchie J T, Thornton P K (1991) CERES-CEREAL model for wheat, maize, sorghum, barley, and pearl millet. Agron Abstracts (1991):78.

Sprague R H Jr and Carlson E H (1982) Building Effective Decision Support Systems. Prentice-Hall, Inc., Englewood Cliffs, NJ.

Thornton P (1991) Application of crop simulation models in agricultural research and development in the Tropics. International Fertilizer Development Center. Muscle Shoals, AL.

Thornton P K, Bowen W T, Ravelo A C, Wilkens P W, Farmer G, Brock J, Brink J E (1997) Estimating millet production for famine early warning: Application of crop simulation modeling using satellite and ground base data in Burkina Faso. Agricultual and Forestry Meteorology 83:95–112.

Thornton P K, Hoogenboom G (1994) A computer program to analyze single-season crop model outputs. Agronomy Journal 86:860–868.

Thornton P K, Hoogenboom G, Wilkens P W, Bowen W T (1995) A computer program to analyze multiple-season crop model outputs. Agronomy Journal 87:131–136.

Thornton P K, Hoogenboom G, Wilkens P W, Jones J W (1994a) Seasonal analysis. Pages 1–66 in Tsuji G Y, Uehara G, Balas S (eds.) DSSAT v3 Vol 3–1. University of Hawaii, Honolulu, HI.

Thornton P K, Wilkens P W, Hoogenboom G, Jones J W (1994b) Sequence analysis. Pages 67–136 in Tsuji G Y, Uehara G, Balas S (eds.) DSSAT v3 Vol 3–2. University of Hawaii, Honolulu, HI.
Tsuji G Y, Uehara G, Balas S (eds.) (1994) DSSAT v3 Vol 1, 2, 3. University of Hawaii, Honolulu, HI.
Uehara G (1989) Technology Transfer in the Tropics. Outlook on Agriculture 18:38–42.
van Kraalingen D W G (1995) The FSE system for crop simulation, version 2.1. Quantitative Approaches in Systems Analysis No. 1, TPE-WAU, Bornsesteeg 47, Wageningen, The Netherlands.
Villalobos, F J, Hall A E, Ritchie J T (1995) OILCROP-Sun V4.1: Code explanation. Research Report, Agronomy Department, University of Buenos Aires, Buenos Aires, Argentina. 77 pp.
Yost R S, Uehara G, Wade M, Sudjadi M, Widjaja-Adhi I P G, Zhi-Cheng Li (1988) Expert systems in agriculture: Determining the lime recommendations for soils of the humid tropics. Research-Extension Series 089–03.88. College of Tropical Agriculture and Human Resources. University of Hawaii, Honolulu, HI.

Modeling and crop improvement

J.W. WHITE
Natural Resources Group, CIMMYT, Lisboa 27, Apartado Postal 6-641, 06600 Mexico, D.F., Mexico

Key words: common bean, breeding, cassava, database, genetic coefficients, genetics, ideotype, peanut, soybean

Abstract
The ability of simulation models to predict growth and development as affected by soil and weather conditions, agronomic practices and cultivar traits makes such models attractive tools for crop improvement. Models have been used to examine the effects on yield of specific traits or suites of traits representing possible ideotypes. By using multi-season data, risk arising from unpredictable water deficit or other effects of weather are easily included. However, the idea of identifying desirable traits through simulation modeling has had a strictly luke-warm reception from plant breeders. Traits such as greater leaf photosynthetic rate may be difficult to select for, or selection for a single trait may cause undesirable changes in other traits. Simulations also may be too imprecise to allow confidence in quantitative results. Although such difficulties should decline as models are improved further, other applications of models may offer more immediate benefits to crop improvement. Models are highly suitable for aiding breeders in understanding genotype by environment interactions, particularly when linked to geographic information systems. Future models should be better integrated with genetic information and should simulate traits related to quality, such as oil and protein content, and to pest and disease resistance mechanisms.

Introduction

Crop simulation models have many applications in crop improvement. Their ability to predict growth and yield as influenced by growing environment, agronomic practices and crop traits immediately suggests the possibility of using models to identify desirable traits or combination of traits potentially leading to the specification of crop ideotypes (Whisler et al., 1986; Boote and Tollenaar, 1994). Similarly, models are highly suitable for studying variation in cultivar response to environment, i.e., the genotype by environment interaction (Shorter et al., 1991; Hunt, 1993; Chapman and Barreto, 1994). In practice, although examples exist of potential application of models to crop improvement, models have not seen wide use by plant breeders. This chapter will first review applications of models to crop improvement and then consider how modeling can be better integrated into crop improvement efforts.

G.Y. Tsuji et al. *(eds): Understanding Options for Agricultural Production, 179–188*

Applications of models to crop improvement

Identification of desirable traits

Selection directly for economic yield is usually perceived as costly and unreliable because of low heritability of yield. Great effort has thus gone to the identification of traits which breeders might select for in order to increase yield indirectly. Although a single trait may be of interest, a suite of traits representing a crop ideotype is often sought (Donald, 1968). The classic examples of ideotypes are found in the erect, dwarf, photoperiod insensitive rice and wheat cultivars, although these were first developed without the use of process-based models.

With simulation models, effects of traits are readily assessed in sensitivity analyses where the coefficients determining traits are varied, and the effects on simulated growth or yield observed. Although not referenced to yield or whole-season growth, de Wit's (1965) simulations of canopy photosynthesis could be considered the first attempt to guide breeding through crop models. De Wit concluded that, for most conditions, an erectophile leaf distribution would lead to little increase in photosynthesis. Various cotton models have been used since 1973 to assess the effect on yield of traits including photosynthetic efficiency, leaf abscission rates and unusual bract types (Whisler et al., 1986). Cock et al. (1979) proposed a cassava ideotype with late branching, large leaves and long leaf life based on a model that used weekly time intervals to simulate leaf development, crop growth and partitioning between shoots and roots.

More recently, Boote and Tollenaar (1994) examined prospects for increasing yield of soybean and maize using the SOYGRO and MAIS models. The soybean results are of particular note because the authors considered known genetic variation, effects of different locations, and possible compensation among traits such as between photosynthetic rate and specific leaf weight. Little potential was found for increasing yields through selection for photosynthesis. At the Florida site, seed yield increased asymptotically with duration of podfilling, and in Ohio, the optimum yield was obtained with a 10-day increase in podfill duration compared with *cv.* Williams. Varying the seed oil and protein contents resulted in large yield changes, which were attributable to costs of biosynthesis. Changing partitioning did not increase yields, leading the authors to suggest that soybean cultivars already partition assimilate very efficiently. Maximum soybean yields were 4870 kg ha^{-1} in Florida and 5760 kg ha^{-1} in Ohio, the difference being attributed to effects of higher temperatures in Florida.

For conditions where water deficits may occur, the number of traits to consider increases, and there is the added complication of considering season-to-season variation increase. Jones and Zur (1984) found that for soybean grown on a sandy soil in Florida, increased root growth was more advantageous than capacity for osmotic adjustment or increased stomatal resistance. By contrast, for dryland cotton farming in Texas, the GOSSYM model predicted that doubling stomatal resistance would lead to a 28% increase in yield, a conclusion subsequently supported by improved cultivars (Whisler et al., 1986).

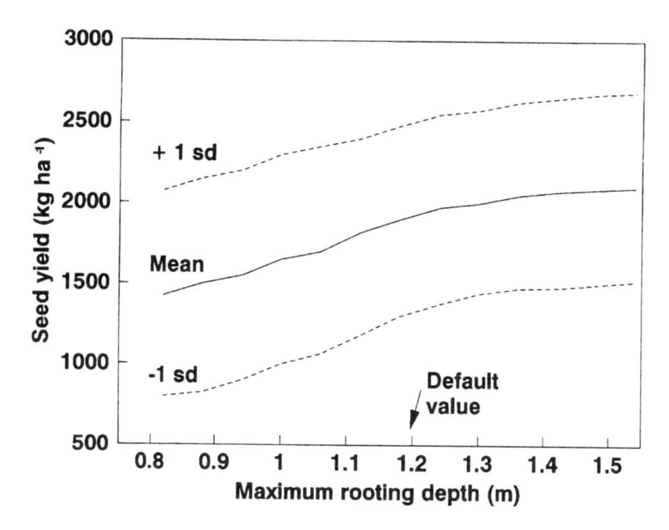

Figure 1. Relation between seed yield of common bean and maximum rooting depth. Simulations are from BEANGRO using *cv.* BAT 477 grown under rainfed conditions at two planting dates over eight years at Palmira, Colombia (Hoogenboom et al., 1988).

For common bean grown at Palmira, Colombia, effects of specific leaf area (SLA), root partitioning, rooting depth and root length weight ratio (RLWR) on seed yield and water use efficiency (WUE) were examined (Hoogenboom et al., 1988). To account for seasonal variation, the study ran simulations with BEANGRO over 8 years using two planting dates per year. Increasing rooting depth from 0.8 to 1.5 m increased yield by over 500 kg ha^{-1} (Figure 1) and WUE, measured as g yield g^{-1} H_2O transpired, by 35%. Yield also increased with increased RLWR or partitioning to roots. For SLA, values over 300 cm^2 g^{-1} gave no increase, while WUE reached a maximum for an SLA of 270 cm^2 g^{-1}. The confidence intervals obtained with multiple simulations run parallel to the line of mean performance (Figure 1, for example), suggesting that while seasonal variation has a large effect on yield, the optimum value for a given trait is relatively insensitive to this variation.

Using PNUTGRO, Boote and Jones (1988) compared the effects of 16 parameters on peanut yield under rainfed conditions over 21 years in Florida. Simulations using default values gave mean yields of 3470 kg ha^{-1} with a range of 1750 to 4380 kg ha^{-1} attributable to the variation in weather conditions. Coefficients from the crop and genetic files were increased to 10% above their standard values for most traits. Increasing canopy photosynthesis and the duration of vegetative and reproductive phases both increased yields over 15%. Coefficients of variation for the 16 traits varied from 17% for photosynthesis to 23% for an increase in reproductive phase.

While sensitivity analyses used to identify opportunities for varietal improvement provide attractive examples of how modeling might guide crop improve-

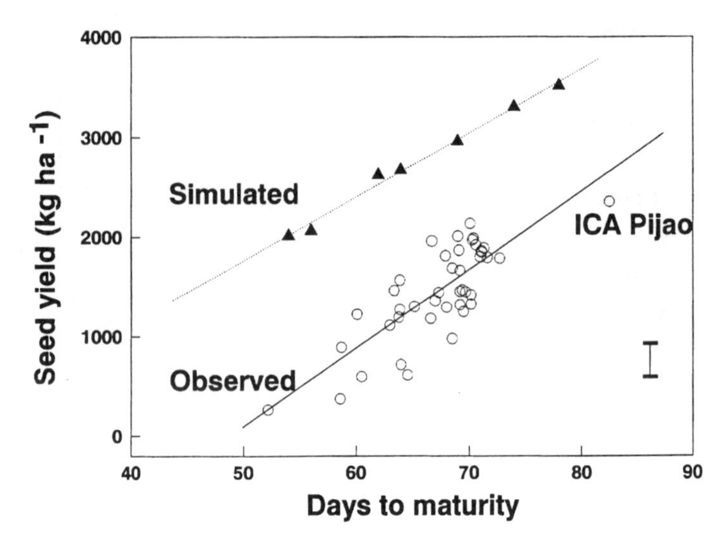

Figure 2. Relation between seed yield of common bean and days to maturity for field data of 42 cultivars and for simulations of seven hypothetical cultivars derived from coefficients of *cv.* Jamapa. Simulations are from BEANGRO V1.01 at Palmira, Colombia (White et al., 1995).

ment, the approach has limitations which are seldom fully confronted. Among these are whether the range of variation tested can be obtained through breeding and whether selection for the trait will result in undesirable changes in other traits. BEANGRO predicts that yield at Palmira, Colombia, will increase linearly with days to maturity in the absence of temperature stress or water deficit (Figure 2). However, it has been difficult to breed lines that mature later than existing cultivars such as *cv.* ICA Pijao. Assuming that increased photosynthetic rates are obtained through greater leaf thickness, one also might expect that selecting for increased photosynthesis would result in a loss of total leaf area that might reduce or negate the expected gain. For CROPGRO applied to common bean using the hedge-row photosynthesis subroutine, the yield increase predicted was less if SLA was varied to account for changes in leaf thickness (Figure 3). However, using the canopy photosynthesis routine, no differences between constant and variable SLA were found (Figure 3). The example thus serves to illustrate further the differences possible among models.

Genotype by environment interactions

In crop improvement, variation in cultivar performance between environments usually forces breeders to evaluate promising materials in multiple environments or over multiple seasons. Recognizing that mean yields across environments may hide important differences in response, many statistical methods have been developed to analyze genotype by environment ($G \times E$) interactions (Lin et al., 1986). One approach is to characterize each environment by the

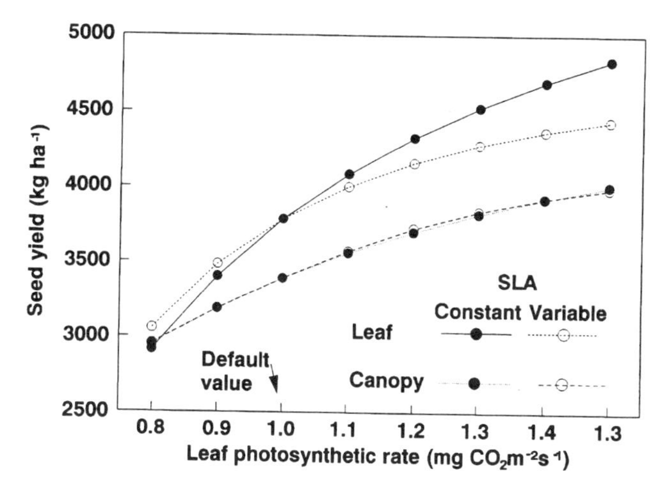

Figure 3. Effect of leaf photosynthetic rate on seed yield of common bean using either the leaf (hedgerow) or canopy subroutines of CROPGRO, and comparing presence or absence of compensatory changes in leaf thickness through variation in specific leaf area (SLA).

mean yield of all materials, and then use the mean as an index of productivity of the site (Finlay and Wilkinson, 1963). Yields of individual cultivars are then regressed against the index, and the slope of the regression line is interpreted as indicating the stability or responsiveness of a given cultivar. However, the approach is often criticized because the index violates assumptions of statistical independence and because it is difficult to relate to environmental factors such as water deficit or temperature stress (Lin et al., 1986). Simulation models could easily provide an independent estimate of site productivity and permit effects of specific environmental factors to be assessed.

A more analytical approach is to use models to examine the physiological basis of $G \times E$ effects. By examining responses to temperature, photoperiod, soil moisture, or nutrients, the mechanisms of adaptation or critical factors determining response may be identified. This information, in turn, can lead to more reliable characterization of growing environments and grouping of cultivars (Chapman and Barreto, 1994). Approaches to understanding problems of adaptation can be as simple as using simulation models to examine water balances for different seasons or locations. In studying the contrasting responses of different bean cultivars in the Mexican highlands, Acosta-Gallegos and White (1995) used BEANGRO to examine the length of growing season at three sites for 10 to 18 years. For two sites, long growth seasons and an early onset of the season were associated with greater probability of adequate rainfall. At the other site, total rainfall was lower and uncorrelated with onset or length of the season. This lead to the proposal that two types of cultivars are needed. A cultivar with a growth cycle that becomes longer with early plantings would suit the first two sites, while a cultivar with a constant, short cycle would be better suited for the third site.

Towards better integration of simulation models into crop improvement

'Real' plant breeding

Perhaps the greatest constraint to the use of models in crop improvement is that few breeding programs focus strictly on selection for increased yield. As Simmonds (1979) noted, 'The plant breeder is working in a complex system, one in which agricultural, technological and socio-economic factors all play parts and are all in a state of more or less continuous change and interaction. There are indeed very few situations in which plant breeding objectives are clear, simple and stable.'

Quality of the harvested product is often as important to growers as final yield. Models such as the CROPGRO series are being modified to simulate changes in seed size and composition. Piper et al. (1993) used SOYGRO to explore the influence of growing temperature on oil and protein content in soybean. Breeders usually seek to combine increased yield with requisite disease or pest resistances, and in many cases increasing resistance may be a more immediate goal than yield improvement. The recent advances in systems for including pest and disease damage in models (Batchelor et al., 1993) should make it easier to assess the effects of damage on crops, and estimate the utility of traits such as indeterminate growth or early maturity as tolerance mechanisms. However, models that examine more complex mechanisms of disease and pest resistance, such as increased lignification or changes in tissue nitrogen content, appear to be lacking.

Strengthening simulation of gene action in models

Simulation models vary greatly in the level of detail used to represent genetic differences among species or cultivars. Since the raw material of crop improvement is genetic variation, failure to represent genetic variation limits the utility of models. From a crop improvement perspective, simulation models can be classified into five levels based on the genetic complexity represented:

(1) Non crop-specific models;
(2) Crop specific models but with no cultivar differences;
(3) Crop models that specify cultivar differences using 'genetic coefficients,' which are usually traceable to traits measured in the field;
(4) Crop models that use genetic coefficients derived from knowledge of actual genotypes of cultivars;
(5) Crop models in which cultivar differences are simulated directly through gene action for specific processes.

Genetic levels 1 and 2 are fairly discrete, but a single model could easily include features corresponding to levels 3 to 5. The IBSNAT models correspond to level 3 and stand out as a group for explicitly addressing cultivar differences. However, it must be acknowledged that their genetic coefficients are problematic. Their values are often difficult to relate directly to measured traits and have to be determined through iterations of comparisons between measured and simulated data. Furthermore, values that produce acceptable simulation results in one environment may not work well in other environments. The underlying problem, however, is not the concept of genetic coefficients but deficiencies in our understanding of how to quantify specific cultivar differences in models.

The only level 4 model appears to be GeneGro (White and Hoogenboom, 1996) developed from BEANGRO V1.01. No models at level 5 exist yet, reflecting the lack of applicable information on gene action at the process level. The levels correspond loosely to a ranking in terms of expected utility to crop improvement efforts. However, a model of level 1 might provide the detail necessary to determine the potential value of traits such as stomatal sensitivity to humidity or strength of root signals for water deficit without reference to a specific crop or cultivar. A model of level 4 or 5 might lack this ability because information on the genetic control of these traits is lacking.

The GeneGro model characterizes cultivars using seven genes for growth habit, phenology, photoperiod response and seed size (White and Hoogenboom, 1996). Cultivar coefficients are specified as linear functions of gene effects. For the calibration set of 30 cultivars grown in various trials at six locations (in Colombia, Guatemala, Mexico and Florida), simulated days to maturity accounted for 85% of variation in observed data, and simulated seed yield, 31% of observed variation (Figure 4). While encouraging, the approach fails to distinguish among cultivars with known differences in response to water deficit or acid soils.

This approach also increases the potential for linking models to genome databases (e.g., Coe, 1994; Lorenzen et al., 1994) and of using molecular biology techniques to characterize genotypes and reduce reliance on field trials to fit model coefficients (Figure 5).

Better software tools

For models to become more accessible to plant breeders, data on soils, weather and cultivars must be readily available. Specifications of minimum data sets (IBSNAT, 1988) and software tools such as DSSAT3 are suitable for analyses of trials using small numbers of lines, but may face limitations in handling large sets of materials that breeders often manage in a single nursery. Software to facilitate determination of cultivar coefficients also show promise (Hunt et al., 1993). However, the ideal would be modeling tools that are fully integrated into plant breeding software systems. Efforts are under way to achieve

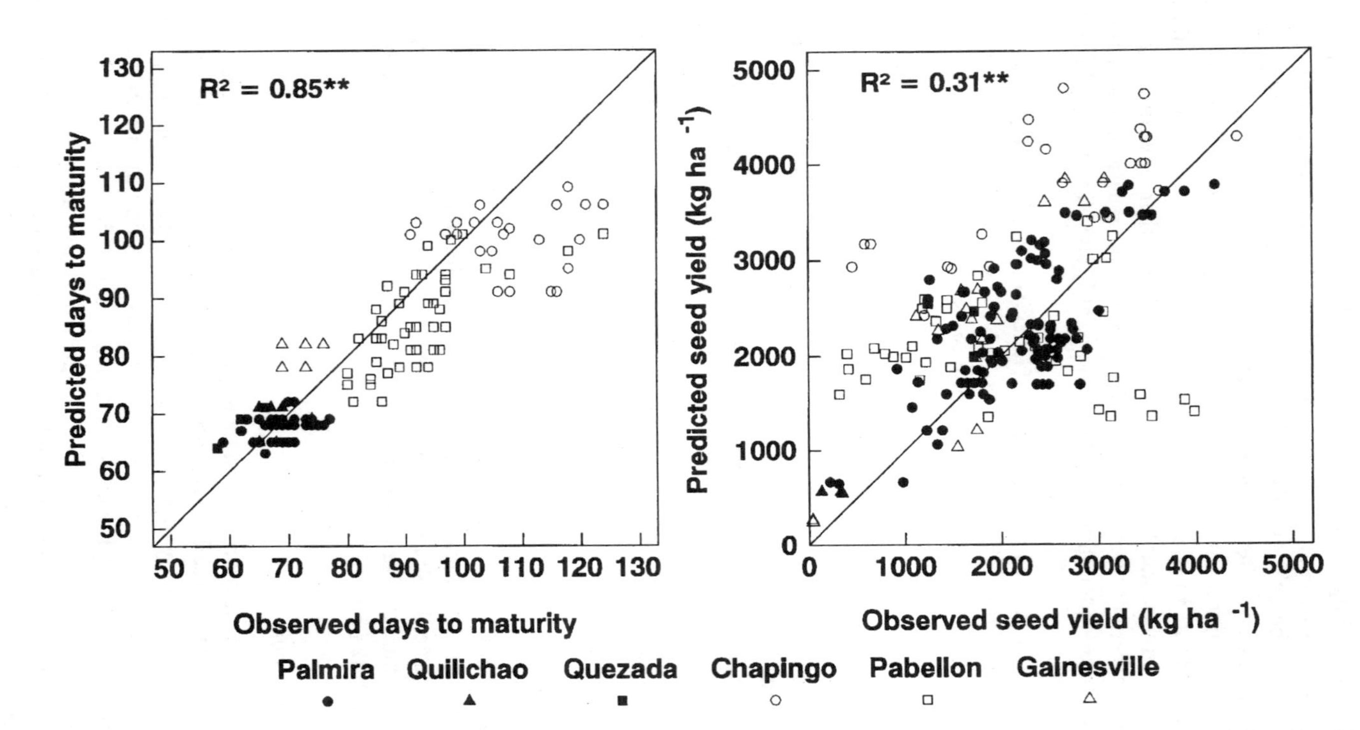

Figure 4. Comparison of observed and simulated days to maturity and seed yield of common bean as predicted by the GeneGro model for 30 cultivars grown at six locations (White and Hoogenboom, 1996).

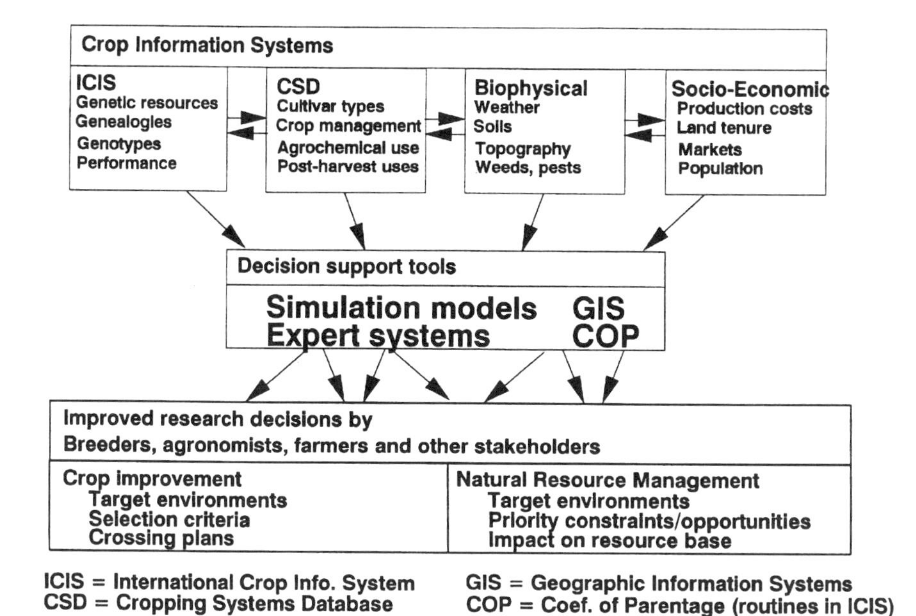

Figure 5. Possible structure for a software system for plant breeding including crop simulation models and GIS.

such integration for the International Wheat Information System (IWIS) (Fox et al., 1996), and more recently, the International Crop Information System (ICIS), which will include versions for wheat, rice, common bean, sorghum, and other major crops.

Conclusions

Although the impact of simulation models in crop improvement is less than one might have expected, the potential benefit is great. Models seem particularly suitable for studies of yield potential and adaptation to rainfed environments. Many models show realistic responses to water deficit, so problems of unpredictable yet predictable precipitation patterns can be examined through simulations over multiple years. As subroutines for simulating other edaphic constraints are improved, models should see increasing use in breeding for a much wider range of conditions. Models that simulate gene action more explicitly seem of particular interest. However, even with major improvements in model design, better sets of base data and software to integrate models with breeding will still be needed.

References

Acosta-Gallegos J A, White J W (1995) Phenological plasticity as an adaptation by common bean to rainfed environments. Crop Science 35:199–204.

Batchelor W D, Jones J W, Boote K J, Pinnschmidt H O (1993) Extending the use of crop models to study pest damage. Transactions of the American Society of Agricultural Engineering 36:551–558.

Boote K J, Jones J W (1988) Applications of, and limitations to, crop growth simulation models to fit crops, cropping systems to semi-arid environments. Pages 63–65 in Bidinger F R, Johansen C (Eds.) Drought Research Priorities for the Dryland Tropics. ICRISAT, Patancheru, India.

Boote K J, Tollenaar M (1994) Modeling yield potential. Pages 553–565 in Physiology, Determination of Crop Yield. American Society of Agronomy, Crop Science Society of America, Soil Science Society of America, Madison, WI, USA.

Chapman S C, Barreto H J (1994) Using simulation models, spatial databases to improve the efficiency of plant breeding programs. Paper presented at ICRISAT $G \times E$ Workshop, Hyderabad, India, 28 Nov.–2 Dec. 1994.

Cock J H, Franklin D, Sandoval G, Juri P (1979) The ideal cassava plant for maximum yield. Crop Science 19:271–279.

Coe E H (1994) The maize genome database: prioritizing. Page 76 in Plant Genome II. Final program, abstracts. The Second International Conference on the Plant Genome, San Diego, CA. 24–27 Jan. 1994. Scherago International Inc., New York, NY.

de Wit, C T (1965) Photosynthesis of leaf canopies. Agricultural Research Reports 663. PUDOC, Wageningen, The Netherlands.

Donald C M (1968) The breeding of crop ideotypes. Euphytica 17:385–403.

Finlay K W, Wilkinson G N (1963) The analysis of adaptation in a plant breeding programme. Australian Journal of Agricultural Research 14:742–754.

Fox P N, Lopez C, Skovmand B, Sanchez H, Herrera R, White J W, Duveiller E, van Ginkel M (1996) International Wheat Information System (IWIS), Version 1. Mexico, D.F.: CIMMYT. (On Compact Disk)

Hoogenboom G, Jones J W, White J W (1988) Use of models in studies of drought. Pages 192–230 in White J W, Hoogenboom G, Ibarra F, Singh S P (Eds.) Research on Drought Tolerance in Common Bean. Documento de Trabajo No. 41. CIAT, Cali, Colombia.

Hunt, L A (1993) Designing improved plant types: a breeder's viewpoint. Pages 3–17 in Penning de Vries F W T, Teng P, Metselaar K (Eds.) System approaches for agricultural development. Kluwer Academic Publishers, Dordrecht, The Netherlands.

Hunt L A, Pararajasingham S, Jones J W, Hoogenboom G, Imamura D T, Ogoshi R M (1993) GENCALC: Software to facilitate the use of crop models for analyzing field experiments. Agronomy Journal 85:1090–1094.

IBSNAT (1988) Technical report 1. Experimental design, data collection procedures for IBSNAT. The minimum data set for systems analysis, crop simulation, 3rd edition revised 1988. Department of Agronomy and Soil Science, University of Hawaii, Honolulu, HI.

Jones J W, Zur B (1984) Simulation of possible adaptive mechanisms in crop subjected to water stress. Irrigation Science 5:251–264.

Lin C S, Binns M R, Lefkovitch L P (1986) Stability analysis: Where do we stand? Crop Science 26:894–900.

Lorenzen L L, Smith P J, Shoemaker R C (1994) Soybase: the building of a genome database. Page 77 in Plant genome II. Final program, abstracts. The Second International Conference on the Plant Genome, San Diego, CA, 24–27 Jan. 1994. Scherago International Inc., New York, NY.

Piper E L, Boote K J, Jones J W (1993) Environmental effects of oil, protein composition of soybean seed. Agronomy Abstracts: 1993 Pages 152–153.

Shorter R, Lawn R J, Hammer G L (1991) Improving genotypic adaptation in crops – a role for breeders, physiologist, modellers. Experimental Agriculture 27:155–175.

Simmonds N W (1979) Principles of crop improvement. Longman Group Ltd., London.

Whisler F D, Acock B, Baker D N, Fye R E, Hodges H F, Lambert R J, Lemmon H E, McKinion J M, Reddy V R (1986) Crop simulations models in agronomic systems. Advances in Agronomy 40:141–208.

White J W, Hoogenboom G (1996) Simulating effects of genes for physiological traits in a process-oriented growth model. Agronomy Journal 88:416–422.

White J W, Hoogenboom G, Jones J W, Boote K J (1995) Evaluation of the dry bean model BEANGRO V1.01 for crop production research in a tropical environment. Experimental Agriculture 31:241–254.

Simulation as a tool for improving nitrogen management

W. T. BOWEN and W.E. BAETHGEN
International Fertilizer Development Center, P.O. Box 2040, Muscle Shoals, Alabama 35662, USA

Key words: simulation, CERES, models, nitrogen, economic, environment

Abstract
Many of the more important processes related to N demand and N supply are described by dynamic simulation techniques. The CERES models, when assembled in the framework of comprehensive crop growth models, are computer-aided tools for exploring different N management options across a wide range of varying cropping practices, soil types, and weather conditions. Simulated outcomes provide estimates of economic and environmental impact of different N management options, including the effects of varying N rates, time of application, placement depth, and source.

Introduction

In modern agriculture, N is applied more frequently and in greater amounts than any other nutrient (Harre and White, 1985). It is also the nutrient that most often limits crop yield. Without added N, growing plants often show a N deficiency characterized by yellow leaves, stunted growth, and lower yields. Because it is an important input, N and factors affecting its availability have been the subject of much investigation. A common purpose of many of these studies has been to develop N recommendation systems that can assist growers in determining the amount of N needed and the best time to apply it. Knowledge of when and how much N to apply is essential not only because N inputs have an economic cost, but also because they have a potential environmental cost if too much N is applied; N in excess of that used by a crop may leak into water supplies and contaminate them by increasing the nitrate load. The amount of N needed, and its ultimate fate, are important issues regardless of the source of N, which may be an organic material or a manufactured fertilizer.

A common approach to estimating the amount of N fertilizer needed by a crop has involved the measurement, in field experiments, of yield response to increasing rates of N fertilizer. In such experiments, the yield response is quantified by fitting a mathematical equation to the data. This equation is then used together with economic information to investigate returns to investments in applied N. Although this approach has proven useful for demonstrating the concept of diminishing returns, it is nothing more than a 'black-box' approach which offers limited information for deriving N recommendations in another

G.Y. Tsuji et al. *(eds): Understanding Options for Agricultural Production, 189–204*

year or at another site. Because N is a dynamic and mobile nutrient, its effect on crop production is rarely the same from year to year. At best, equations from variable N rate experiments describe only historical relationships in the data, and as such offer little insight into the processes that must be understood to better manage N inputs.

More informative approaches to making N recommendations have followed from a better understanding of the processes that affect crop N demand as well as soil N availability. Although N recommendation systems often vary in how they are implemented, most are based on the same principles. Generally, the amount of additional N needed by a crop can be estimated from both the crop N requirement and the soil N supply. While the crop N requirement is set by the internal N needed to obtain a specific yield, the soil N supply is set by the properties of the soil, particularly the organic matter content and other properties that affect the extent and rate of microbial decomposition. For most high yielding crops in most soils, the crop N requirement will exceed the soil N supply, and it is the difference between these two that approximates the amount of supplemental N needed by a growing crop.

Most N recommendation systems calculate the amount of fertilizer N to apply from estimates of the soil N supply, a target yield level, and an expected fertilizer efficiency; the latter since not all of the applied N is usually recovered by the crop (Dahnke and Johnson, 1990; Stanford and Legg, 1984). To estimate these components, measurements of one or more of the following variables are usually needed: plant N content required to reach a target yield level, mineral N released during soil incubation (potentially mineralizable N; Cabrera et al., 1994), total N content of the soil, mineral N present in the soil profile at planting, and leaf N concentration or chlorophyll content at a specific growth stage. Relationships between any of these variables and the estimated amount of extra N to apply are usually defined in field calibration studies.

Field calibration studies provide information on which to base N recommendations given site-specific circumstances, and as such need to be conducted across a broad range of soil-crop-climate conditions. Nevertheless, it can easily be argued that most N recommendation systems are based on too few calibration studies, with an insufficient range of conditions represented. Undoubtedly, as greater accuracy is sought in making N fertilizer recommendations, more comprehensive field calibration will be necessary (Meisinger, 1984). There is, however, a practical limit to the number of soil-crop-climate conditions that can be included in field calibration studies.

An additional tool now available for evaluating N management practices and to assist in deriving N recommendations is one based on computer simulation. Using dynamic simulation techniques, systems scientists have been able to construct computer models capable of simulating many of the more important processes related to N demand and N supply. When assembled within the framework of comprehensive crop growth models, the dynamic nature of these models makes them valuable instruments for exploring different N management options across a wide range of cropping practices, soil types, and weather conditions.

It must be stressed at the beginning that simulation models are not a substitute for standard testing and monitoring of the N status in soils and plants. Rather, they provide a tool to extend the value of such measurements by facilitating a systematic analysis of the way plant, soil, weather, and management factors interact to affect N dynamics. In this chapter, we use the CERES simulation model to conduct a systematic study of some of these factors, which will hopefully demonstrate how such a model might be used to gain insight into improved N management. The version of CERES used in this exercise is that released as version 3.1 (Hoogenboom et al., 1994; Bowen et al., 1995), which was built upon earlier versions of the model (Jones and Kiniry, 1986; Ritchie et al., 1989).

Major components of a systematic analysis

A general approach to estimating the amount of N fertilizer to apply to a crop (N_f) may be based on the simple calculation

$$N_f = (N_y - N_s)/E_f$$

where N_y is the internal plant N required to attain the expected yield, N_s is the N supplied by the soil, and E_f is the expected efficiency or fraction of applied N the crop is expected to recover. This approach, described in more detail by Stanford (1973), provides a framework for examining the main factors that must be considered when managing N inputs. Within this framework, the N balance in agricultural fields can be characterized by defining three major components: (i) the crop N demand, (ii) the N supplied by the soil, and (iii) the N added in organic or mineral fertilizer. To optimize N management for a specific situation, quantitative estimates of each of these components are needed along with economic information.

The CERES model provides a useful tool for analyzing the quantitative effect that controlled factors (management), uncontrolled factors (weather), and site-specific soil properties may have on the major components of the N balance. The model is comprehensive in that it simulates most major processes associated with both crop N demand and plant available N (Table 1). It also simulates the availability of different fertilizer N materials as well as the decomposition of plant residue sources of N. The capability to simulate the complex interaction of these major processes also makes it possible to estimate the likely efficiency of different N sources and management practices. The model provides a balanced approach in the level of detail used to describe soil and plant processes; all processes listed in Table 1 are simulated using a daily time step. The comprehensive and dynamic nature of the model means it can be used to evaluate, on a site-specific basis, many alternative management practices against

Table 1. Major processes simulated and environmental factors affecting those processes in the CERES N model.

Process simulated	Main factors influencing process
CROP N DEMAND	
Growth	Solar radiation, temperature
Development	Photoperiod, temperature
PLANT AVAILABLE N	
Mineralization/immobilization	Soil, temperature, soil water, C/N ratio
Nitrification	Soil temperature, soil water, soil pH, NH_4^+ concentration
Denitrification	Soil temperature, soil water, soil pH, soil C
NO_3^- leaching	Drainage
Volatilization[a]	Soil temperature, soil pH, surface evaporation, NH_3 concentration
Urea hydrolysis	Soil temperature, soil water, soil pH, soil C
Uptake	Soil water, inorganic N, crop demand, root length density

[a] Presently simulated in only CERES-Rice for flooded conditions.

the uncertainties of weather. Simulation results, when based on reliable input data, can then provide critical information for defining the best N management practices from both an economic and environmental perspective.

Crop N demand

Characterization of crop N demand at maximum or near maximum yield is useful for establishing the time course and quantity of N needed by the crop. Generally, the maximum or attainable yield of a crop is associated with a minimum amount of plant N; this minimum is often referred to as the internal N requirement (Stanford and Legg 1984). Although a growing crop may take up more than this minimum amount, the extra nitrogen (luxury consumption) does not result in any yield benefit. Therefore, to optimize N management and avoid its inefficient use, it is important to know what is the expected attainable yield and its associated internal N requirement.

'Attainable yield' will be defined here as the yield obtained when all nutrients are nonlimiting and the crop is free of diseases, pests, and weeds. Growth may, however, be constrained by water availability if rainfall or irrigation management are not optimal. Note that this definition differs from the commonly used 'potential yield' definition which assumes water is also nonlimiting (Penning de Vries et al., 1989). Accordingly, the attainable yield of a specific crop at a specific site will be determined largely by the genetic potential of the cultivar grown, the weather conditions, and the water-holding characteristics of the soil. For the same cultivar, attainable yield will be expected to vary from site to site and year to year as a consequence of the interaction of genetic traits with photoperiod, temperature, solar radiation, and available water.

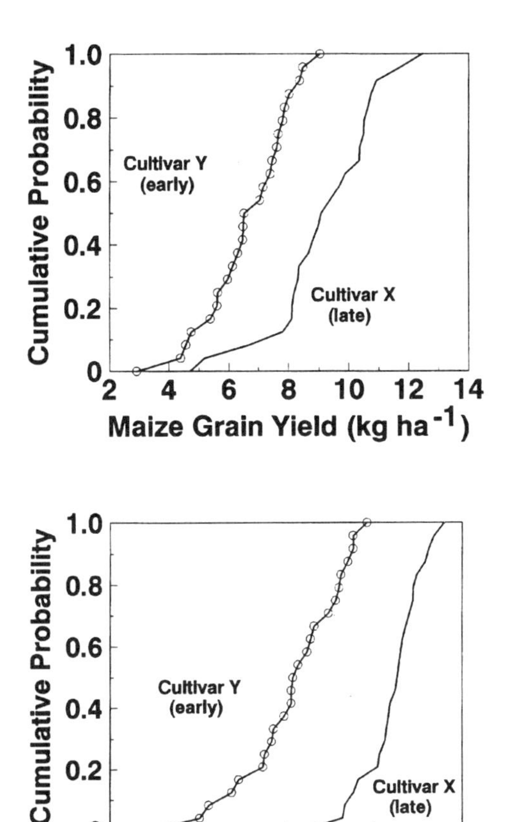

Figure 1. Simulated cumulative probability curves showing (a) maize grain yield and (b) maize aboveground N for two cultivars using 25 years of weather from south central Brazil.

An illustration of how attainable yields for maize and the associated aboveground N content (grain N plus stover N in the aboveground dry matter) might vary between cultivars and across seasons is shown in Figure 1. These simulated values were obtained using soil data and 25 years of historical daily weather records for a site in south central Brazil, and reflect expected outcomes for rainfed crops with an optimum supply of nutrients. Clearly, at least at this site, cultivar X would be expected to perform better than cultivar Y; in most years, cultivar X would yield between 8 and 11 t ha^{-1} while cultivar Y would yield between 5 and 8 t ha^{-1}. To obtain yields between 8 and 11 t ha^{-1}, cultivar X needs 200–240 kg N ha^{-1} in the aboveground dry matter, substantially more than the 130–180 kg N ha^{-1} needed by the lower yielding cultivar Y. For both cultivars, the internal N requirement is about the same, averaging 12.0 g N kg^{-1}

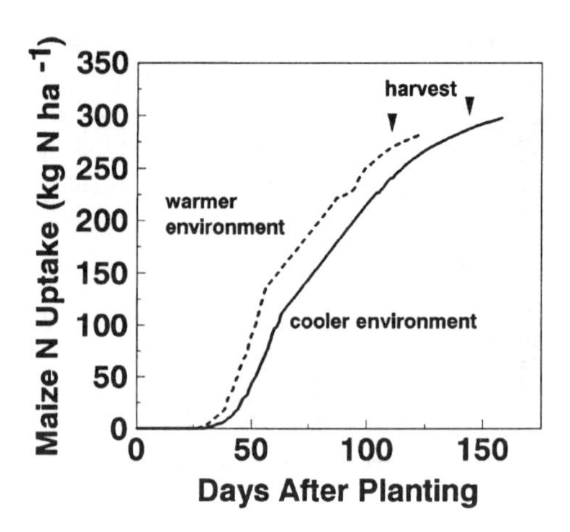

Figure 2. Simulated maize N uptake for the same cultivar grown in two environments differing only in daily temperature.

of aboveground dry matter produced at maturity. This number matches the internal N requirement Jones (1983) and Stanford and Legg (1984) found when they analyzed maize yield data from several field experiments.

Much of the difference in simulated yields shown for the two cultivars in Figure 1 can be attributed to differences in time to physiological maturity. Whereas, on average, cultivar Y reached maturity at about 90 days after planting, cultivar X usually reached maturity at about 120 days after planting. Clearly, the longer duration of growth translated into greater yields for cultivar X at this site.

Simulation can also be used to estimate the time course or pattern of N uptake by a growing crop. Such information can be used to define the period of most active N uptake, with a management goal being to have sufficient mineral N available in the soil during this period. Typical patterns of N uptake simulated for two different maize crops are illustrated in Figure 2. For this example, the same cultivar was grown at two different locations varying in temperature regimes, i.e., the CERES-maize model was run for the same cultivar, management, and soil type but with two different weather files. The warmer environment produced faster initial growth and earlier demands on N, while the cooler environment supported a longer growing period which resulted in slightly greater total N uptake. Similar simulation studies can easily be conducted to define N uptake patterns and amounts for any number of controlled and uncontrolled factors, e.g., the effect of different planting dates, planting densities, varieties, soil types, or weather conditions. For a specific site, this information can then be used to seek better synchrony between crop N demand and N supply.

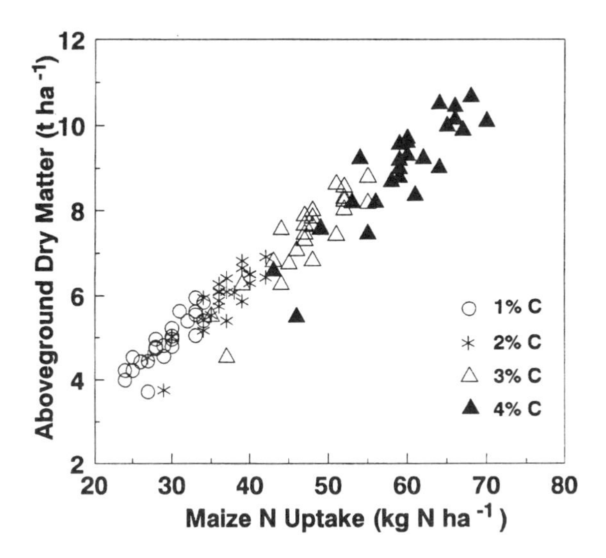

Figure 3. Simulated relationship between maize N uptake and aboveground dry matter, using 25 years of weather, for four soils differing only in the level of soil C.

Soil N supply

The N supplied by the soil can come from two major sources: (i) mineralization of soil organic N during the growing season, and (ii) mineral N present in the soil profile at planting. Both sources should be considered when estimating the amount of supplemental N needed by a growing crop. Their relative importance, however, will depend primarily on previous management, soil organic matter, and rainfall.

The mineralization of soil organic N is a biological process, which means the amount of N this process makes available is greatly influenced by the level of microbial activity and the amount of C substrate available. For most mineral soils, the proportion of C and N in the soil organic matter is fairly constant with an average C:N ratio of 10:1. Such a ratio favors net mineralization, or a net release of N to available mineral forms if organic materials with higher C:N ratios are not added. Immobilization of some N occurs as part of the internal turnover of N through mineralization-immobilization, but as long as C:N ratios remain below about 25 the net result is a release of mineral N (Stevenson, 1986). When C:N ratios reach over 25 there is usually a loss of plant available N as it becomes tied up through net immobilization.

In general, when all other factors are equal (weather, crop management, soil texture), plant available N increases as the level of organic C in a soil increases. To illustrate, Figure 3 shows the simulated relationship between maize N uptake and aboveground dry matter for typical soil C levels ranging from 10 to 40 g C kg^{-1} of soil (C:N ratios of 10:1); simulations were run using 25 years

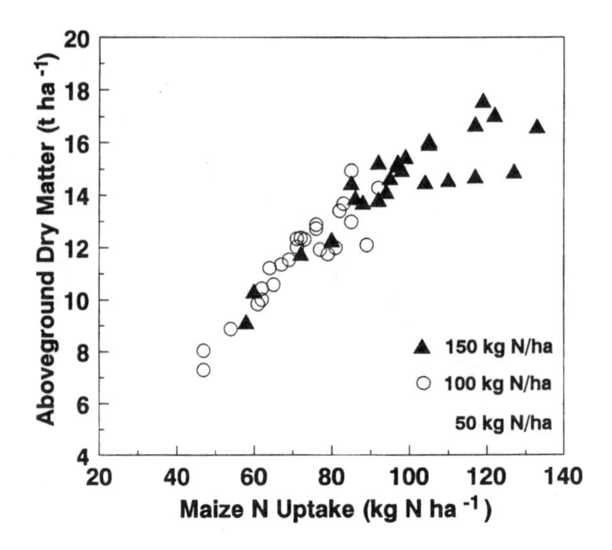

Figure 4. Simulated relationship between maize N uptake and aboveground dry matter, using 25 years of weather, for the same soil differing only in the amount of mineral N present in the soil profile at the start of the simulation.

of weather for each of four soils differing only in the level of C. For a given soil C level, the scatter in points for a given maize N uptake is a reflection of the effect year-to-year variability in weather has on crop growth and N related processes.

The amount of mineral N present in the soil profile at planting (initial mineral N) often has a substantial impact on the need for supplemental N, particularly in less humid environments. Initial mineral N may vary across sites and years due to previous crop effects, management and type of crop residues, and residual N left from external sources applied to earlier crops. If rainfall is not excessive, much of the initial mineral N can remain available to a crop throughout the growing season.

Once again it is easy to demonstrate the potential importance of initial mineral N with a simulation example. Figure 4 shows the relationship between maize N uptake and aboveground dry matter for 25 model runs (25 years of weather) for each of three levels of initial mineral N (50, 100, and 150 kg N ha^{-1}). These results show that this source of N can be significant if the mineral N remains in the profile long enough for it to be tapped by growing roots. Similar results have been obtained in real world assessments (Vanotti and Bundy, 1994), and for this reason soil testing programs strongly suggest sampling the crop root zone for initial mineral N when deriving N recommendations (Hergert 1987).

The amount of initial mineral N a growing crop might use is highly dependent on rainfall and rooting depth. For example, Figure 5a shows the relationship between rainfall and maize N uptake obtained from simulating 200 growing seasons where rainfall ranged from 100 to 1700 mm, and where each simulation started with an initial mineral N content of 100 kg N ha^{-1} in a rooting profile 1.2 m deep. For the conditions of this simulation (i.e., soil type, crop variety, and management), N uptake would likely decrease as seasonal rainfall increases above about 500 mm, mainly because leaching losses would increase as rainfall increases above 500 mm (Figure 5b).

With appropriate soil, weather, and crop information, these same types of simulation studies can be performed on a site-specific basis, with any number of possible scenarios examined. For example, changing the cultivar, management, or water-holding characteristics of the soils used to derive Figures 3–5 would likely result in slightly different but comparable relationships. Likewise, running the same simulation examples for sites with different weather patterns should provide different but analogous outcomes.

Sources of supplemental N

The simulation examples cited earlier illustrate how the CERES model can be used to estimate crop N requirements and soil N supply for specific situations. Particularly important is the capability to quantify these components, on a site-specific basis, across a wide range of likely weather conditions. A quantitative understanding of how the crop N requirement and soil N supply change in response to weather and management can then be used to better estimate the likely benefit of supplemental N. Supplemental N in the CERES model can be added as plant residue or manufactured fertilizer N. Algorithms dealing with the transformation of animal manure have not yet been included, although Hoffman and Ritchie (1993) have proposed a model which could be adapted.

Since crop N requirement and soil N supply are likely to change with management, soil type, and weather conditions, the amount of supplemental N needed to attain a specific yield would be expected to vary as well. For this reason N response functions so often vary from year to year and site to site. Although the model might be used to draw N response functions for many different conditions, it could also be used to estimate the need for supplemental N more directly as the difference between crop N requirement and soil N supply. For example, Figure 6 shows how the need for extra N, expressed as the difference between crop N requirement and soil N supply, might vary for maize using 25 years of weather. Also illustrated in this figure is the likely effect of initial mineral N assuming the amount present in the profile at planting is either 50 or 100 kg N ha^{-1}. The need for supplemental N is expected to vary according to weather conditions, but it will also depend on other factors such as the attainable yield level and N supplied by mineralization of the soil organic matter or available as residual mineral N.

Were it possible to know future weather with certainty, and were it possible to recover all applied N with unit efficiency, it would then be easy to use the

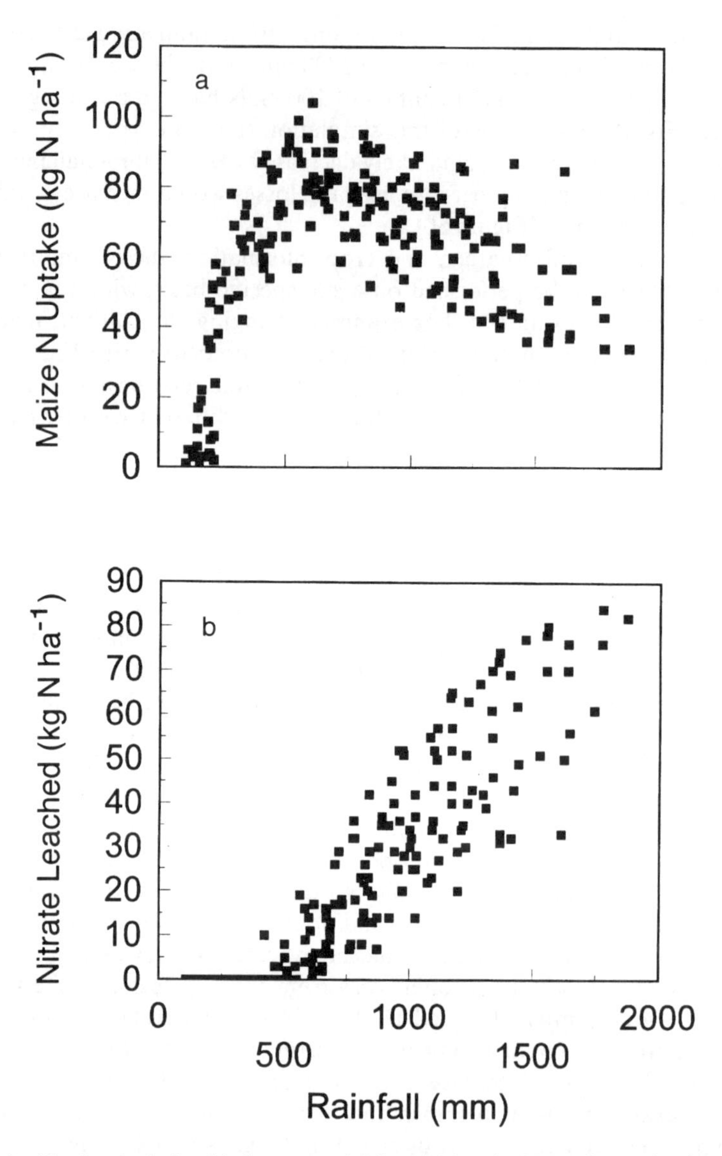

Figure 5. Simulated relationship between rainfall and (a) maize N uptake or (b) nitrate leached below 1.2 m for 200 different growing seasons with rainfall ranging from 100 to 1700 mm.

type of simulation results shown in Figure 6 to prescribe an exact optimum rate of N. The real world, however, is not so conveniently predictable or efficient. Weather conditions cannot be forecast beyond a few days, and the recovery of applied N is seldom complete due to the many interacting N processes and management factors that affect its efficiency. Nevertheless, by

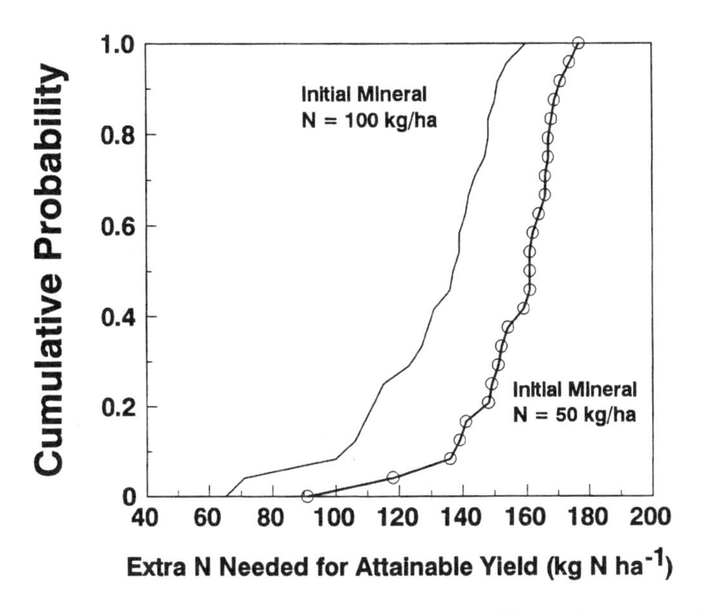

Figure 6. Simulated cumulative probability curves showing the difference between crop N requirement and soil N supply for maize grown in the same soil with two different amounts of initial mineral N, for 25 different seasons.

employing models that simulate many of these complex interactions, combined with the use of appropriate soil, crop, and historical weather data, it should be possible to estimate the likely impact of different N management practices.

With the CERES model, N management practices that can easily be examined include the effect of varying N rates, time of application, placement depth, and source. In addition to crop yield response, other factors to be examined might include economic returns and the potential for excessive nitrate leaching. By running the model with a proposed N management practice for many different weather years, one can estimate the risks involved from both an economic and environmental perspective. An analysis program that greatly facilitates the analysis of such risks is distributed with the DSSAT v3 software (Thornton et al., 1994), for which Thornton and Hoogenboom (1994) offer application examples.

An example of how an increase in applied N might affect both yield and nitrate leaching potential on two different soil types is shown in Figure 7. These data were obtained by running the maize model with N fertilizer rates increasing from 25 to 300 kg N ha^{-1}, assuming the given rate was applied at planting to maize grown on either a shallow sand or a deep loam. The points on the graph, each of which represents the mean value obtained using 25 years of weather, show the relationship between yield and nitrate leaching for N rates as they increase in increments of 25 kg N ha^{-1} from left to right starting at 25 kg N ha^{-1} and ending at 300 kg N ha^{-1}.

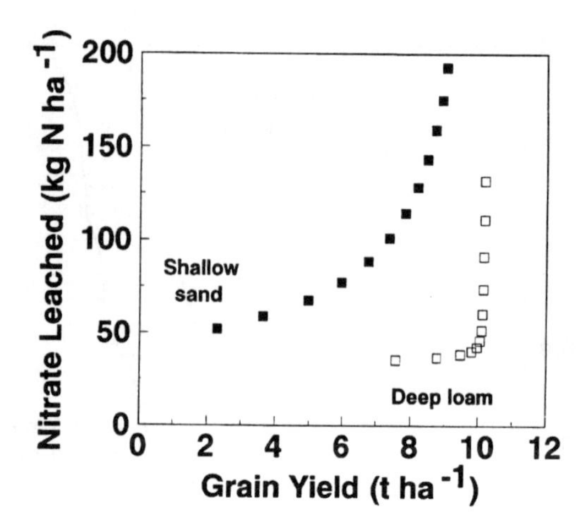

Figure 7. Simulated tradeoff curves showing the relationship between maize grain yield and nitrate leached for two different soils.

The presentation of simulated data as shown in Figure 7 helps to depict the potential tradeoffs for a proposed N management practice, which in this case represents the tradeoff between a benefit and an environmental cost, i.e., greater yield due to applied N versus potential leaching of nitrate to groundwater. According to these results, there is a distinctly different tradeoff curve for each soil type. Whereas both soils show a yield response to applied N, leaching losses are likely to be much greater on the shallow sand than the deep loam.

If one were to define a critical value for nitrate beyond which leaching losses were not permitted, then such tradeoff curves could be used as a guide to determine maximum allowed N rates. For example, a critical value of 100 kg N ha^{-1} for nitrate leaching would mean that no more than 150 kg N ha^{-1} should be applied to the shallow sand. For the deep loam, however, near maximum yields would be obtained with the application of less N with minimum losses due to leaching. The same sort of analysis could be repeated with economic returns used as a measure of benefit rather than yield.

The simulation examples presented are simple illustrations of the way the CERES model might be used to estimate the likely outcome of different N management practices. The next section is intended to further demonstrate its value for addressing issues related to N management by citing published studies designed to evaluate and use the model in practical ways.

Model evaluation and application examples

There is a vast literature devoted to both the testing and application of the CERES model, with only a small proportion of it directed, however, towards the testing and application of the N balance component. One of the earliest studies on the capability of the model to realistically predict response to fertilizer N applications was conducted for winter and spring wheats by Godwin and Vlek (1985). In this study, they were able to demonstrate that the model performed well in predicting not only grain yield but also plant N uptake across a wide range of N application rates for wheat grown in different environments including Syria, Australia, North America, and Europe. They then illustrated how the model could be run with many years of generated weather data to compare different N management strategies for improving fertilizer efficiency given the uncertainty of weather.

In a later study with maize, Jones and Kiniry (1986) provided evidence that the CERES model realistically predicted grain yield response to increasing rates of N fertilizer. They also demonstrated the sensitivity of the model to the accurate specification of mineral N present in the soil profile at the start of the simulation. If initial mineral N is not measured accurately, then the reliability of the simulated response to N must be questioned. Data and model testing results from these same studies plus others were also presented by Godwin and Jones (1991) in their detailed description of the N balance model.

A particularly good example of how the CERES model might be used together with site-specific data to examine N management strategies was offered by Keating et al. (1991, 1993) for maize in semiarid Kenya. Using the model as a quantitative tool, they were able to demonstrate that moderate improvements in management, mostly related to planting density, fertilizer N application, and residue management, would likely result in greater earnings in most years even though there is a risk of crop failure in some years due to drought. Their study involved first an evaluation of the model for its accuracy in simulating the interaction of water and N supply and management under local conditions, which pointed to weaknesses in the model regarding the simulation of plant response to severe water and N stress. After modifying the model to better account for extreme water and N stress, they ran it with various management scenarios using long-term weather data to quantify the economic risks of modest changes in plant population, N management, and crop residue management.

A similar evaluation and application of the maize model was done for Malawi by Singh et al. (1993) to quantify weather-related risks for different N management options. Thornton et al. (1995) then took this application of the model one step further by linking it to a geographical information system (GIS) with spatial databases of soils and weather. Such a linkage permitted the analysis of the impact of N management on both crop yield and nitrate leaching potential at the regional level, by taking into account not only differences in soil types and weather across the region but also year-to-year variability in weather. Particularly useful was the capability to then map the variability in maize yields and nitrate leaching potential due to the interaction of management, soil, and weather factors for the region.

Thornton and MacRobert (1994) presented a simulation study with the CERES model designed to quantify the economic value of perfect weather information for scheduling N fertilizer applications. Their study showed the potential for using the model in a forecasting mode along with historical weather data and current season weather to better estimate the most cost effective management of N fertilizer applications as the season progresses. In another study, Bowen and Papajorgji (1992) used the CERES model to estimate the expected benefit from providing emergency supplies of N fertilizer to Albania for its winter wheat crop at a time when the country had insufficient fertilizer and wheat stocks.

More critical evaluations of specific components of the CERES N balance model were conducted by Bowen et al. (1993), Campbell et al. (1994), and Quemada and Cabrera (1995). Although Bowen et al. (1993) showed that the model provided realistic simulations of legume green manure decomposition and N release, they also demonstrated the inability of the model to account for retarded nitrate leaching in the subsoil of variable charge soils such as Oxisols. A slight modification, however, to the leaching subroutine showed that retarded movement due to net positive charge in the subsoil could be realistically simulated; this modification only required that an adsorption coefficient for nitrate be calculated for each soil layer.

Campbell et al. (1994) were interested in using the CERES model to estimate soil N supply in long-term cropping systems based on a fallow-wheat rotation. To obtain more realistic estimates of N supplied by the mineralization of soil organic matter with time, they proposed a modification to the model that split the soil humus pool into a slow N release fraction and a rapid N release fraction, rather than the one-pool humus model originally incorporated into CERES. The modified model performed better when simulating different rotation systems in years with varying precipitation and soil water availability.

Quemada and Cabrera (1995) conducted laboratory incubation experiments designed to improve the capability of the CERES model to simulate N release from crop residues left to decompose on the soil surface. Since the original decay rate constants used in the model were for plant residue incorporated into the soil, they found it was necessary to decrease the size of these constants for residue placed on the surface. They also found better results were obtained when the relative size of the carbohydrate, cellulose, and lignin plant residue pools could be modified by the user using measured data, rather than assuming all plant residue had a fixed proportion of 20% carbohydrate, 70% cellulose, and 10% lignin.

Conclusions

Nitrogen management can be improved upon through the insight provided by comprehensive crop simulation models such as the CERES model. As a continually evolving tool, such models have the potential to help both researchers and farmers better understand how soil, crop, weather, and management factors interact to affect crop N demand, soil N supply, and fertilizer use efficiency on a site-specific basis. With the model and appropriate input data, any number of management scenarios can be examined and compared for their impact on not only economic returns but also the potential for excessive leaching of nitrates. Economic and environmental (nitrate leaching) risks due to uncertain weather can also be quantified.

References

Baethgen W E, Christianson C B, Lamothe A G(1995) Nitrogen fertilizer effects on growth, grain yield, and yield components of malting barley. Field Crops Res 43:87–101.

Bowen W T, Papajorgji P (1992) DSSAT estimated wheat productivity following late-season nitrogen application in Albania. Agrotechnology Transfer 16:9–12.

Bowen W T, Jones J W, Carsky R J, Quintana J O (1993) Evaluation of the nitrogen submodel of CERES-Maize following legume green manure incorporation. Agronomy Journal 85:153–159.

Bowen W T, Wilkens P W, Singh U, Thornton P K, Ritchie J T (1995) A Generic CERES Cereal Model. Agronomy Abstracts p 62, American Society of Agronomy, Madison, WI, USA.

Cabrera M L, Vigil M F, Kissel D E (1994) Potential nitrogen mineralization: Laboratory and field evaluation. Pages 15–30 in Havlin J L, Jacobsen J S(ed.) Soil testing: Prospects for improving nutrient recommendations. Soil Science Society of America Special Publication 40. Soil Science Society of America, Madison, WI, USA.

Campbell C A, Jame Y W, Akinremi O O, Beckie H J (1994) Evaluating potential nitrogen mineralization for predicting fertilizer nitrogen requirements of long-term field experiments. Pages 81–100 in Havlin J L, Jacobsen J S(ed.) Soil testing: Prospects for improving nutrient recommendations. Soil Science Society of America Special Publication 40. Soil Science Society of America, Madison, WI, USA.

Dahnke W C, Johnson G V (1990) Testing soils for available nitrogen. Pages 127–139 in Westerman R L (ed.) Soil testing and plant analysis, 3rd ed. Soil Science Society of America, Madison, WI, USA.

Godwin D C, Jones J W (1991) Nitrogen dynamics in soil-crop systems. Pages 287–321 in Hanks R J, Ritchie J T (Eds.) Modeling plant and soil systems. Agronomy Monograph #31, American Society of Agronomy, Madison, WI, USA.

Godwin D C, Vlek P L G (1985) Simulation of nitrogen dynamics in wheat cropping systems. Pages 311–332 in Day W, Atkins R K (eds.) Wheat growth and modeling. Plenum Press, NY, USA.

Harre E A, White W C (1985) Fertilizer Market Profile. Pages 1–24 in Engelstad O P (ed.) Fertilizer technology and use, 3rd ed. Soil Science Society of America, Madison, WI, USA.

Hergert G W (1987) Status of residual nitrate-nitrogen soil tests in the United States of America. Pages 73–88 in Soil testing: sampling, correlation, calibration, and interpretation. Soil Science Society of America, Special Publication 21, Madison, WI, USA.

Hoffmann F, Ritchie J T (1993) Model for slurry and manure in CERES and similar models. Journal of Agronomy and Crop Science 170:330–340.

Hoogenboom G, Jones J W, Wilkens P W, Batchelor W D, Bowen W T, Hunt L A, Pickering N B, Singh U, Godwin D C, Baer B, Boote K J, Ritchie J T, White J W (1994) Crop models. Pages 95–244 in Tsuji G Y, Uehara G, Balas S (eds.) DSSAT v3. Vol. 2. University of Hawaii, Honolulu, HI, USA.

Jones C A, Kiniry J A (eds.) (1986) CERES-Maize: a simulation model of maize growth and development. Texas A&M University Press, College Station, Texas, USA.

Jones C A (1983) A survey of the variability in tissue nitrogen and phosphorus concentrations in maize and grain sorghum. Field Crops Research 6:133–147.

Keating B A, Godwin D C, Watiki J M (1991) Optimising nitrogen inputs in response to climatic risk. Pages 329–358 in Muchow R C Bellamy J A (eds.) Climatic risk in crop production: Models and management for the semiarid tropics and subtropics. CAB International, Wallingford, Oxford, United Kingdom.

Keating B A, McCown R L, Wafula B M (1993) Adjustment of nitrogen inputs in response to a seasonal forecast in a region of high climatic risk. Pages 233–252 in Penning de Vries F W T, Teng P S, Metselaar K (eds.) Systems approaches for agricultural development. Kluwer Academic Publishers, Dordrecht, The Netherlands.

Meisinger J J (1984) Evaluating plant-available nitrogen in soil-crop systems. Pages 391–416 in Hauck R D (ed.) Nitrogen in crop production. American Society of Agronomy-Crop Science Society of America-Soil Science Society of America, Madison, WI, USA.

Penning de Vries F W T, Jansen D M, ten Berge H F M, Bakema A (1989) Simulation of ecophysical processes of growth in several annual crops. Simulation Monograph Series, PUDOC, Wageningen, The Netherlands and IRRI, Los Banos, The Philippines.

Quemada M, Cabrera M L (1995) CERES-N model predictions of nitrogen mineralized from cover crop residues. Agronomy Journal 59:1059–1065.

Ritchie J T, Singh U, Godwin D C, Hunt L A (1989) A user's guide to CERES maize – V2.10. International Fertilizer Development Center, Muscle Shoals, AL, USA.

Singh U, Thornton P K, Saka A R, Dent J B (1993) Maize modelling in Malawi: A tool for soil fertility research and development. Pages 253–273 in Penning de Vries F W T, Teng P S, Metselaar K (eds.) Systems approaches for agricultural development. Kluwer Academic Publishers, Dordrecht, The Netherlands.

Stanford G, Legg J O (1984) Nitrogen and yield potential. Pages 263–272 in Hauck R D (ed.) Nitrogen in crop production. American Society of Agronomy-Crop Science Society of America-Soil Science Society of America, Madison, WI, USA.

Stanford G (1973) Rationale for optimum nitrogen fertilization in corn production. Journal of Environmental Quality 2:159–166.

Stevenson F J (1986) Cycles of soil: carbon, nitrogen, phosphorus, sulfur, micronutrients. John Wiley and Sons, New York, NY USA.

Thornton P K, Hoogenboom G (1994) A computer program to analyze single-season crop model outputs. Agronomy Journal 86:860–868.

Thornton P K, MacRobert J F (1994) The value of information concerning near-optimal nitrogen fertilizer scheduling. Agricultural Systems 45:315–330.

Thornton P K, Hoogenboom G, Wilkens P W, Jones J W (1994) Seasonal analysis. Pages1–66 in Tsuji G Y, Uehara G, Balas S (eds.) DSSAT v3. Vol. 3. University of Hawaii, Honolulu, HI USA.

Thornton P K, Saka A R, Singh U, Kumwenda J D T, Brink J E, Dent J B (1995) Application of a maize crop simulation model in the central region of Malawi Exp. Agriculture 31:213–226.

Vanotti M B, Bundy L G (1994) Frequency of nitrogen fertilizer carryover in the humid midwest. Agronomy Journal 86:881–886.

The use of a crop simulation model for planning wheat irrigation in Zimbabwe

J.F. MACROBERT[1] and M.J. SAVAGE[2]

[1] *Formerly Research Manager, Agricultural Research Trust, P.O. Box MP 84, Harare, Zimbabwe. Current affiliation: Elim Mission, Private Bag 2007, Nyanga, Zimbabwe*

[2] *Professor and Head, Department of Agronomy, University of Natal, Pietermaritzburg, South Africa*

Key words: wheat, deficit irrigation, Zimbabwe, CERES, ZIMWHEAT, management, gross margin

Abstract

Wheat is grown in Zimbabwe during the relatively dry, cool winter with irrigation. On most large-scale farms, land resources exceed irrigation water resources. Consequently, the efficient use of water is of prime concern. This has led to the development and adoption of deficit irrigation techniques, with the aim of maximizing net financial returns per unit of applied water rather than per unit land area. This often means that less water is applied than that required for maximum yields, and water deficits are allowed to develop in the crop. This chapter describes an interactive computer program, based on a modification of CERES-Wheat version 2.1, that searches for the intraseasonal irrigation regime that maximizes the total gross margin for a particular soil, cultural and weather scenario, within the constraints of land and water availability. The optimum irrigation solution generated by the program provides a basis on which a farmer can plan irrigation management strategies.

Introduction

Spring wheat is an important crop in Zimbabwe. In 1990, 325,454 t were produced at an average yield of 5.92 t ha^{-1} (Smith and Gasela, 1991). The crop is grown during the relatively dry, cool winter with hand-moved sprinkler irrigation using water from limited stocks. Consequently, the efficient use of water is of prime concern.

Land resources are usually abundant during the winter months, whereas water is not. This has led to the development and adoption of deficit irrigation techniques (James, 1988), with the aim of maximizing net financial returns per unit of applied water rather than per unit land area. This often requires that less water be applied than that required for maximum yields, which means that water deficits may be allowed to develop in the soil and crop.

The problems of deficit irrigation management are of three types: first, in deciding the land area to be irrigated and the minimum return interval that can be employed with the existing irrigation infrastructure; second, in determining the intraseasonal sequence and amounts of irrigation to apply; and third, in evaluating the financial returns and risk associated with allowing the develop-

G.Y. Tsuji et al. *(eds): Understanding Options for Agricultural Production, 205–220*

ment of soil water deficits. Analyses beyond conventional irrigation management decisions are thus required for deficit irrigation, and Heermann et al. (1990) concluded that one cannot give general recommendations for deficit irrigation because each situation is unique and depends on a number of factors. Stegman et al. (1983) advocated the use of computer simulation to evaluate deficit irrigation options, and some attempts have been made in this direction (e.g., Howell and Hiler, 1975; MacRobert, 1994; Martin, 1984; Martin et al., 1989).

This chapter describes an interactive computer irrigation optimization program that was developed to assist farmers in Zimbabwe decide on an irrigation strategy that would maximize the total gross margin of their wheat. The program is based on CERES-Wheat (Ritchie and Otter, 1984; Godwin et al., 1989) and incorporates irrigation and economic criteria to predict the intraseasonal irrigation regime that maximizes the total gross margin for a particular soil, management, and weather scenario, within the constraints of land and water availability. The program provides an example of the practical use of crop simulation models in assisting with farm decision making.

Program description

The wheat irrigation optimization program, named WIRROPT7, was written in Microsoft's QuickBASIC 4.00 and is useable on IBM AT-compatible computers operating under MS-DOS 3.1 or higher. WIRROPT7 uses a modified version of the CERES-Wheat crop simulation model called ZIMWHEAT (MacRobert, 1994). Many changes were made to the phenological and growth subroutines of CERES-Wheat version 2.10. Some changes consisted of minor calibrations of empirical equations, whilst others involved significant modification to various procedures. Details of these changes are available in MacRobert (1994) and are summarized below.

The initial validation of CERES-Wheat under Zimbabwean conditions showed that the model gave biased and imprecise predictions of phenological development, particularly under deficit-irrigated conditions. The simulation of tillering was poor and the model tended to over-predict dry matter accumulation and under-predict leaf area indices. The yield component and grain yield predictions were also generally imprecise. On the other hand, for most data sets, the simulated soil water contents were similar to measured soil water contents. These inconsistencies prompted a revision of the phenological and growth subroutines of the model. In the phenological subroutine, new thermal time durations and base temperatures for all growth phases were determined from regressions of the rate of phasic development on mean air temperatures. For growth phases one, two and three, a base temperature of 4°C was established, whereas for growth phases four and five, a base temperature of 3°C was used. The revised model included the prediction of leaf emergence (as apposed to leaf appearance) and first node appearance. In order to hasten plant development under conditions of soil water deficit stress, daily thermal time was made to increase whenever the actual root water uptake declined below 1.5 times the potential plant evaporation. These changes improved the prediction of crop phasic development.

Changes were also made to the growth subroutine, including the following: (1) The extinction coefficient in the exponential photo-synthetically active radiation (PAR) interception equation was reduced from 0.85 to 0.45. (2) An allowance was made for the interception of PAR by the wheat ears during growth phases four and five. (3) The area to mass ratio of leaves was increased from 115 $cm^2\ g^{-1}$ to 125 $cm^2\ g^{-1}$ during growth phase two and this was allowed to decrease under conditions of water deficit stress. (4) Tiller production during growth phase one was made a function of daily thermal time, total daily solar radiant density and plant density, moderated by high air temperatures and a new soil water deficit factor that takes the dryness of the surface soil layer into account. (5) A cold temperature routine was added to reduce kernel numbers whenever the exposed minimum air temperature decreased below 0 C during the period ear emergence to the start of the linear kernel growth phase (cold temperatures during anthesis occasionally cause reductions to kernel numbers in Zimbabwe). (6) The kernel growth rate was gradually increased during growth phase four, and the rate of kernel growth was increased under conditions of water deficit stress during growth phase five. Taken together, the modifications made to CERES-Wheat improved predictions of tillering, ear density, growth duration (Figure 1a), and yield under conditions of deficit irrigation in Zimbabwe (Figure 1b).

The irrigation optimization program aims to answer the following pre-season questions in the context of maximizing net financial returns:

(1) With a given water and land resource, what is the area of wheat that should be grown to maximize the total gross margin?
(2) With a given irrigation equipment infrastructure, what is the minimum return interval of irrigation to be employed, and when should intraseasonal irrigation be applied?
(3) For a given optimal area of wheat, what will be the expected costs of production and returns?

WIRROPT7 is modular, the main program supporting 20 subroutines. The computational flow of the program is shown in Figure 2. The program contains a main menu by which the user interacts with six input menus; these specify the identity, irrigation constraints, economic criteria and the soil, weather and cultural conditions appropriate to the farm in question. Once these parameters have been chosen, the main program calls ZIMWHEAT iteratively to simulate irrigation application dates, phenology and yield. On output, irrigation dates and amounts and yield are converted into financial indices to determine the best irrigation regimes for the farm in question. These are listed on the computer

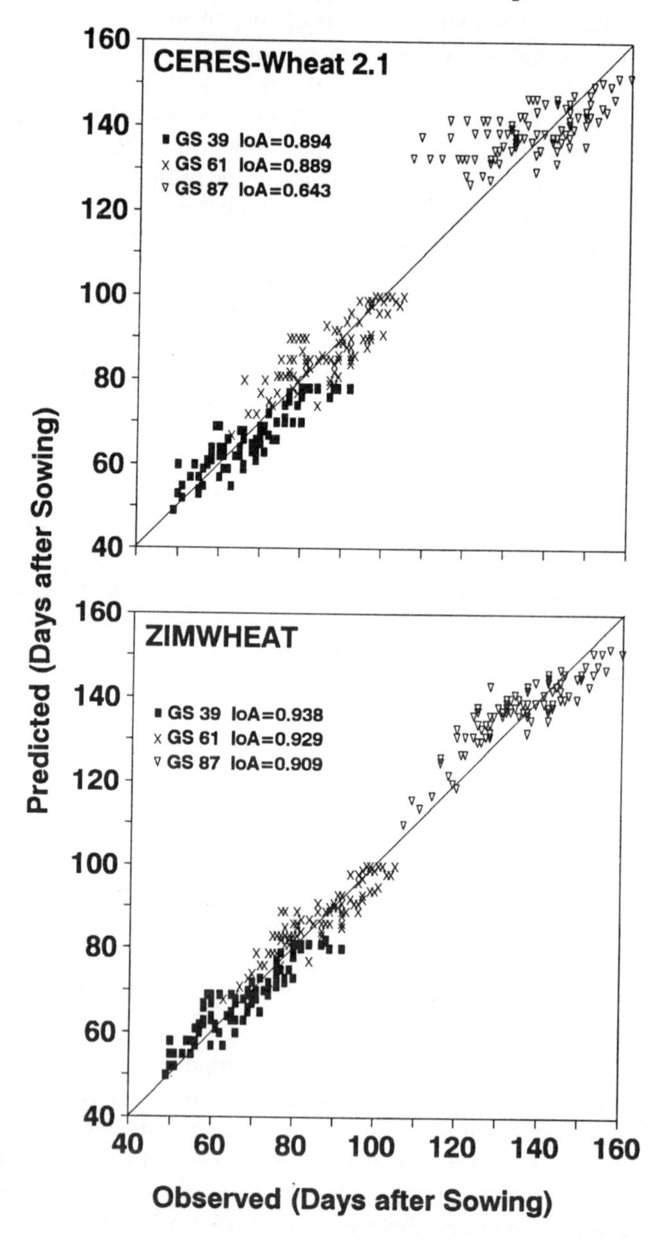

Figure 1. Comparison of the predictions by CERES-Wheat V2.1 and the modified version (ZIMWHEAT) with observed data from 97 independent data sets for (A) phasic development and (B) field yield of spring wheat in Zimbabwe. Growth stages are according to the scale of Zadoks et al. (1974), *IoA* is the index of agreement (Wilmott, 1982), and *r* is Pearson's correlation coefficient.

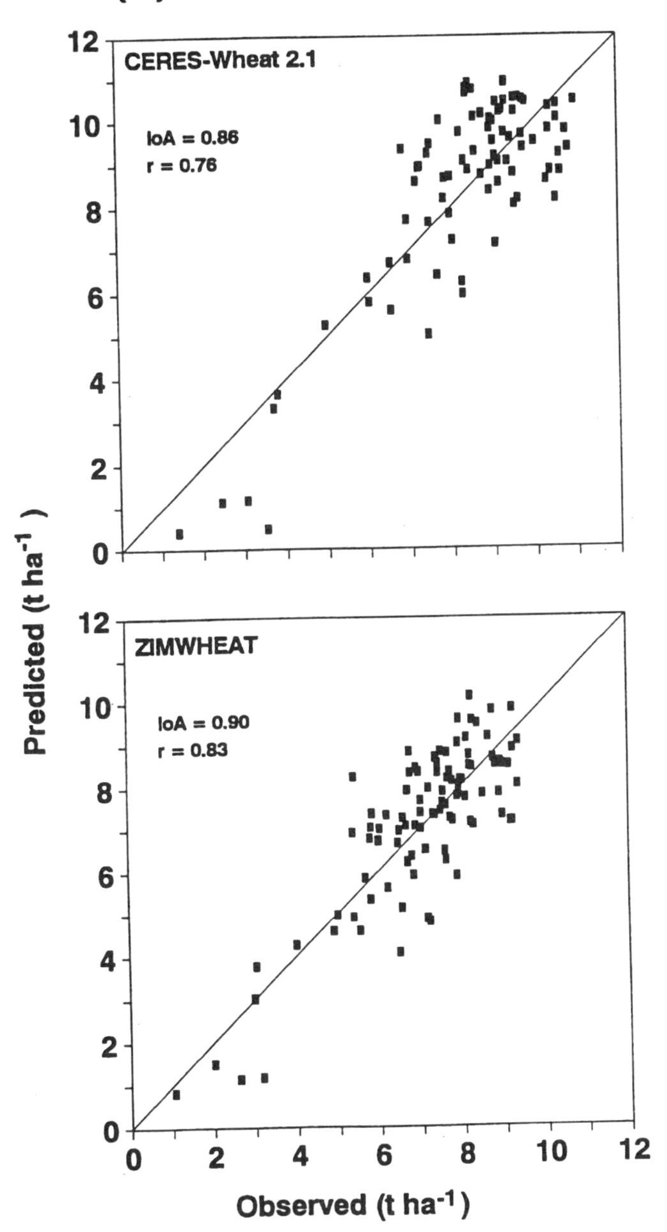

Figure 1. (*Continued.*)

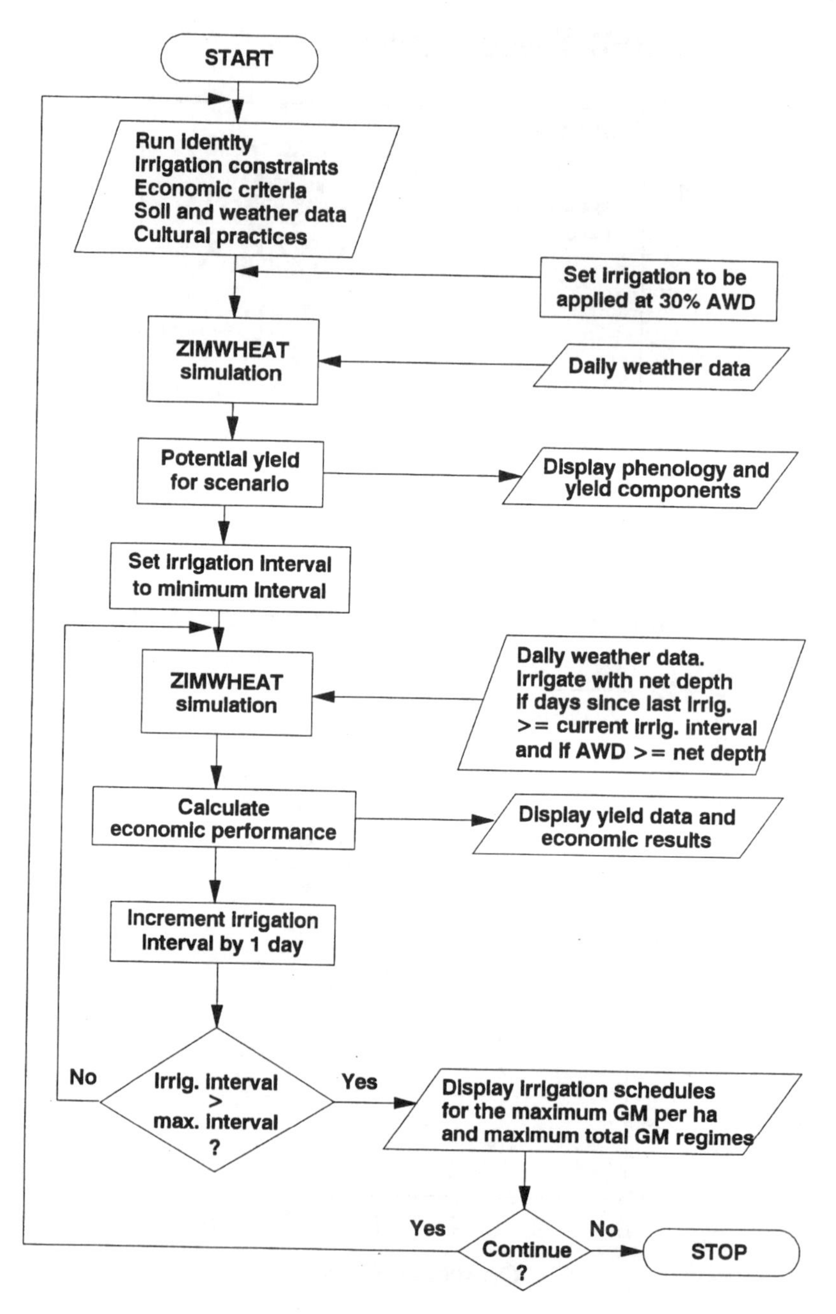

Figure 2. Diagram of the computational flow of the irrigation optimization program WIRROPT7. *AWD* is the percent depletion of plant available water in the soil profile, and *GM* is the gross margin.

screen for evaluation. The program also outputs the simulated dates of important growth stages, including the projected harvest date. The period from physiological maturity to harvest was simulated using the algorithm developed by Chen and McClendon (1984). Although their model was used on soybeans, Parsch et al. (1991) found it adequate for the prediction of grain drydown of wheat.

Input routines

(a) *Farm and run identification.* Each optimization encounter with the model is preceded by an opportunity to identify the simulations by specifying the farm name, a four-character code for the farm, and a two-digit year code.

(b) *Irrigation constraints.* Most irrigation farms in Zimbabwe have more arable land than irrigable land, as dictated by the irrigation water resources. The majority of irrigation schemes consist of hand-moved, overhead sprinkler systems. Such schemes are flexible in that they can be extended relatively easily to cover a larger area than the original design. However, there is usually a maximum area to which the scheme may be expanded, as determined by underground or portable main lines and/or pump capacity. Once the field is laid out, the irrigation scheme becomes fairly rigid in terms of the frequency with which irrigation can be applied. Farmers generally lay out their irrigation network to cater for a minimum return interval, which is decided upon before the season starts by taking into account factors such as the soil water holding capacity, crop water demand, and potential evaporation.

The amount of irrigation that may be applied in one setting is dependent on the nozzle discharge, sprinkler spacing, and set duration. Most irrigation schemes have sprinkler spacings of 12 m by 18 m or 18 m by 18 m, and nozzle discharges of around 1 $m^3 h^{-1}$. Water application rates therefore range between 4 and 7 mm h^{-1}. The set duration is usually between six and 12 hours, with shorter durations applied on the sandier soils with lower water holding capacity. Twelve-hour set durations are generally favored, however, because they enable two moves per day during daylight hours. More frequent irrigation pipe changes require night moves, which can be difficult. On a 12 h set time (usually taken as 11.5 h to account for the time taken in moving the pipes), gross irrigation application amounts range between 45 and 80 mm (a common standard is 55 mm on soils of high water holding capacity). Irrigation application efficiencies vary greatly, but a figure of 85% is commonly used.

Water used in wheat production is derived from river flow, surface storage, and underground reserves. Farmers relying on underground reserves have the least knowledge of the total quantity of water available, but they generally know the hourly pump capacity of a well. Farmers extracting water from dams and rivers usually have defined water rights, which specify the seasonal quantities of water that may be pumped. From wherever a farmer obtains his water, he can estimate the total quantity of water available to him for a given season.

The irrigation constraints facing a farmer, and required as inputs by the program, are therefore as follows:

(1) The amount of potentially irrigable land available (*POTAREA*, ha).
(2) The total seasonal amount of water available (*WATRES*, ml).
(3) The minimum return interval (*MinInterval*, d).
(4) The maximum return interval (*MaxInterval*, d).
(5) The normal, full set irrigation application amount (*IrrApp*, mm).
(6) The irrigation efficiency (*EffIrr*, %).

(c) *Economic criteria.* The principal wheat production cost variables are irrigation and fertilizer. In the program, irrigation costs are considered the dominant variable costs. Consequently, the program requires the following production costs:

(1) The cost of irrigation water (*IrrCost1*, ZW\$ mm^{-1} where ZW\$ refers to Zimbabwean dollars). This represents the cost of delivering one mm of water to the field and may comprise energy costs, repairs and maintenance, depreciation on equipment, and labor.
(2) The cost of applying one cycle of irrigation (*IrrCost2*, ZW\$ $cycle^{-1}$). Because the program considers both the total amount of irrigation applied and the frequency of irrigation, it is necessary to include a cost component related to irrigation frequency. As irrigation schemes in Zimbabwe are mainly hand-moved sprinkler systems, this cost component consists mostly of labor costs.
(3) The harvest and transport cost (*HarvCost*, ZW\$ t^{-1}). Irrigation amount affects yield, and so harvest and transport costs need to be included. These costs account for the harvesting operation and the transport of grain to market.
(4) The constant costs of production (*ConstCost*, ZW\$ ha^{-1}). Certain costs of wheat production are essentially independent of irrigation amount. These include the costs of land preparation, seed and fertilizer. Although fertilizer (especially N fertilizer) is linked to yield potential, it was considered unnecessary to include it specifically in the model, particularly as the model assumes that all nutrients are non-limiting. Furthermore, research has indicated that wheat responds similarly to nitrogen fertilizer application over a range of water application amounts (Agricultural Research Trust, 1988).

In order to calculate the gross margin per hectare (*GMha*, ZW\$ ha^{-1}) and other financial indices, the program also prompts for the net price of wheat grain (*NetPrice*, ZW\$ t^{-1}), which is the grain price less marketing costs.

(d) *Weather file selection.* Representative weather files are included with the program. The names of these files are stored in a weather listing file. This file, together with the historical weather data files, conforms to the DSSAT version 2 file conventions (IBSNAT, 1986). Weather conditions during the winter period (1 May to 30 September) in Zimbabwe do not vary greatly, and so the use of a representative weather file is considered adequate for irrigation planning purposes. The user has the opportunity to test irrigation scenarios using other weather files if required.

(e) *Soil profile selection.* Representative soil profiles are listed for user selection. The program utilizes the IBSNAT (1986) standard file for defining soil profiles. Additional soil pedons may be added at the user's discretion.

The actual initial soil water content of each soil layer may be specified by the user at the start of the simulation, if known. Alternatively, the user may simply specify the percent depletion of the profile in terms of available water content. If this latter option is chosen, the program adjusts the soil water content of each layer to give the initial soil water content required for appropriate functioning of the soil water balance routines.

(f) *Crop cultural practices.* The program presents the user with a choice of cultivars, which have genetic coefficients appropriate to ZIMWHEAT. Additional cultivars may be added to the list at the user's discretion. Following cultivar selection, the user is prompted to enter the sowing date, plant density (plants m^{-2}), and seeding depth (cm).

Irrigation optimization routines

The program seeks to find the intraseasonal irrigation regime that maximizes the total gross margin within the constraints of land and water availability. The process for doing this is straightforward. First, all irrigation regimes have the same establishment procedure. This procedure applies irrigation water on the day of sowing, five days after sowing (DAS), if necessary, and at 21 DAS in quantities that ensure that the whole soil profile is at or near the upper limit of plant available water by 21 DAS. The only constraint is that the amount of irrigation applied with each irrigation must be equal to either a full or a half application of the normal full set irrigation (*IrrApp*) as specified by the user. This procedure conforms to the establishment principles given by MacRobert (1992), and they are practical in terms of a farmer's capability.

Second, the program loops once through ZIMWHEAT to simulate the potential yield by irrigating whenever 30% of the profile available soil water has been depleted. This irrigation triggering level of soil water depletion has been cited as one that will achieve maximum yields (James, 1988). No regard is taken of the values of *IrrApp* or *MinInterval* specified by the user, since the aim of this iteration is to simulate the maximum potential yield.

From the maximum yield and the total irrigation amount applied, the gross margin per hectare (*GMha*, ZW\$ ha^{-1}) is calculated as

$$GMha = (YIELD \times NetPrice) - (IrrApp \times IrrCost1) - (Nirr \times IRRCost2) - (YIELD \times HarvCost) - ConstCost,$$

where *YIELD* is the predicted field yield, t ha^{-1}. The area of wheat that could be grown (*AREA*, ha) is calculated as *WATRES* divided by the total irrigation application (*TotIrr*, mm ha^{-1}), multiplied by 100 to convert mm ha^{-1} to ml ha^{-1}. The gross margin per unit of applied water (*GMmm*, ZW\$ mm^{-1}) is calculated by dividing *GMha* by *IrrApp*. The total costs (*TOTCOST*, ZW\$) and the total gross margin (*TotGM*, ZW\$) are computed from the product of the *AREA* and the sum of the costs per ha and from the product of *AREA* and *GMha*, respectively. These indices provide the starting point from which the program begins to seek the irrigation regime that will maximize total gross margin.

The program continues to loop through the ZIMWHEAT crop simulation model, incrementing the minimum irrigation return interval by one day with each loop, starting with *MinInterval* and ending with *MaxInterval*. Irrigation after establishment (i.e., at 21 DAS) is applied only if the number of days from the last irrigation is greater than or equal to the current irrigation interval and if the profile soil water content has been depleted by at least the amount $IrrApp \times EffIrr$. On completion of each loop, the gross margin per hectare, the gross margin per unit of irrigation water, the total gross margin, and the total costs are calculated and displayed. The program selects the irrigation regime that produces the maximum gross margin per hectare and the maximum total gross margin and displays the irrigation schedule and plant characteristics of each for evaluation by the user.

An example of irrigation optimization

As an example of the use of WIRROPT7, the program was run on two soil types differing in water holding capacity: a low water holding capacity soil from Kutsaga, located some 25 km from Harare (a sand with a mean available water content of 0.093 m^3 m^{-3} and a drainage coefficient of 0.7 day^{-1}), and a high water holding capacity soil located on the Agricultural Research Trust Farm, Harare (ART 1988), a sandy clay with a mean profile available water content of 0.150 m^3 m^{-3} and a drainage coefficient of 0.2 day^{-1}. The same weather file and cultural conditions were applied to both soils. The irrigation application amounts and the minimum and maximum return intervals differed according to soil type. A high water application rate (55 mm per set) and longer return intervals were applied on the ART 1991 soil, whereas a lower water application rate (40 mm per set) and shorter return intervals were applied on the Kutsaga soil. These application rates are common on such soil types in

Table 1. Input specifications used in the irrigation optimization example on two soils types.

Parameter	ART 1991	Kutsaga
Identification		
Farm name	ART Farm	Kutsaga
Farm code	ARTF	KUTS
Field code	01	02
Year	93	93
Run code	ARTF9301	KUTS9302
Irrigation constraints		
Area of land available, ha	100	100
Quantity of water available, ml	400	400
Maximum return interval, d	17	15
Minimum return interval, d	10	8
Normal full set application, mm	55	40
Irrigation efficiency, %	85	85
Economic criteria		
Net price of wheat, ZW$ t^{-1}	1450	1450
Irrigation water cost, ZW$ mm^{-1}	8	8
Irrigation application cost, ZW$ $cycle^{-1}$	12	12
Harvesting and transport cost, ZW$ t^{-1}	30	30
Constant costs of production, ZW$ ha^{-1}	2000	1600
Soil profile characteristics		
Name	ART 1991	Kutsaga
Available water depletion, %	30	40
Cultural conditions		
Cultivar	W170/84	W170/84
Sowing date	12 May	12 May
Plant density, plants m^2	200	200
Seeding depth, cm	2	2

Zimbabwe. The constant costs of production were set lower on the Kutsaga soil than the ART soil in anticipation of lower yields on the Kutsaga soil.

Details of the input parameters for these runs are given in Table 1. The program was executed from within the QuickBASIC 4.00 environment on an IBM-compatible 486 computer running at 25 MHZ under MS-DOS 6. The optimization time for each soil type was between four and five minutes (the duration of the optimization process depends partly on the time taken for the user to interact with the input routines).

The first output screen (Table 2 for the ART soil) presents the phenological development, yield and yield components of the first optimization loop. This loop establishes the potential yield for the site, sowing date and cultivar. The simulated potential yields for the two soil types were equal. This is principally because irrigation is applied to ensure a non-limiting soil water balance. However, because the application rates for the two soils were different, the attainment of potential yield on the ART soil required a minimum return interval of seven days, whereas the minimum return interval for potential yields on the Kutsaga soil was five days (Tables 3 and 5, respectively). Nevertheless,

Table 2. The first output screen generated by WIRROPT7 during the simulated optimization of irrigation of wheat on the ART soil. *Price* is the net price of wheat grain, *Irr* is the irrigation water cost, *App* is the irrigation application cost, *Harv* is the harvesting cost, *Const* are the constant costs of production, and *DAS* is the days after sowing.

Simulation of potential yield
Code: ARTF9301 Weather: KUTS0112.RND Soil: 4 (16) Cv: W170/84
Price = \$1450/t Irr = \$8.00/mm ha App = \$12.00/cycle Harv = \$30/t Const = \$2000/ha

Event	Date	DAS
Sowing date:	12 May	0
Germination:	13 May	1
Emergence:	18 May	6
Terminal spikelet:	30 Jun	49
First node:	9 Jul	58
Flag leaf emergence:	30 Jul	79
Flowering started:	22 Aug	102
Begin grain-filling:	1 Sep	112
Physiological maturity:	12 Oct	153
Harvest maturity:	21 Oct	162

Yield = 10.08 t/ha; Ear density = 451 per m^{-2}; Kernel mass = 43.1 mg
Kernels per plant = 138; Kernels per square meter = 27520

Table 3. The second output screen from the simulated optimization of irrigation of wheat on the ART soil, showing the effect of incrementing the minimum irrigation interval (*Irr Int*) from *MinInterval* to *MaxInterval* on the total irrigation applied (*Tot Irr*), yield, gross margin per unit area (*GM*, \$/ha), gross margin per unit of applied water (*GM*, \$/mm), the area of wheat that could be grown within the constraints of land and water availability, the total cost (*Tot cost*) and the total gross margin (*Tot GM*).

Searching for maximum total gross margin
Code: ARTF9301; Weather: KUTS0112.RND; Soil: 4 (16); Cv: W170/84
Price = \$1450/t; Irr = \$8.00/mm ha; App = \$12.00/cycle; Harv = \$30/t; Const = \$2000/ha

Irr Int days	Tot Irr mm	Yield t/ha	GM \$/ha	GM \$/mm	Area ha	Tot Cost \$	Tot GM \$
7	670	10.08	6788	10.1	60	467536	405384
10	578	10.08	7562	13.1	69	488617	523790
11	578	10.08	7562	13.1	69	488617	523790
12	578	10.08	7562	13.1	69	488617	523790
13	523	9.73	7511	14.4	77	504634	575002
14	523	9.71	7491	14.3	77	504601	573465
15	468	8.91	6809	14.6	86	523243	582586
16	468	8.77	6603	14.1	86	522871	564997
17	413	7.82	5714	13.9	97	546009	554076

Maximum gross margin per ha on a 10 d interval: GM/ha = \$7562
Maximum gross margin per mm on a 15 d interval: GM/mm = \$14.56
Maximum total gross margin on a 15 d interval: MaxTotGM = \$582586
Optimization process took 3.8 minutes

the total water application and potential crop areas were similar for each soil type.

The simulated effects of lengthening the irrigation interval from the minimum to the maximum intervals are given in Table 3 for the ART soil and in Table 5 for the Kutsaga soil. The optimization procedure on the ART soil indicated that potential yields and maximum gross margins per unit area could be achieved with either a ten, 11 or 12 day minimum irrigation interval. The use of any of these cycle lengths would enable the production of 69 ha with the available water resources. However, with a lengthening of the irrigation interval beyond 12 days, less water per unit area was required, thereby enabling a greater area of production, and although the yield and gross margin per unit area declined, the gross margin per unit of water increased, as did the total gross margin. The maximum total gross margin was achieved with a minimum irrigation interval of 15 days.

There are thus two options on the ART soil. The hypothetical farmer may aim for maximum gross margin per unit area with a frequent irrigation schedule. Alternatively, he could reduce the water application per unit area by irrigating less frequently, and thereby grow a larger area and achieve greater total profits. To help distinguish between these two alternatives, the program presents the total cost and total gross margin data, together with the irrigation schedule and crop development information (Table 4). The farmer's choice may depend on factors such as attitude to risk, bearing in mind that the long term variability of yield and total gross margin tends to increase with a decrease in water application (MacRobert, 1994), the relationship between total costs and returns, and the practicality of the proposed irrigation schedule. For example, if the farmer grew 86 ha with a 15-day minimum irrigation interval, his costs would be ZW$ 34,626 greater than if he grew 69 ha on a 10-day interval, but his total gross margin would be increased by ZW$ 58,796. However, if the increased cost incurred added finance charges, there may not be any advantage in the 15-day interval over the 10-day interval.

On the Kutsaga soil, both the maximum gross margin per unit area and the maximum total gross margin were achieved with a minimum return interval of eight days (Table 5). With a lower frequency of irrigation, a greater area of wheat could be grown, but total costs increased and the total gross margin decreased. Thus, for this particular case, the choice of irrigation management is simple, and the simulated irrigation schedule and crop development information is presented (Table 6) to assist the user in deciding the intra-seasonal allocation of water.

Conclusions

The WIRROPT7 irrigation optimization program is a relatively simple and 'friendly' program to work with, and provides a means by which a farmer or advisor may rapidly and conveniently evaluate irrigation management options on wheat. The user may loop through the program *ad infinitum*, changing any of the input variables to assess new options and alternatives.

However, WIRROPT7 can only be viewed as a management aid within the context of the limitations of the modified CERES-Wheat model. Although this

Table 4. Output screens generated during the simulated optimization of irrigation of wheat on the ART soil, showing details of the irrigation regime that produced (A) the maximum gross margin per unit area and (B) the maximum gross margin per unit of applied water.

A. The irrigation regime and characteristics of the 'maximum GM/ha regime'

Code: ARTF9301; Weather: KUTS0112.RND; Soil: 4 (16); Cv: W170/84
Price = $1450/t; Irr = $8.00/mm ha; App = $12.00/cycle; Harv = $30/t; Const = $2000/ha

Irrigation application			Interval	AWD			
Date	DAS	mm	days	%	Growth stage	Date	DAS
12 May	0	55	–	30	Leaf three	2 Jun	21
2 Jun	21	28	21	15	First node	9 Jul	58
22 Jun	41	55	20	36	Flag leaf	30 Jul	79
7 Jul	56	55	15	38	Anthesis	22 Aug	102
20 Jul	69	55	13	38	P. maturity	12 Oct	153
1 Aug	81	55	12	36	Harvest	21 Oct	162
17 Aug	97	55	16	39			
28 Aug	108	55	11	39	Yield = 10.08 t/ha		
7 Sep	118	55	10	38	Area = 69 ha		
17 Sep	128	55	10	41			
27 Sep	138	55	10	48	GM per ha = $7562/ha		
					per mm = $13.09/mm		
		578 mm gross					
					Total GM = $523790		
		491 mm net			Total Cost = $488617		

B. The irrigation regime and characteristics of the 'maximum GM/mm regime':
Code: ARTF9301; Weather: KUTS0112.RND; Soil: 4 (16); Cv: W170/84
Price = $1450/t; Irr = $8.00/mm ha; App = $12.00/cycle; Harv = $30/t;Const = $2000/ha

	Irrigation application		Interval	AWD			
Date	DAS	mm	days	%	Growth stage	Date	DAS
12 May	0	55	–	30	Leaf three	2 Jun	21
2 Jun	21	28	21	15	First node	9 Jul	58
22 Jun	41	55	20	36	Flag leaf	30 Jul	79
7 Jul	56	55	15	38	Anthesis	22 Aug	102
20 Jul	71	55	15	45	P. maturity	8 Oct	149
6 Aug	86	55	15	52	Harvest	19 Oct	160
21 Aug	101	55	15	52			
5 Sep	116	55	15	68	Yield = 8.91 t/ha		
20 Sep	131	55	15	89	Area = 86 ha		
		468 mm gross			GM per ha = $6809/ha		
					GM per mm = $14.56/mm		
					Total GM = $582586		
		397 mm net			Total Cost = $523243		

model has shown reasonable accuracy in the prediction of wheat yield under irrigated conditions in Zimbabwe, it is by no means a perfect predictor of yield, and suffers from the inadequacy of not simulating nutrient dynamics or pest problems. WIRROPT7 also uses historical weather to predict a likely outcome from a particular set of irrigation, soil, cultivar and economic conditions for use in a future situation. In cognizance of future uncertainty, the simulated

Table 5. The second output screen from the simulated optimization of irrigation of wheat on the Kutsaga soil, showing the effect of incrementing the minimum irrigation interval (*Irr Int*) from *MinInterval* to *MaxInterval* on the total irrigation applied (*Tot Irr*), yield, gross margin per unit area (*GM*, $/ha), gross margin per unit of applied water (*GM*, $/mm), the area of wheat that could be grown within the constraints of land and water availability, the total cost (*Tot cost*) and the total gross margin (*Tot GM*).

Searching for maximum total gross margin
Code: KUTS9302; Weather: KUTS0112.RND; Soil: 11 (31); Cv: W170/84
Price = $1450/t; Irr = $8.00/mm ha; App = $12.00/cycle; Harv = $30/t; Const = $1600/ha

Irr Int days	Tot Irr mm	Yield t/ha	GM $/ha	GM $/mm	Area ha	Tot Cost $	Tot GM $
5	666	10.08	7142	10.7	60	448582	428667
8	600	10.09	7739	12.9	67	459651	515938
9	560	9.45	7156	12.8	71	467389	511176
10	520	8.89	6697	12.9	77	476517	515118
11	480	8.16	5995	12.5	83	486740	499561
12	440	7.26	5051	11.5	91	498357	459205
13	400	6.54	4361	10.9	100	512832	436065
14	400	6.16	3820	9.6	100	511691	382026
15	360	5.48	3177	8.8	100	476430	317705

Maximum gross margin per ha on a 8 d interval: GM/ha = $7739
Maximum gross margin per mm on a 8 d interval: GM/mm = $12.90
Maximum total gross margin on a 8 d interval: MaxTotGM = $515938
Optimization process took 5.1 minutes.

Table 6. Output screen generated during the simulated optimization of irrigation of wheat on the Kutsaga soil, showing details of the irrigation regime that produced the maximum gross margin per unit area. The irrigation regime and characteristics of the 'maximum GM/ha regime':

Code: KUTS9302; Weather: KUTS0112.RND; Soil: 11 (31); Cv: W170/84
Price = $1450/t; Irr = $8.00/mm ha; App = $12.00/cycle; Harv = $30/t; Const = $1600/ha

Irrigation application			Interval	AWD			
Date	DAS	mm	days	%	Growth stage	Date	DAS
12 May	0	40	–	40	Leaf three	2 Jun	21
16 May	4	20	4	11	First node	9 Jul	58
2 Jun	21	20	17	19	Flag leaf	30 Jul	79
16 Jun	35	40	14	38	Anthesis	22 Aug	102
28 Jun	47	40	12	41	P. maturity	12 Oct	153
9 Jul	58	40	11	42	Harvest	21 Oct	162
18 Jul	67	40	9	40			
26 Jul	75	40	8	39	Yield = 10.09 t/ha		
5 Aug	85	40	10	41	Area = 67 ha		
17 Aug	97	40	12	41			
25 Aug	105	40	8	41	GM per ha = $7739/ha		
2 Sep	113	40	8	45	GM per mm = $12.90/mm		
10 Sep	121	40	8	43			
18 Sep	129	40	8	57	Total GM = $515938		
26 Sep	137	40	8	67	Total Cost = $449651		
4 Oct	145	40	8	79			
		600 mm gross					
		510 mm net					

output of WIRROPT7 must be used with caution. The best use of the program would be to give guidelines on the area of wheat to grow and the minimum return interval that should be planned for. The actual intra-seasonal allocation of water should be carried out using real-time information of crop water use, as dictated by weather, soil, crop and management conditions.

References

Agricultural Research Trust (1988) ART winter report. Supplement to The Farmer including Tobacco News, 23 March, 1989.

Chen L H, McClendon R W (1984) Selection of planting schedule for soybeans via simulation. Transactions of the American Society of Agricultural Engineers 27:92–32.

Godwin D C, Ritchie J T, Singh U, Hunt L A (1989) A user's guide to CERES-Wheat v.2.10. International Fertilizer Development Center, Muscle Shoals, Alabama, USA.

Heermann D F, Martin D L, Jackson R D, Stegman E C (1990) Irrigation scheduling controls and techniques. Pages 509–535 in Stewart B A, Nielsen D A (Eds.) Irrigation of agricultural crops. Agronomy Monograph #30. American Society of Agronomy-Crop Science Society of America–Soil Science Society of America, American Society of Agronomy, Madison, WI, USA.

Howell T A, Hiler E A (1975) Optimization of water use efficiency under high frequency irrigation. 1. Evapo-transpiration and yield relationship. Transactions of the American Society of Agricultural Engineers 18:873–878.

IBSNAT (1986) Decision Support System for Agrotechnology Transfer (DSSAT). Technical Report 5. International Benchmark Sites Network for Agrotechnology Transfer, University of Hawaii.

James L G (1988) Principles of farm irrigation system design. John Wiley and Sons, New York.

MacRobert J F (1992) Irrigation management for efficient use of water. 2. Deficit irrigation of wheat. The Farmer, 23 April, 1992.

MacRobert J F (1994) Modelling deficit irrigation of wheat in Zimbabwe. Unpublished Ph.D. dissertation, University of Natal, Pietermaritzburg, South Africa.

Martin D L (1984) Using crop yield models in optimal irrigation scheduling. Unpublished Ph.D. dissertation, Colorado State University, Fort Collins, CO, USA.

Martin D L, Gilley J R, Supalla R J (1989) Evaluation of irrigation planning decisions. Journal of the Irrigation and Drainage Division of the American Society of Civil Engineers 115:58–77.

Parsch L D, Cochran M J, Trice K L, Scott H D (1991) Biophysical simulation of wheat and soybean to assess the impact of timelines on double-cropping economics. Pages 511–534 in Hanks J, Ritchie J T (Eds.) Modeling plant and soil systems. Agronomy Monograph 31. American Society of Agronomy–Crop Science Society of America- Soil Science Society of America, American Society of Agronomy, Madison, WI.

Ritchie J T, Otter S (1984) CERES-Wheat: A user oriented wheat yield model. Preliminary documentation. United States Department of Agriculture, Agricultural Research Service, ARS 38: 159–175. AgRISTARS Publication No. YM-U3-04442-JSC-18892.

Smith B, Gasela R (1991) Wheat Strategy Paper. Zimbabwe Cereal Producers' Association in conjunction with the Grain Marketing Board, Harare, Zimbabwe.

Stegman E C, Musick J T, Stewart J I (1983) Irrigation water management. Pages 763–816 in Jensen M E (Ed.) Design and operation of farm irrigation systems. American Society of Agricultural Engineers Monograph 3, St. Joseph, MI, USA.

Willmott C J (1982) Some comments on the evaluation of model performance. Bulletin of the American Meteorological Society 63:1309–1313.

Zadoks J C, Chang T T, Konzak C F (1974) A decimal code for the growth stages of cereals. Weed Research 14:415–421.

Simulation of pest effects on crops using coupled pest-crop models: the potential for decision support

P.S. TENG[1], W.D. BATCHELOR[2], H.O. PINNSCHMIDT[3], G.G. WILKERSON[4]

[1]*Division of Entomology and Plant Pathology, International Rice Research Institute, P.O. Box 933, Manila, The Philippines*

[2]*Department of Agricultural and Biosystems Engineering, Iowa State University, Ames, Iowa 50011–3080, USA*

[3]*Tropeninstitut, Abt. Phytopathologie und Angew. Entomologie, Justus-Liebig-Universitaet, Bismarckstr. 16, D–35390 Giessen, Germany*

[4]*Department of Crop Sciences, North Carolina State University, Raleigh, NC 27695–2647, USA*

Key words: coupled pest-crop models, CERES-Rice, CROPGRO, SOYWEED, multiple pests, physiological coupling points, pest damage, pest effects, crop loss, simulation, decision support system

Abstract

Pest management decision support systems have evolved from rudimentary single decision rules to multiple criteria optimization software. In its simplest form, a decision support tool could be a pest management threshold calculated using empirical relations and field data on a calculator. A sophisticated form would be interactive computer systems that utilize simulation models, databases, and decision algorithms, in an integrated manner, to address normative problems. Central to the decision making process in pest (insect, disease, weed) management is information on the effect that a particular pest population has on the economic output of the crop. This effect depends on crop development stage, the prevailing environment, and the crop genotype's yield potential and ability to compensate for pest injury. In this paper, we present a conceptual framework for linking pest effects to crop models, and detail the coupling techniques used in linking pest and crop models and demonstrate, with examples, how this provides output for decision support. The crop models belonging to the CERES and CROPGRO families are used to exemplify situations for linking pest effects to crop growth and development via twenty-one links (CROPGRO) and twenty for CERES-RICE. Methods are described for representing pest dynamics, since these affect the pest-crop interaction, and the kind of pest data required for input into pest-crop combination models. Five basic methods of quantifying pest dynamics are proposed – (a) Field assessment, (b) *A priori* assumptions, (c) Analytic modeling, (d) Pest simulation models, and (e) Use of pest simulation models interlinked with crop models. The concept and techniques for using common coupling points in multiple-pest situations are described.

Introduction

Modern approaches and techniques for pest management increasingly rely on timely information to make strategic and tactical decisions. With rapid enhancements in microcomputer hardware and software, and their improved availability and user-friendliness, it is now possible to advance the decision making process in ways previously considered impractical. These advances have been

G.Y. Tsuji et al. *(eds): Understanding Options for Agricultural Production, 221–266*

significant (cf. reviews by Teng and Savary, 1992 and Teng and Rouse, 1984), but would not have been possible had there not been corresponding research done to generate the social, economic and biological knowledge needed to improve pest management decision-making. In the context of developing countries, all pest management research may even be viewed as ultimately contributing to the development of decision toolkits for specific pests or pest complexes in a single crop or cropping system (Teng, et al., 1994a). (In this paper, the term 'pest' collectively refers to insects, diseases and weeds.)

Decision support systems have evolved from rudimentary single decision rules to multiple criteria optimization software. In its simplest form, a decision support tool could be a pest management threshold calculated using empirical relations and field data on a calculator. A sophisticated form would be interactive computer systems that utilize simulation models, databases, and decision algorithms, in an integrated manner, to address normative problems. The latter is illustrated by the DSSAT (Decision Support System for Agrotechnology Transfer) developed by the IBSNAT (International Benchmark Sites Network for Agrotechnology Transfer) project of a consortium of U.S. universities (Jones, 1993). Decision support systems, in whatever form should produce decision rules for intervening in a situation either directly or indirectly. A distinction is made between strategic decisions, with a longer time step for decision making, e.g., what crop cultivar should be grown in the upcoming season in response to a pest outbreak in the past season, and tactical decisions, e.g., when and is it economical to spray a particular fungicide? This distinction is needed to consider the timeliness and directness of interventions for pest management. Indeed, the time step for decision making during a rice growing season is so small with some pathogens that it may even be necessary to use automated microclimate monitoring systems to ensure timeliness of a spray recommendation (Teng, 1994b).

Decision tools in pest management have broadly been categorized into knowledge, physical, communication and policy tools (Teng, 1994a). Knowledge tools are rules or guidelines used for decision making (such as a threshold for intervention); physical tools directly affect the system structure or dynamics (e.g., seed), communication tools are different media for conveying a pest management message (e.g., bulletin boards, radio, etc.) and policy tools are regulations or laws aimed at influencing the use of knowledge, physical or communication tools (e.g., law to ban specific insecticides). A common element in the use of many tools is requirement for information on pest status, crop status and the anticipated effect of pest status on crop productivity in the presence of various management options. Pest management decision processes do not differ much from other management cycles and possess the elements of objective setting, planning with management options, monitoring and evaluation of performance, and correction of performance to meet objectives (Dent, 1978). In this paper, we will discuss methods to obtain data on pest and crop status, and on pest-caused yield effects.

Central to the decision making process in pest management is information on the effect that a particular pest population has on the economic output of the crop. This effect has to be premised on crop development stage, the prevailing environment, and the crop genotype's yield potential and ability to compensate for pest injury.

The effect of pest populations on crop yield has historically been quantified as damage functions (Pinnschmidt et al., 1994), and until crop physiological models executing on microcomputers became a reality, most damage functions were empirical regression equations specific to the cultivar and location of the experiment. Such empirical equations were simple models reflecting pest effects on crops, but could not be used when weather, crop cultivar, soil type or management practices changed, and were thus limited in their extrapolation value. The literature shows that until about the mid 1980s, empirical models and thresholds were being developed for many pests (Zadoks, 1985), and there was relatively little effort to link pest models to crop models, even though the need to do so, and the knowledge for the linking, was articulated by many workers (Rouse, 1988; Teng, 1988).

Our objectives in this paper, therefore, are a) to present a conceptual framework for linking pest effects to crop models, and b) to detail the coupling techniques used in linking pest and crop models and demonstrate, with examples, how this provides output for decision support. To achieve objective (a) also requires that we discuss ways to represent pest dynamics, since these affect the pest-crop interaction, and pragmatically, what kind of pest data is available for input into pest-crop combination models. Throughout the paper, the crop models referred to belong to the CERES and CROPGRO families endorsed by the IBSNAT project.

Concepts and the biology for linking pest effects to crop growth and development

Pests, whether they are insects, pathogens or weeds, occur in populations which exhibit their own dynamic and spatial characteristics. To capture these dynamics requires that techniques be available for quantifying the pest population and for modeling the dynamics. Some common measures of pest population are numbers of a specific pest per unit area (for insects and weeds), and proportion of injured tissue such as percent leaf defoliation or leaf severity (for diseases). Sampling techniques furthermore have to be developed to match the spatial characteristics of each pest population. In the context of estimating pest effects on crop yield, the growth stages at and during which infestation or infections occur is important, hence population curves are needed. These are in turn either quantified or simulated using models which range from single equations to systems models (Teng, 1987).

Table 1. Example of a pest progress curve file to be used as input file for the multiple pest-coupled CERES-Rice model, containing a time axis (*DAS* = days after seeding), pest identifiers (PID), and pest level values for multiple pests over time. *LB* = % leaf blast, *SHBs* = % sheath blight on the sheath, *SHBl* = *% sheath blight on the leaf, DEF* = % defoliation, *DTILL* = absolute detillering rate/m^2, *PB* = % panicle blast. From: Pinnschmidt et al., 1995.

		Pest 1	Pest 2	Pest 3	Pest 4	Pest 5	etc.	Pest 10
	PID:	LB	SHBs	SHBl	DEF	DTILL	etc.	PB
DAS:								
35		0.112	0.000	0.000	0.004	2.888	.	0.000
42		1.453	1.493	0.325	2.198	9.570	.	0.000
49		11.360	2.665	3.813	7.353	19.315	.	0.000
56		7.450	9.123	4.568	10.579	39.434	.	0.000
63		2.280	23.533	6.439	7.707	69.773	.	0.000
70		1.360	26.632	4.475	6.406	35.202	.	0.000
77		0.873	33.560	5.580	9.054	19.973	.	0.451
84		0.431	38.790	7.480	8.307	11.101	.	1.223
91		0.116	42.546	8.530	9.709	6.097	.	12.560
⋮		⋮	⋮	⋮	⋮	⋮	⋮	⋮
etc.		etc.	etc.	etc.	etc.	etc.	etc.	etc.

Quantifying pest dynamics for crop-pest modeling

Five basic methods of quantifying pest dynamics are described in the following sections.

Field assessment

Using appropriate sampling and measuring techniques (Gaunt, 1990, Walker 1990; Shepard and Ferrer, 1990; Pinnschmidt et al., 1994), field scouting activities on a per field or per plot basis result in data on pest levels over time. The resulting pest progress curves may then be stored on input files to be read by a crop model before executing the crop simulation for the respective field or plot. Such input files will typically contain field or plot identifiers, a time axis with corresponding observed pest levels, and identifiers or names (PID) for each pest progress curve (Table 1). Since the interval of observation will usually be at least one week or longer, pest levels have to be interpolated between observation dates to obtain data that are compatible with the crop models' time step. This methodology has been employed by Batchelor et al. (1993); Pinnschmidt and Teng (1994), and Pinnschmidt et al. (1994, 1995).

A priori assumptions

Based on historical data or experience gained in previous field studies, educated guesses may be made about the probable progress of the severity level of a pest during the course of a growing season. Pest progress curves thus quantified can be processed as described in the previous paragraph. Examples of this approach can be found in Pinnschmidt and Teng (1994) and Pinnschmidt et al. (1994, 1995).

Analytic modeling

Differential equations and their integrated forms can be used to produce pest progress curves that resemble those typically observed in field pest populations. This approach is most suitable for pests whose usual patterns of population dynamics are somewhat regular, e.g., linear, exponential, asymptotic and/or sigmoidal, or unimodal. There are many growth functions to choose from in this context. Mathematical descriptions of disease progress curves and their differentials, with illustrated examples, are provided by Waggoner (1986) and Campbell and Madden (1990). An exponential function will commonly produce a progress curve where the increment per time step increases over time. Its differential will resemble a power function. A monomolecular function will yield a progress curve that steadily increases, with diminishing increments, towards an asymptote K. Its differentiated form will be an asymptotically declining curve. Logistic, log-logistic, and Gompertz functions will produce sigmoidal growth curves approaching K. But while a logistic curve is symmetrical, i.e., the point of inflection is at $K/2$, the latter curves are asymmetrical with their point of inflection less than or equal to $K/2$, the Gompertz curve is asymmetirical with its point of inflection as $0.37K$. Consequently, their differentiated forms will usually produce unimodal curves that are bell-shaped for the logistic and usually skewed to the right for the log-logistic and the Gompertz function. The so-called beta-function can also be used to produce unimodal curves. While the asymptotic functions contain only one or two parameters and their shapes are quite predetermined, the Bertalanffy–Richards function contains an additional shape parameter. It is therefore known for its flexibility and can produce curve shapes of any of the aforementioned functions. Another growth function, the Weibull model, contains a parameter that represents disease onset time (t_0), and thus combines the flexibility of the Bertalanffy–Richards model with the ability to handle a variable t_0.

While the curve parameters take care of the basic curve characteristics such as inflection point, skewness, rate of change, and shape, one might additionally want to affix the given curve maximum (X_m) to a desired maximum level (X_{max}) and adjust the pest levels (X_t, where $X =$ pest level and $t =$ time in relation to X_{max}). This can be done by standardizing X as follows:

$$X'_t = X_t(X_{max}/X_m) \tag{1}$$

where $X'_t = X$ after standardization. This was done for the curves shown in Figure 1 that resulted from a Gompertz growth function and its differential. The approach has been employed by Boote et al. (1993), Pinnschmidt and Teng (1994), and Pinnschmidt et al. (1995). These workers also wanted to control the pest onset time (t_0) in the simulation, i.e., the first time when pest levels assume values greater than zero. This can easily be done by assuming that:

$$X = f(t - t_0) \tag{2}$$

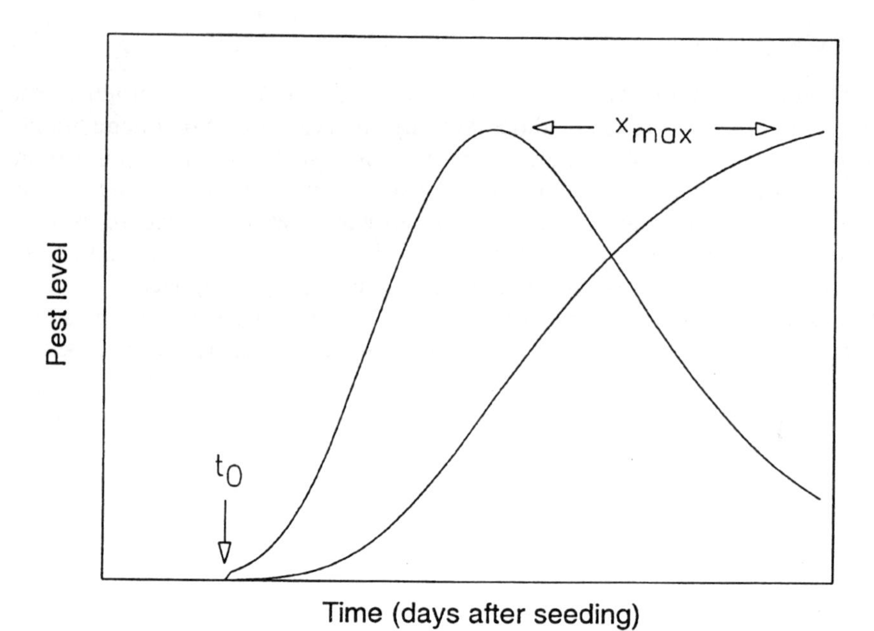

Figure 1. Example of a sigmoidal pest progress curve and its differential derived from the Gompertz function. Both curves were standardized on their asymptote respectively maximum (x_{max}). Variable t_0 denotes the time of pest onset, i.e., the first time when the pest level assumes a value >0.

Subsequent plotting of X versus t will result in a parallel shift of magnitude t_0 of the initial curve $f(t)$ along the time axis. It should be mentioned that another growth function, the Weibull model, contains a parameter that represents disease onset time. It thus combines the flexibility of the Bertalanffy–Richards model with the ability to handle a variable t_0. For mathematical description, visual examples, and author references of disease progress curves and their differentials, we refer to the exhaustive reviews by Waggoner (1986) and Campbell and Madden (1990).

Pest simulation models

In this approach, a pest population model would have to be run prior to executing the crop model. Simulation approaches to modeling pest dynamics, and more specifically disease dynamics, were reviewed by Rabbinge and Carter (1983) and Teng (1985). Useful hints on the construction of simulation models of pest population dynamics can be found in Rabbinge et al. (1989). The structure and *modus operandi* of such models is usually quite complicated and takes the effects of driving variables, such as weather factors, on various aspects and stages of the pests' life cycles into account. However, for the sake of simplicity we might assume that pest dynamics are made up of two basic processes: growth dx/dt (e.g., newly hatched insect pest individuals, new infections, lesion growth, etc.) and removal dx_s/dt (e.g., death of insect pest individuals, senesced lesions, etc.) of a pest population. Both processes determine the current size x of a pest population that will usually be the one exerting damage effects on the crop:

$$dx/dt = rx(1 - x) \tag{3}$$

$$dx_s/dt = r_s x \tag{4}$$

$$x = \sum (dx/dt) - \sum (dx_s/dt) \tag{5}$$

where r = intrinsic rate of pest growth and r_s = intrinsic rate of pest removal. In pest simulators, each of the terms in the equations above might actually consist of many sequentially solved equations that are needed to model specific aspects and subprocesses as affected by driving variables. For example, the term x can be split into classes representing lesions or larvae of different age, parameters r and r_s might be functions of environmental, host, and pathogen factors, and dx/dt might be subdivided into new infections and growth of existing lesions. Other important subprocesses to be modeled are oviposition, hatching, survival, search behavior, and migration for insect pests and sporulation, spore dispersal, spore deposition, and latency for diseases. The pest levels x computed by the pest model can be stored as progress curves on a file similar to the one shown in Table 1. The information on this file then has to be read and processed by the crop model prior to a crop simulation run in analogy to the approach described under field assessment and *a priori* assumptions.

Use of pest simulation models interlinked with crop models

In the four previous paragraphs we discussed ways to quantify pest levels and/or pest dynamics independently from the development and current status of the crop. While this is a 'one-way' approach because the pest affects the crop without the reverse being true, pest simulators interlinked with crop models represent a 'two-way' approach: pest model variables drive the crop model and vice versa. Such models have been called 'true' pest-crop models (Pinnschmidt and Teng, 1994; Pinnschmidt et al., 1990a, 1995) and were constructed for a variety of crops and pests e.g., by Gutierrez et al. (1975, 1977, 1983); Benigno et al. (1988); Pinnschmidt(1997); Pinnschmidt et al. (1990b); Graf et al. (1991); and Luo et al. (1997). In these models, pest growth and removal will commonly depend on the availability and quality of suitable host resources. In the case of diseases, removal might also directly be tied to host senescence. In turn, the current pest population will consume and/or occupy host resources and alter physiological processes of the host which will affect host development. The latter aspect will be detailed in the next two sections. To incorporate host effects on pest dynamics, the equations (3), (4) and (5) may be modified as follows:

$$dx/dt = rx(1 - x/K)f_1(t) \quad (6)$$

$$dx_s/dt = x(1 - r_s f_2(t))(1 - (K_s/dt)/K)) \quad (7)$$

$$x = \sum (dx/dt) - \sum (dx_s/dt) \quad (8)$$

with K = total pest-exploitable host resource (e.g., total leaf area as a crop model variable), $f_1(t)$ = relative suitability (e.g., susceptibility) of host resources with respect to dx/dt as a function of time, respectively, host age, $f_2(t)$ = relative effect of host age on pest removal as a function of time, and dK_s/dt = removal rate of K. In practice, pest-crop models that realistically simulate population dynamics in the field, would have a higher level of detail than the equations presented here.

Generic coupling points and damage computation

Pests may either consume plant tissue or disrupt plant processes. For instance, corn earworm consumes small and large seed, and leaf tissue, and sheath blight in rice reduces the rate of photosynthesis, transpiration, and respiration. Conceptually, pest damage can be simulated in crop models by subtracting biomass from the pools being consumed, or by reducing rates of processes affected by pest damage. In this model, biomass is stored in leaf, stem, seed, shell, and root pools which represent growth of these crop components. Biomass can be lost from each pool through pest damage. For instance, leaf mass (dL/dt) is added to leaf mass pool (L) daily during the vegetative growth period for soybean. Insect damage can then reduce the biomass in the leaf pool according to the daily pest consumption (dLc/dt).

The CROPGRO and CERES models are a series of continuous, process-oriented models that predict daily increases in vegetative and reproductive growth and development. Daily increases in biomass are computed by partitioning daily photosynthate to vegetative, reproductive and root components. Development and growth can be affected by temperature, photoperiod, and limitations in nitrogen and water. Phenology is predicted using a photothermal time concept in which development time can be modified by temperature and photoperiod. These models compute growth of plant organs (i.e., leaves, stems, seeds, shells, and roots) using first order differential equations which are solved using a daily time step. Daily photosynthesis is computed based upon light interception and modified by sub-optimum environmental conditions (temperature, nitrogen, etc.). The carbon fixed by photosynthesis is partitioned to each plant component according to the development stage of the crop. Biomass is maintained in separate pools for each major plant component.

CROPGRO pest effects. In the crop models, the variables that represent pest damage (i.e., dLc/dt term in Figure 2) are called coupling points. They allow

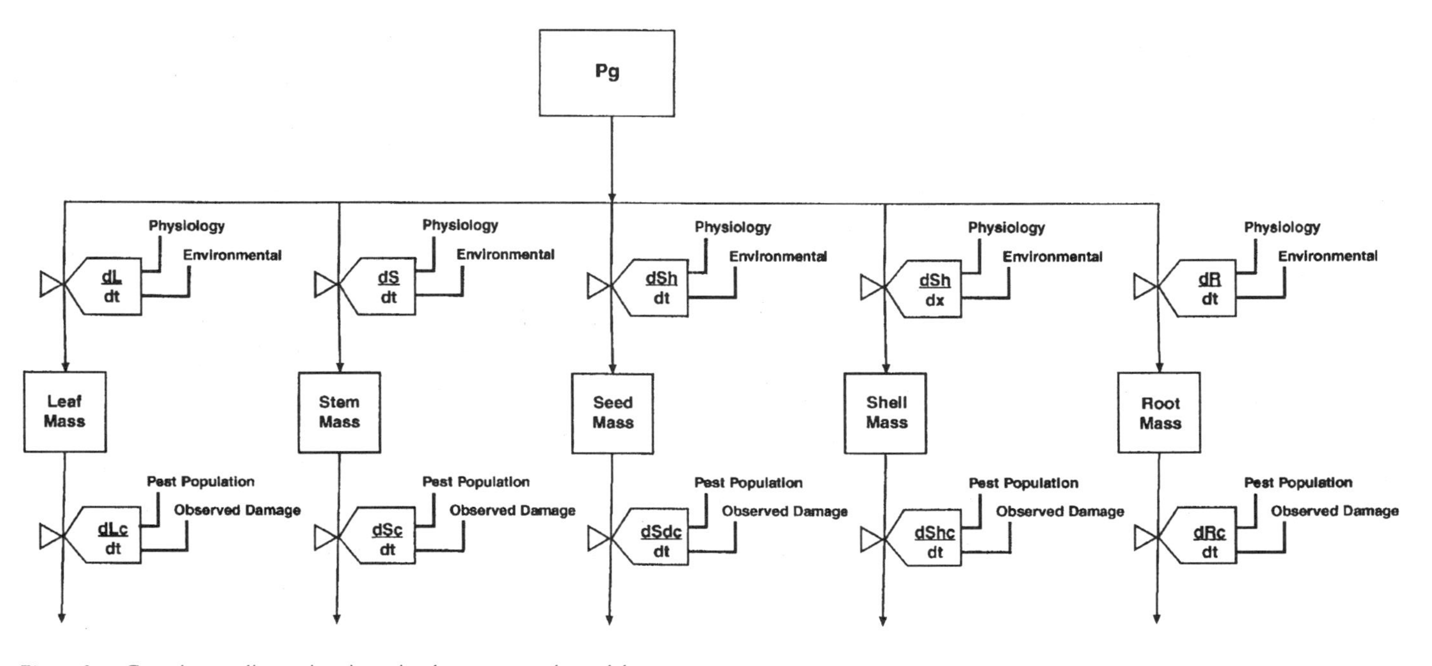

Figure 2. Generic coupling points in a simple crop growth model.

the model to simulate biomass losses resulting from observed pest populations or damage levels. Mathematically, the application of damage to crop model variables can be expressed by

$$X_{it} = X_{it}^{*} - \sum_{p=1}^{m} D_{ipt} \tag{9}$$

where,

X_{it} = state (or other model) variable i on day t, after damage has been applied,
X_{it}^{*} = state (or other model) variable i on day t before applying pest damage,
D_{ipt} = amount of damage applied to state (or other models) variable i on day t by pest type p, and
m = total number of pest types feeding on variable i.

Using this approach, multiple pest damage to plant components can also be simulated by summing damage from each pest and subtracting it from the biomass pool being consumed. The D_{ipt} term indicates that pest damage to each crop component is summed over all pest types and subtracted from the model variable. Note that the units on D_{ipt} must be the same as the units on the crop model variable X_{it}. A list of generic coupling points is shown in Table 2 for CROPGRO and Table 3 for CERES-RICE.

The real challenge in this approach is to compute daily values of D_{ipt} for pest populations or observed damage in the field. In some cases, pest population on particular dates are reported, while in other cases, damage to the crop is reported in absolute or relative terms on different dates. In the simple case when a pest population on N_{pt} (pests m^{-2} ground area) is reported for pest p on day t, and this pest population damages only one coupling point (i), then D_{ipt} can be computed by

$$D_{ipt} = N_{pt} \times C_{ip} \tag{10}$$

In this equation, N_{pt} would be the input (observed) pest population on day t for pest p and C_{ip} is a 'pest coefficient' that represents the amount of coupling point i damage per individual pest per day. We refer to this method of computing D_{ipt} as the 'absolute damage rate' method.

If the availability of the crop component is small relative to the amount being consumed in a day, the damage to variable i may be lower than that computed by equation (10) because of search/encounter frequencies or because of intraspecific competition for food sites. In the CROPGRO model, a method of applying damage was developed that will allow for competition for sites and for a pest switching to alternate coupling points for feeding based on work by Frazer and Gilbert (1976). This method is referred to as 'daily absolute damage rate with competition and preference'. If this method is used to define damage

Table 2. Coupling points and damage types used to apply damage in the CROPGRO crop models.

Coupling point	Units[1]	Available damage types[2]	Coupling point damage variable
Leaf area index	$m^2\ m^{-2}$	1,2,3,4	LAD
Leaf mass	$g\ m^{-2}$	1,2,3,4	LMD
Stem mass	$g\ m^{-2}$	1,3,4	SMD
Root mass	$g\ m^{-2}$	1,3,4	RMD
Root length	$cm\ cm^{-2}$	1,3,4	RLF
Root length volume	$cm\ cm^{-3}$	1,3,4	RLV
Small seed number	no. m^{-2}	1,3,4	SDNS
Large seed number	no. m^{-2}	1,3,4	SDNL
Mature seed number	no. m^{-2}	1,3,4	SDNM
Small seed mass	$g\ m^{-2}$	1,3,4	SDMS
Large seed mass	$g\ m^{-2}$	1,3,4	SDML
Mature seed mass	$g\ m^{-2}$	1,3,4	SDMM
Small shell number	no. m^{-2}	1,3,4	SHNS
Large shell number	no. m^{-2}	1,3,4	SHNL
Mature shell number	no. m^{-2}	1,3,4	SHNM
Small shell mass	$g\ m^{-2}$	1,3,4	SHMS
Large shell mass	$g\ m^{-2}$	1,3,4	SHML
Mature shell mass	$g\ m^{-2}$	1,3,4	SHMM
Whole plant	no. m^{-2}	1,3,4	WPD
Assimilate	$g\ m^{-2}$	1,3	ASM
Necrotic leaf area index	$cm^2\ cm^{-2}$	1,3	PDLA

[1] Per unit ground area.
[2] Where damage type are: (1) daily absolute damage rate, (2) percent observed damage, (3) daily percent damage rate, and (4) daily absolute damage rate with pest competition and food preference effects.

by a pest, daily damage is computed by

$$D_{ipt} = X_{it}^{*} \times (1 - e^{-N_{pt}C_{ip}/X_{it}^{*}}) \tag{11}$$

When X_{it}^{*} is large compared to the demand for the site $(N_{pt}C_{ip})$, then D_{ipt} approaches $(N_{pt}C_{ip})$, as in equation (10). When X_{it}^{*} is small, then equation (3) ensures that D_{ipt} remains below both X_{it}^{*} and $(N_{pt}C_{ip})$. This results in lower consumption than specified by the value of C_{ip}, so we allow damage to a secondary food source, *j*, when it has been defined for a pest. Demand for the second food preference *j* at time $t(FE_{jpt})$ remaining after feeding on the first food preference is computed by

$$FE_{jpt} = (N_{pt} \times C_{ip} - D_{ipt}) \times C_{jp}/C_{ip} \tag{12}$$

and the damage rate to the secondary food preference is computed by

$$D_{jpt} = X_{jt}^{*} \times (1 - e^{-FE_{jpt}/X_{jt}^{*}}) \tag{13}$$

A third preference for food can be defined in a similar manner using corn earworm feeding on soybean as an example. Assume that a population (N_{pt}) of corn earworm was measured at a density of 1.5 larva m^{-2} on June 17. Corn earworm prefer to feed on small seed (C_{ip}) at a rate of 10 seed larva^{-1} d^{-1} (Table 4). They feed on the secondary food source, large seed (C_{jp}), at a rate

Table 3. Physiological coupling points available in the CERES pest module (see Figure 3).

Code	Description of pest damage effect	Crop variables primarily affected
CP1A	Leaf consumption	Leaf area Leaf weight
CP1B	Root consumption	Root weight Root length volume
CP1C	Stem consumption	Stem weight
CP1D	Panicle and grain consumption	Grain weight Panicle weight Grain number
CP2	Leaf covering	Photosynthetic potential/day
CP3	Photosynthesis reduction	Assimilates/day
CP4	Assimilate consumption	Assimilates/day
CP5	Respiration increase	Assimilates/day
CP6	Phloem blockage	Assimilates/day
CP7A	Leaf growth rate reduction	Leaf growth/day
CP7B	Root growth rate reduction	Root growth/day
CP7C	Stem growth rate reduction	Stem growth/day
CP7D	Panicle and grain growth rate reduction	Grain growth/day
CP8	Light competition	Assimilates/day Leaf senescence/day
CP9	Leaf senescence acceleration	Leaf senescence/day
CP10	Xylem blockage	Rroot water uptake/day
CP11	Altered transpiration	Transpiration/day
CP12	Stand reduction	Plant population

Table 4. Typical time series file (file T) containing examples of pest progress data for six pest and damage types for soybean. The pest name corresponding to each pest header is: CEW6 – 6th instar corn earworm; SL6 – 6th instar soybean looper; VC5 – 5th instar velvetbean caterpillar; VC6 – 6th instar velvetbean caterpillar; SHAD – percent canopy shading by shade cloth; GC5 – 5th instar green cloverworm. Each pest and damage is defined in the pest coefficient file.

*EXPERIMENTAL DATA (T): UFBI9101 SB Drew field drought study

@TRNO	DATE	CEW6	SL6	VC5	VC6	SHAD	GC5
1	91147	0.0	1.5	0.25	0.4	30.0	0.0
1	91154	0.5	2.8	0.25	0.6	30.0	0.6
1	91161	0.9	3.3	0.1	1.2	0.0	0.1
1	91168	1.5	6.4	0.0	1.6	-99	0.0
1	91175	0.2	0.5	0.0	1.8	-99	0.0
1	91181	0.0	1.2	0.0	1.0	-99	0.0
1	91188	-99	1.5	-99	0.2	-99	0.0
1	91195	-99	3.3	-99	0.0	-99	0.0

of 2.5 seed larva^{-1} d^{-1}. At time t, the plant contains 30.0 small seed m^{-2} larva^{-1} d^{-1} (X^*_{it}) and 400.0 large seed m^{-2} (X^*_{jt}). Ignoring pest density and competition effects, damage to small seed would be 15 seed m^{-2} d^{-1} by equation (12). However, using equation (11), the effects of density and competition for the primary feeding site results in a computed D_{ipt} of 11.8 small seed m^{-2} d^{-1}. The demand not satisfied by the primary food source (15.0 minus 11.8 small seed m^{-2} d^{-1}) can be converted into the equivalent of 0.80 large seed m^{-2} d^{-1} of the secondary food source (Fe_{jpt}) according to equation (12).

The damage to be applied to large seed can then be computed using equation (15). This results in damage of 0.799 large seed m^{-2} d^{-1}. The unmet demand can again be computed as 0.001 large seed m^{-2} d^{-1} (0.80 minus 0.799 m^{-2} d^{-1}) and the same procedure can be used to convert this damage to equivalent units of the third food source preference for corn earworm, which is leaf mass.

Scouting reports often describe damage as daily percent damage to a coupling point (e.g., defoliation due to hail). Thus, a third method of defining damage is when field observations report a percent loss to coupling point i at time t. We prefer to do this as the 'percent damage' method. This method is selected, D_{ipt}, is computed by

$$D_{ipt} = (R_{ipt}/100) \times X^*_{it} \tag{14}$$

where R_{ipt} is the percent damage to coupling point i caused by pest p (or in this case, reported damage) on day t.

In some experimental plots, it is more convenient to report relative damage throughout the season. In this case, damage to coupling point i is measured by comparison to the same coupling point of a different treatment. We refer to this as the 'percent observed damage' method. In this case, the total amount of damage is known for observation dates throughout the season, however, neither the daily percent nor absolute damage rate is known. The total amount of damage on a given day would have been accumulated over previous days, and this reporting method does not contain information on when the damage occurred. Therefore, in CROPGRO, equations were defined to estimate the daily absolute damage that would be required to simulate the percent observed damage for leaf mass, leaf area, and stem mass coupling points. If damage is reported by this method, D_{ipt} is computed by first writing an expression for the percent observed damage (Pit) in terms of absolute damage (D_{ipt});

$$P_{it} = 100 \times \left(1 - \frac{X^*_{it} - D_{ipt}}{Xt_{it} - Xs_{it}}\right) \tag{15}$$

or, by rearranging

$$D_{ipt} = X^*_{it} - \left(1 - \frac{P_{it}}{100}\right)(Xt_{it} - Xs_{it}) \tag{16}$$

In these equations, P_{it} is the percent difference in the amount of coupling point i in two treatments at time t (i.e., one treatment that has not been damaged and the other that has accumulated damage). The variable Xt_{it} is the cumulative amount of coupling point i produced by the crop model up to time t without any damage or senescence and Xs_{it} is the cumulative senescence of coupling point i up to time t. The variable X^*_{it} is the value of the coupling point before damage is applied at time t. In the field, Xt_it and Xs_{it} are not known, but in crop models, they can easily be computed so that the observed P_{it} values can be used to compute absolute damage rates as defined by equation (16).

Damage routines may be structured to provide flexibility in collecting damage data. Two different methods can be used to describe damage information. In typical farm operations, pest population data can be collected through field scouting and damage to crop components can be computed if pest feeding rates are known. In other farm or experimental plots, the actual amount of damage can be measured. This approach is useful when pest populations are difficult to measure, or when the source of damage is unknown. Two data files are used to define the pest linkage. A field-specific record of observed pest or damage levels is contained in the time series file (file T) associated with each experiment. A typical file T with recorded pest populations is shown in Table 4. The first two columns in this file contain the treatment number and observation date. Each subsequent column has a 3 or 4 character header that is unique for each pest. The measured level of pest and/or damage type is input in these columns with an appropriate 3 or 4 character header abbreviation for the respective pest or damage type. The model uses linear interpolation between observation dates to compute daily pest populations or damage levels. This approach for recording observed pest and damage levels follows a typical field scouting record format.

A crop-specific pest coefficient file defines individual pests or damage in terms of coupling points and feeding rates. Feeding rate coefficients provide a means of converting from pest population to damage to coupling points. Observed pest levels or damage reported in the pest progress file must have a corresponding definition in the pest coefficient file. A typical pest coefficient file for soybean is shown in Table 5. The first column is the pest number, *LN*, that is used internally for record keeping in the crop model. The second column contains a 4 character pest identifier (PID) that links a pest in the progress file with a definition of a pest in the pest coefficient file. The next column contains a 21 character description of the pest or damage (*PNAME*). The fourth column contains the damage characterization method (*PCTID*). This is a code corresponding to the method used to apply damage to the coupling point variable in the model. Next, up to 6 coupling point damage variables (*PCPID*) can be listed for each pest definition. Damage will be applied to corresponding coupling points through these damage variables, which are defined in Table 2. The last column contains a damage coefficient (*PDCF1*). In the crop models, the damage coefficient is multiplied by the damage level in the pest progress file to determine the amount of damage to apply. Thus, for pests, this coefficient should be the feeding rate that converts population to daily feeding rate in terms of mass consumption per day. The amount of damage caused by the population can be computed by multiplying pest population by *PDCF1*. For damage types such as manual defoliation, this coefficient should be 1.0, indicating that percent defoliation is recorded in the pest progress file.

Four different damage characterization methods were defined to describe observed damage data according to equations (9) through (16):

Table 5. A typical pest coefficient file for soybean consists of the following columns: (1) pest number (N); (2) pest abbreviation (PID); (3) pest name (PNAME); 40 damage characterization method (PCTID); (5) coupling point identifier (PCPID) defined in Table 1; (6) feeding rate coefficient (PDCF1); (7) additional coefficient not used in CROPGRO (PDCF2), (8) units of the feeding rate coefficient (Units), and (9) source for damage rates (Source).

LN	PID[1]	PNAME	PCTID	PCPID	PDCF1	PFCF2	Units	Source
01	CEW6	Corn Earworm[2]	4	SDNS	10.00000000	0.0000	no./larva/d	Batchelor et al., 1989
				SDNL	2.50000000	0.0000	no./larva/d	Szmedra et al., 1988
				LAD	0.00505000	0.0000	m^2/larva/d	Szmedra et al., 1988
02	VBC5	5 Instar Velevetbean[3]	1	LAD	0.00081000	0.0000	m^2/larva/d	Reid, 1975
03	VBC6	6 Instar Velvetbean	1	LAD	0.00144000	0.0000	m^2/larva/d	Reid, 1975
04	SL4	Soybean Looper[4]	1	LAD	0.0044000	0.0000	m^2/larva/d	Reid and Green, 1975
05	SL5	Soybean Looper	1	LAD	0.00071000	0.0000	m^2/larva/d	Reid and Green, 1975
06	SL6	Soybean Looper	1	LAD	0.00124000	0.0000	m^2/larva/d	Reid and Green, 1975
07	SGSB	Stinkbug[5]	4	SDNS	15.00000000	0.0000	no./m^2/d	Batchelor et al., 1989
				SDNM	5.00000000	0.0000	no./m^2/d	Batchelor et al., 1989
08	FAW	Fall Armyworm	1	LMD	2.00000000	0.0000	g/larva/day	estimated
09	RTWM	rootworm	1	RLV	1.00000000	0.0000	cm/cm^2 /lar/d	estimated
10	PCLA	Obs.% defoliation	2	LAD	1.00000000	0.0000	%	
11	PSTM	Obs.% Stem damage	2	SMD	1.00000000	0.0000	%	
12	PDLA	% Diseased Leaf Area	3	PDLA	1.00000000	0.0000	%/day	
13	PRP	% Reduction in Photo	3	ASM	1.00000000	0.0000	%/day	
14	PLAI	% daily LAI dest.	3	LAD	1.00000000	0.0000	%/day	
15	PLM	% daily Leaf Mass	3	LMD	1.00000000	0.0000	%/day	
16	PWP	% Whole Plants	3	WPD	1.00000000	0.0000	%/day	
17	PSDN	% All Seed Dest.	3	SDNL	1.00000000	0.0000	%/day	
				SDNS	1.00000000	0.0000	%/day	
				SDNM	1.00000000	0.0000	%/day	
18	PSHN	% All Shell Dest.	3	SHNL	1.00000000	0.0000	%/day	
				SHNL	1.00000000	0.0000	%/day	
				SHNL	1.00000000	0.0000	%/day	
19	PPDN	% All Pod dest.	3	PPDN	1.00000000	0.0000	%/day	
10	PRTM	% Root mass dest.	3	RMD	1.00000000	0.0000	%/day	

[1] Pest identifier or abbreviation for the pest or damage type.
[2] Corn Earworm (*Heliothis Zea*).
[3] Velvetbean Caterpillar (*Anticarsia gemmatalis*).
[4] Soybean Looper (*Upseudoplusia includens*).
[5] Southern Green Stinkbug (*Nezara virdula L.*).

(1) daily absolute damage rate,
(2) percent observed damage,
(3) daily percent damage rate, and
(4) daily absolute damage rate with pest competition and food preference effects.

Tables 2 and 3 show the allowable damage types available for each coupling point. The coupling point damage variable contains the amount of damage and the damage type descriptor tells CROPGRO how to apply the damage to the coupling point. These damage types provide four ways to describe damage using the coupling point damage variables.

The selection of the method for applying damage depends upon the type of damage data that can be collected. Absolute daily damage is useful when pest population data can be collected. If pest populations and feeding rates are known, daily damage to coupling points can be computed in units of mass per unit area. Percent observed damage typically occurs when observations of plant components are compared between some scientific treatment and control. Defoliation can occur in one treatment, resulting in a percent difference in leaf mass between a treatment and control. In this case, the source of damage may not be known, however, time series measurements describing the percent reduction between treatment and control is known. Daily percent damage rate is useful when damage can be measured as percent on a daily basis. For instance, in a manual defoliation study, 33% and 66% defoliation may be applied on a particular day during the season. In this case, damage is applied as a percent daily damage. Another example would be the application of a shade cloth that blocks 80% of incoming light, resulting in an 80% reduction in daily assimilate production during the period of shading. The last damage type is intended to be used for insects that compete for feeding sites. If insect demand for the primary food source is high relative to supply, some damage is partitioned to secondary food sources. The approach used in CERES-Rice and CROPGRO is different (see Batchelor et al., 1993; Pinnschmidt et al., 1994).

CERES-RICE pest effects. In crop models such as CERES-Rice (Alocilja and Ritchie, 1988), the accumulated biomass or yield at harvest is the cumulative outcome of complex interactions among physiological processes M_m under the influence of environmental and crop management inputs, as well as varietal characteristics. The various M_m's represent state variables, such as leaf area index and ontogenetic stage, and rate variables, such as absorbed photosynthetically active radiation, photosynthesis, and leaf senescence that are linked in order to drive the daily crop growth rate. Following the concept of Boote et al. (1983), each M_m can be considered as a potential physiological coupling point (= *CP*, Figure 3) where pests can affect the crop. For example, a defoliating insect will affect the leaf mass state variable, while a disease that induces leaf senescence will affect the daily leaf senescence rate. Some 20 coupling points

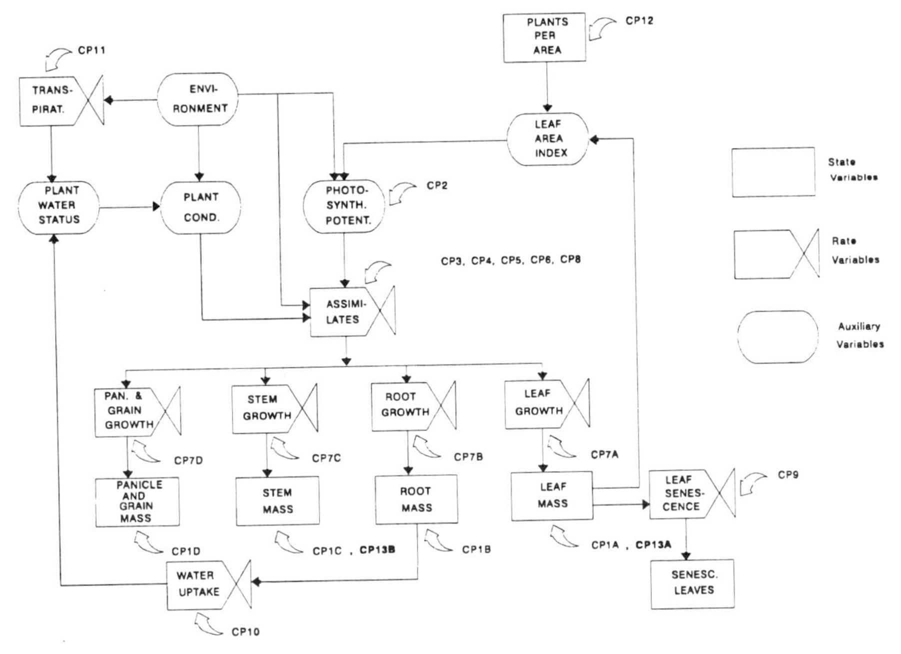

Figure 3. Relational diagram of CERES-Rice, showing the physiological coupling points (*CP*, see arrows) established in the model. Description of coupling points: *CP1A* = leaf consumption, *CP1B* = root consumption, *CP1C* = stem consumption, *CP1D* = panicle and grain consumption, *CP2* = leaf coverage, *CP3* = photosynthesis reduction, *CP4* = assimilate consumption, *CP5* = respiration increase, *CP6* = phloem blockage/reduction of assimilate translocation, *CP7A* = leaf growth rate reduction, *CP7D* = grain growth rate reduction/removal of grain sinks, *CP8* = light competition, *CP9* = leaf senescence acceleration, *CP10* = *xylem blockage/reduction of root water uptake*, *CP11* = altered transpiration, *CP12* = stand reduction, *CP13A* = leaf reserve blockage, *CP13B* = stem reserve blockage. From: Pinnschmidt et al., 1995.

were thus identified in the source code of the CERES-Rice model for potential application of pest damage (Pinnschmidt and Teng, 1994; Pinnschmidt et al., 1995). Using this approach, it is possible to simulate multiple pest damage effects mechanistically.

In addition to pest levels over time (see section on Quantifying pest dynamics for crop-pest modeling), a second type of information must be 'known' to the crop model if pest damage effects are to be included in a simulation run. It consists of coupling point identifiers, damage coefficients, and damage equation specifyers that are needed to interpret pest levels and convert them into damage to be applied to the correct physiological coupling points of the crop model. This information can be either contained in the model code (Pinnschmidt, 1991, Pinnschmidt et al., 1990), or be supplied from external files (Batchelor et al., 1993; Boote et al., 1993; Pinnschmidt and Teng, 1994; Pinnschmidt et al., 1994, 1995). Table 6 shows such a file that contains multiple pest identifiers (*PID*), pest descriptions, coupling point identifiers (PCPID) for individual pests

Table 6. Example of a pest specification file to be used as an input file for the multiple pest-coupled CERES-Rice model. It contains pest identifiers (PID), pest progress curve type identifiers (PCTID), pest coupling point identifiers (PCPED), and damage coefficients (*a*, *be*). *LB* = % leaf blast, *CB* = % collar blast, *PB* = % panicle blast, *SHB*s = % sheath blight on the sheath, *SHB1* = % sheath blith on the leaf, *BLB* = % bacterial leaf blight, *OLD* = other leaf diseases, *LF* = no. Leaf folder larvae/m² *DEF* = % other defoliation, *R* = rat damage, *DTILL* = absolute detillering rate/m², *DHWH* = % stem borer dead hearts and white heads, *WEED* = % ground cover by weeds. From: Pinnschmidt et al., 1994a.

Pest no.	PID	Pest description	PCTID	PCPID	a	b
1	LB	Leaf blast (%)	2	CP2	1.00	1.00[1]
				CP3	1.00	2.04[1]
				CP9	0.00	0.99
2	CB	Collar blast (%)	2	CP3	1.00	1.00
				CP6	1.00	1.00
				CP9	0.00	0.00
				CP10	1.00	1.00
3	PB	Pan. blast sev. (%)	2	CP7D	0.80	1.00
4	SHB1	sh.bl., leaf (%)	2	CP2	1.00	1.00
				CP3	1.00	1.00
				CP9	0.00	0.00
5	SHBS	sh.bl., sheath (%)	2	CP1D	1.00	1.00
				CP3	1.00	1.00
				CP9	0.00	0.00
				CP10	1.00	1.00
6	LF	Leaf folder larvae	1	CP1A	20.00	1.00
7	DEF	Defoliation (%)	2	CP1A	1.00	1.00
8	DHWH	d.heart + w.head (%)	2	CP3	1.00	1.00
				CP12	1.00	1.00
9	R	Rat damage (%)	2	CP12	1.00	1.00
10	OLD	oth. Leaf dis. (%)	2	CP2	1.00	1.00
				CP9	0.00	0.00
11	BLB	bact. 1.blight (%)	2	CP2	1.00	1.00
				CP9	0.00	0.00
12	WEED	Weeds (% cover)	2	CP8	1.00	0.70
13	DTILL	Absolute detillering	4	CP12	1.00	1.00

[1] Bastiaans, 1991; all other coefficients are based on educated guesses; for weeds, b = light extinction coefficient of weed canopy.

p and corresponding damage coefficients (a_{pm}, b_{pm}), and progress curve type identifiers (*PCTID*) for each pest. The PCPIDs define which coupling points M_m are to be affected by a given pest. The PCTIDs indicate whether pest levels will be interpreted as absolute damage rates or population number counts of pest individuals by considering inter- and intraspecific competition among pests ($PCTID = 1$), disease or damage percentages ($PCTID = 2$), or absolute damage rates or population number counts of pest individuals without considering inter- and intraspecific pest competition ($PCTID = 4$). Prior to a simulation run, the crop model has to read this file and match its information with the pest level data supplied by a file such as shown in Table 1. The pest identifiers (PID) on both files serve as the matching criteria. During a simulation run, damage is then applied to each coupling point identified for a given set of pests using the approach generalized by Pinnschmidt and Teng (1994) and

Pinnschmidt et al. (1995): a physiological process or variable (M_m) can be affected at any given time t during the simulation by multiplying it by a damage factor (F_m) that expresses the relative degree of damage imposed on M_m through pests:

$$M'_{mt} = M_{mt} F_{mt} \tag{17}$$

where $M'_{mt} = M_{mt}$ adjusted for pest damage. F_{mt} is a function of the combined potential damage caused by all pests that affect M_{mt}. It represents the fraction of the respective crop variable remaining after daily damage is applied and will thus usually be e[0,1]. The computation of F_{mt} depends on pest levels, on whether these are defined as counts of the number of pest individuals or as percentage of diseased or damaged crop, on pest damage and/or feeding coefficients, and on whether the M_m to be affected represents a crop physiological rate or state variable. If M_m represents a state variable, such as plant leaf area, number of tillers, or biomass, F_{mt} is computed as:

$$F_{mt} = F1_{mt} F2a_{mt} \tag{18}$$

while, if M_m represents a crop physiological rate variable such as the daily photosynthetic potential, photosynthesis, or carbohydrates partitioned to different plant organs, F_{mt} is computed as:

$$F_{mt} = F1_{mt} F2b_{mt} \tag{19}$$

$F1_{mt}$ denotes the fraction of M_{mt} remaining after accounting for damage effects of pests whose infestation levels are defined as number counts of pest individuals. It is computed using an equation derived from the functional response model proposed by Frazer and Gilbert (1976):

$$F1_{mt} = e^{-[(a_{1m}N_{1t})/K_{1t} + \ldots (a_{pm}N_{pt})/K_{pt}]} \tag{20}$$

with N_{pt} = number of individuals of pest pm^{-2}, a $\bar{p}_m$ potential feeding or damage rate per N_p, measured in units of K_{pt}, but with regards to M_{mt}, and K_{pt} = food or space resource for pest p on a given day t. Usually, K_p will represent crop variables such as green leaf area, number of tillers, daily assimilates, etc. Note that K_p can be identical to M_m.

$F2a_{mt}$ and $F2b_{tm}$ both denote the fraction of M_{mt} remaining after accounting for damage effects of pests whose infestation levels are defined as the percentage of the crop diseased or damaged. But while $F2a_{mt}$ is used if M_{mt} refers to crop state variables such as tissue biomass, $F2b_{mt}$ is used if M_{mt} refers to a crop physiological process or rate variable and it is computed as:

$$F2b_{mt} = (1 - a_{1m}/P_{1t}/100)^{b_{1m}} \ldots (1 - a_{pm}P_{pt}/100)^{b_{pm}} \tag{21}$$

where $P_{pt}\, e[0, 100]$ = percentage of host organs diseased or damaged on day t, a_{pm} = coefficient that determines the damage equivalency, and b_{pm} = coefficient that describes the proportionality of the damage effect (Bastiaans, 1991). The computation of $F2a_{mt}$ is complicated by the fact that P_{pt} does not reflect the actual damage rate that occurred on crop state variables between two observation points in time, but only indicates the current relative damage status of the system as it was observed at a given time. Moreover, actual net damage rates can be 'masked' by growth and senescence of the host. For example, P_{pt} can decrease over time due to growth and senescence of the host, even though the absolute net damage rates are greater than zero. In order to estimate the actual amount of damage that has to be applied daily to the respective crop state variables, each P_{pt} has to be converted into a net rate of new daily damage P'_{pt} by accounting for growth and senescence of host organs:

$$P'_{pt} = \left\{ P_{pt} K_{pt}/100 - \sum_{t=e}^{t} [(P_{pt-1} K_{pt-1}/100) - (P_{pt-2} K_{pt-2}/100)] + \sum_{t=e}^{t} [(P_{pt-1} K_{pt-1}/100) \times S_{pt}] \right\} / K_{pt} \tag{22}$$

where $P_{pt} K_{pt}/100$ converts percent damage into absolute damage $e\,[0, M_{mt}]$, the first summation term represents cumulative changes of $P_{pt} K_{pt}/100$ over time, and the second summation term keeps track of the $P_{pt} K_{pt}/100$ that has cumulatively been removed through host senescence S_{pt}. $S_{pt}\, e[0, 1]$ is defined as the relative amount of K_{pt} that senesces between day $t-1$ and t and is derived from the actual leaf senescence rate computed by the crop model. $F2a_{mt}$ can now be computed as:

$$F2a_{mt} = 1 - P'_{1t} - \dots P'_{pt} \tag{23}$$

with $P'_{1t} + \dots P'_{pt}\, e[0, 1]$.

There are a few special cases where this approach has been modified. If a pest effect consists of leaf senescence acceleration (CP9), M'_{mt} is directly computed as:

$$M'_{mt} = M_{xt}[1 - (1 - M_{mt}/M_{st}) F1_{mt} F2b_{mt}] \tag{24}$$

for a given day t. Note that in this case, M_{mt} = senescence rate of green leaf area without pest stress, $M'_{mt} = M_{mt}$ corrected for pest stress, and M_{xt} = green leaf area on day t. Light competition (CP8) caused by weeds is modeled using an equation that assumes that weeds are uniformly distributed within the crop and have the same height and canopy architecture as the crop (after Spitters, 1989):

$$M'_{mt} = a_c L_{ct}/(a_c L_{ct} + \dots a_{wp} L_{wpt})(1 - e^{-[a_c L_{ct} + a_{w1} L_{w1t} + \dots a_{wp} L_{wpt}]}) R_t E \tag{25}$$

where a_c, a_{wp} = light extinction factor of the crop and weeds p, respectively; L_{ct}, L_{wpt} = crop leaf area, and leaf area of weed p, respectively, at time t; E = light use efficiency, and R_t = photosynthetic active radiation. Here, M'_{mt} represents the potential photosynthesis of the crop after accounting for weed light competition. If the total leaf area index (crop plus weeds) exceeds 4, an additional effect of weed light competition on the daily rate of crop leaf senescence is assumed:

$$M'_{mt} = L_{ct}0.004(L_{ct} + L_{w1t} + \dots L_{wpt} - 4) \qquad (26)$$

where M_{mt} = green leaf area that senesces on that day.

In the above sections, we have described in detail the techniques to quantify pest dynamics and then the relationship of these to crop growth and productivity in two models. Insects and diseases have common parasitic effects on crops while weeds, because they are commonly higher plant species, tend to have different mechanisms to affect crops.

Weed competition

Weeds may reduce crop yields through competition for limited resources: primarily light, water, and nutrients. In some cases weeds may also release chemicals which can suppress germination or growth of the crop (allelopathy). It is not yet possible in a field situation to separate the relative effects of the two processes (Cousens, 1992). It is also difficult to design experiments to determine whether below- or above-ground interference is more important. Perhaps because of these difficulties, many experiments have concentrated on quantifying the reduction in final yield which can be attributed to a particular weed population, without regard to the mechanisms through which that yield reduction occurred. Zimdahl (1980) summarized much of the literature on weed crop competition. Since his review, many more studies have been completed. Numerous studies have looked at the effect of weed density (e.g., Barrentine, 1974; Bloomberg et al., 1982; Stoller et al., 1987; Legere and Schreiber, 1989; Baysinger and Sims, 1991; Bridges et al., 1992; Fellows and Roeth, 1992). Studies have also quantified the importance of duration of competition (Baysinger and Sims, 1991; Bridges et al., 1992; Fellows and Roeth, 1992), and weed versus crop emergence date (Bloomberg et al., 1982; Baysinger and Sims, 1991; Qasem, 1992). Other factors which have been demonstrated to be of importance include crop variety (McWhorter and Hartwig, 1972; Fiebig et al., 1991), crop stand density (Marwat, 1988; Legere and Schreiber, 1989), and rainfall patterns (Geddes et al., 1979; Mortensen and Coble, 1989).

Because weed control decisions can be very complex, a number of computerized decision aids have been developed to help pest managers decide if a herbicide application is necessary, and if so, which one to use. Some of these programs use a minimal level of herbicide efficacy as the primary means of determining which treatments are appropriate for a particular weed population. Other programs estimate yield loss which is likely to occur without treatment and with each possible herbicide, and make a recommendation based on maximizing expected net return. Mortensen and Coble (1991) reviewed the available programs in each category. Programs which utilize an economic threshold approach estimate the combined effects of a multi-species weed complex on crop yield in a number of different ways. Aarts and de Visser (1985) and Wilkerson et al. (1991) consider weed species and density in determining potential damage. Keisling et al. (1984) also consider amount of time from weed emergence to the decision date when estimating the amount of damage which will occur during the rest of the growing season. Kropff and Lotz (1992) suggest using relative leaf area of weeds (leaf area of weeds/leaf area crop + weeds) shortly after crop emergence to estimate yield loss. The relative leaf area gives an indication not only of weed density, but also of weed size relative to crop size.

To date, the decision models consider only a small subset of the factors which may affect crop yield loss. Although weed scientists know that crop variety, planting date, row spacing, and crop stand will affect a weed's ability to compete with the crop, determining how to modify yield loss estimates as all these factors change is not really feasible through field research, given the large number of weed species which may occur in any given field, the tremendous diversity in growth habits of weed species, and the expense and difficulty of conducting multi-factor experiments. Extrapolating experimental results from one location to another is complicated by the fact that many important weed species are photoperiodic, as are many crop plants. Competitive ability may change during the course of the growing season as partitioning coefficients for root and leaf growth change as the plants shift from vegetative to reproductive growth.

One method which can be used to interpret experimental results and investigate the possible mechanisms of competition is the development of dynamic models of crop and weed growth and competition. A number of models of weed/crop competition have been developed. Rimmington (1984) has developed a model for the effect of light competition on daily dry matter production in a binary mixture. The canopy is divided into horizontal mixed and unmixed layers according to the heights and canopy structures of the component species.

Spitters and Aerts (1983); Kropff (1988); and Kropff and Lotz (1992) simulated competition for light and water between crops and weeds. They considered the vertical distribution of leaf area and root mass in the determination of light interception and water uptake by the competing species. Leaf area of both species is assumed to be homogeneous. The percentage of light penetrating a layer of crop canopy is described by an exponential equation (Spitters and Aerts, 1983). Light interception by each species is calculated from its share of the total leaf area weighted by the light extinction coefficient (Kropff, 1988). Kropff et al. (1992) found good agreement between field data and simulation output using this model, but point out that the assumption of horizontal homogeneity can cause deviations at low weed density or when row spacing of the crop is large.

Kiniry et al. (1992) simulate competition for light, water, and nutrients between two competing plant species. They have extended the approach to modeling utilized in EPIC (Williams et al., 1984). Light interception by the two competing species is calculated using the system of Spitters and Aerts (1983).

Graf et al. (1990) simulate competition of rice with weeds for nitrogen and light. They allow simulation of multi-species complexes by dividing weed flora into six groups based on differences in leaf shape, growth form, height and phenology. They calculate a potential nitrogen uptake rate for each plant group, then apportion available nitrogen according to the proportion of root space explored by each group. Plant height and leaf area distribution are used to estimate light interception by each plant group according to a method similar to that used by Spitters and Aerts (1983). Simulation results compared well with field data. The model slightly overestimated rice biomass in weedy plots, and underestimated it in early-weeded plots.

All of the models discussed above assume a homogeneous horizontal distribution of crop and weed leaf area. A number of studies (Barrentine and Oliver, 1977; Gunsolus, 1986; Monks and Oliver, 1988) have shown that crop yield increases as distance from a weed increases. This can be explained in large part by the fact that the closer a crop plant is to a weed, the longer that plant will be in direct competition with the weed for light, nutrients, and water. Gunsolus (1986) found that canopy diameter of common cocklebur (*Xanthium strumarium L.*) grown in competition with soybean increased from 16.2 cm at 4 weeks to 129.2 cm at 18 weeks after planting.

Coupling insect pest, weed and disease effects to crop models for decision support

Accurate identification of the damage mechanisms associated with a particular pest's effects on crop growth and development is vital for subsequent use of the models to estimate yield loss. The damage mechanisms are translated for simulation into coupling points that link pest to crop. Because there are sufficient differences in approach for the three main pest groups, pest-crop coupling for decision support is discussed here separately for insects and weeds, and then diseases as part of a multiple pest scenario.

Insect pests

Two examples are detailed here to illustrate different types of pest damage.

Corn earworm in soybean. Corn earworm can cause severe damage to soybean in the southeastern United States, and it is often necessary to determine how irrigated and non-irrigated soybean production systems respond to corn earworm damage. In our example (Table 7), treatments 3 and 4 contained irrigated and non-irrigated Bragg soybean subjected to a hypothetical corn earworm population (Table 7). These observed population levels were placed in the time series file for treatments 3 and 4. A 4 character identification (CEW6) was selected to identify and define the sixth instar corn earworm populations. Next,

Table 7. Corn earworm population data collected from field scouting for the soybean experiment UFGA7802.

Day of year	Corn earworm population, no. m^{-2}
222	0.0
229	0.5
236	1.0
243	2.5
250	5.5
257	11.0
264	14.1
271	13.1
278	0.5
285	0.8
292	0.9

the 'daily absolute damage rate with pest competition and food preference effects' damage characterization method (PCTID) was assigned because corn earworm can damage multiple crop components depending upon preference and competition for food sites. Three feeding sites can be damaged by corn earworm depending upon preference and competition. Thus, three different coupling points were used to describe the pest. Corn earworm prefer to feed on small seed, so the coupling point damage variable SDNS (Table 2) was placed first in the pest coefficient file. Because of limited availability for small seed, CEW6 can feed on large seed. Thus the coupling point damage variable for large seed (SDNL) was placed second in the hierarchy. Finally, corn earworm will feed on leaf mass if seeds are not readily available. The leaf area index coupling point (LAD) was placed last in the food hierarchy for CEW6. Damage coefficients (PDCF1) were then found from the literature to describe the feeding rate of CEW6 on the three coupling points. Because the damage type is set to 4, damage is applied preferentially to small seed first at a rate of 10.0 small seed $larva^{-1}\ d^{-1}$. If the food demand of the population cannot be satisfied by small seed, some of the damage will be partitioned to large seed at a damage rate of 2.5 seed $larva^{-1}\ d^{-1}$. Damage can also be partitioned to leaf mass at a rate of 0.00505 $m^2\ larva^{-1}\ d^{-1}$ if the population is large enough (Table 5).

Figures 4 and 5 show the resulting leaf and seed mass for each of the four treatments, respectively. The leaf and seed mass were higher in the irrigated than in the non-irrigated treatment, demonstrating the model's response to water stress. In both the irrigated + CEW6 and non-irrigated + CEW6 treatments, corn earworm began consuming leaf mass before seed growth was initiated. Once seed growth began, however, CEW6 damage shifted from leaf mass to seed mass based on CEW6's hierarchy of food preference for small seed, large seed, then leaf mass. Seed growth was initiated slightly later in the non-irrigated treatment. In the irrigated + CEW6 treatment, the CEW6 population reduced the seed mass by approximately 40% compared to the irrigated control. In the non-irrigated + CEW6 treatment, the corn earworm population

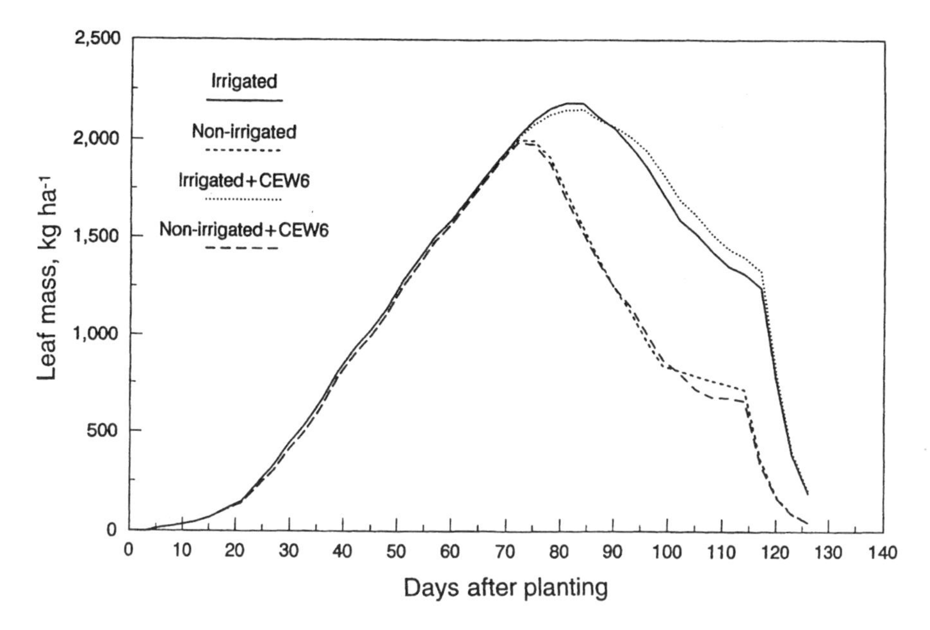

Figure 4. Effect of corn earworm damage on leaf mass for the soybean experiment.

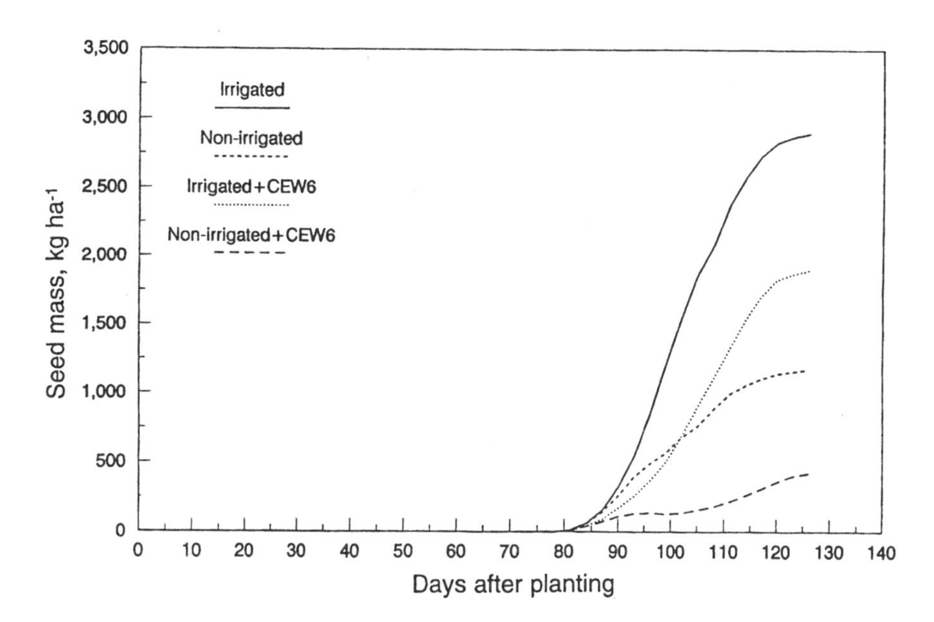

Figure 5. Effect of corn earworm damage on soybean seed mass.

Table 8. Percent defoliation measured at different observation dates for the soybean experiment UFQU7902. Defoliation occurred due to velvetbean caterpillar population, however, the population level was unknown.

Day of year	Observed defoliation %
241	0.0
248	22.6
255	56.1
262	50.0
271	52.5
276	57.1
283	57.9
288	61.7
292	53.3
295	71.4
303	100.0

reduced yields by 70% compared to the non-irrigated control treatment. This was primarily due to consumption of seed mass rather than limitations of photosynthesis due to leaf area reduction from consumption. Thus, in the low input system (non-irrigated, no insecticides), the losses due to insects is much more severe than in the irrigated system.

Defoliation of soybean by velvetbean caterpillar (subsubhead)

In many research experiments, defoliation may occur and either the source or the population level may be unknown. This can result in different observed values of leaf mass for a treatment and a control plot. This example illustrates use of the percent observed damage to simulate effects of defoliation from an unknown source on soybean yield. In this experiment, an insecticide was used in the control treatment while none was used for the second treatment. Periodic measurements of leaf area index in the untreated treatment showed severe defoliation due to velvetbean caterpillar larva. In this case, pest population data were not collected, but percent defoliation for the damaged treatment was measured and is shown in Table 8. In this experiment, defoliation was simulated by specifying the percent observed defoliation over time.

The coefficients defining an observed defoliation damage type were added in the pest coefficient file (Table 5). The four character identifier for percent cumulative leaf area damage was PCLA. The coupling point was leaf area index, indicated by the CROPGRO leaf area index damage variable, LAD (Table 2). The damage characterization method selected was 2, which indicates that damage will be entered in the time series file as observed defoliation. The crop model will compute the daily damage required to obtain this level of defoliation on the observation dates. The damage coefficient is 1.0, so that damage recorded in the time series file can be directly applied to the crop model. This coefficient is typically used to convert insect populations to feeding rates. When a damage type, such as this one, is defined, direct measures of damage are typically recorded in the time series file and the damage coefficient is set to 1.0.

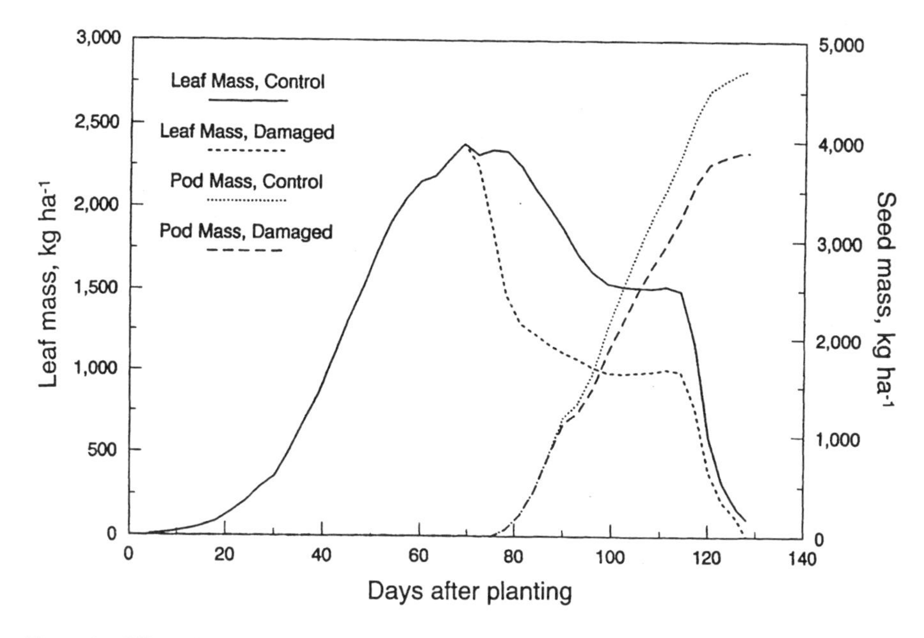

Figure 6. Effect of velvetleaf caterpillar damage on leaf and seed mass for the experiment.

A column with the header PCLA was entered in the time series file UFQU7902.SBT. Percent damage on each observation date was entered for treatment 2. The crop model linearly interpolates between observations to obtain daily levels of observed damage. Note that an entry of −99 indicates that data were not available. Thus, CROPGRO skips that entry for linear interpolation of daily damage.

In this example, damage was applied to leaf area index. The resulting leaf area index for control and defoliation treatments is shown in Figure 6. The crop model computed the daily leaf mass damage required to obtain the defoliation observed on the observation dates. The effect of defoliation on seed yield is also shown. Seed yield was reduced due to a reduction of daily photosynthesis.

This approach for applying damage to crop models was first outlined by Batchelor et al. (1992) and Pinnschmidt et al. (1990a). It was applied to the SOYGRO (Jones et al., 1988) and PNUTGRO (Boote et al. 1986) models by Batchelor et al., (1993). Pinnschmidt et al. (1990b) also used a similar approach to apply pest damage to the CERES-Rice model. Batchelor et al. (1993) tested the response of SOYGRO to experiments containing manual defoliation and defoliation from velvetbean caterpillar. The simulated plant response was similar to the observed plant response. They tested the response of the PNUTGRO model to leafspot disease by applying measured levels of defoliation, diseased leaf area, and pod detachment. The response of the model was similar to field observations. Finally, they conducted a sensitivity analysis of yield loss for several other damage types and compared the predicted yield loss to losses reported in the literature.

Weeds

A weed competition model (SOYWEED/LTCOMP) linked SOYGRO V5.41 (Jones et al., 1988) and described by Wilkerson et al. (1990) and Wiles and Wilkerson (1991) is similar in many respects to the models described previously. It was designed to simulate low weed populations near economic thresholds and therefore considers the differential competition of a weed with crop plants at varying distances.

In this model all interference between soybean and weeds is attributed to competition for light and water. It is assumed that the crop is planted in conventional, widely-spaced rows (approximately 76 to 106 cm row spacing) and that the weeds are uniformly distributed throughout the field. The approach taken to modeling weed growth in the absence of competition is very similar to that used in modeling soybean growth (Wilkerson et al., 1983). The processes simulated include photosynthesis, respiration, dry matter partitioning, and leaf senescence. Equations for all processes are presented by Wilkerson et al. (1990). The light interception model is described by Wiles and Wilkerson (1991). Although this model originally utilized SOYGRO V5.4 with the soil-water model replaced by a simpler two-compartment model, the weed model has now been linked to SOYGRO V5.41, and it utilizes the SOYGRO V5.41 soil-water model.

Competition for light. Competition for light is modeled by defining an 'area of influence' for each weed. The weed competes for light with the crop only within this area. The area of influence expands as the weed grows until it meets the area of influence of another weed. The expansion of the area of influence depends upon simulated weed leaf area (Figure 7). Each day of the simulation, the proportion of the soybean field within the area of influence of all weeds is updated based on the prior days's increase in weed leaf area.

Light interception by soybean and weed within the area of influence is determined from daily radiation, the amount of weed and crop leaf area present, and weed and crop canopy structure (Wiles and Wilkerson, 1991). An approach similar was used by Spitters and Aerts (1983). The canopy structure of the competing species within the area of influence is divided into three layers. The top layer is the leaf area of the taller component. The remainder of the canopy is divided into two equal parts. If the components are almost equal in height, then the top layer consists of the top 1/4 of the height, and each of the bottom layers 3/8 of the height.

Crop leaf area is distributed vertically according to a triangular function (Pereira and Shaw, 1980; Norman, 1979). Common cocklebur leaf area is

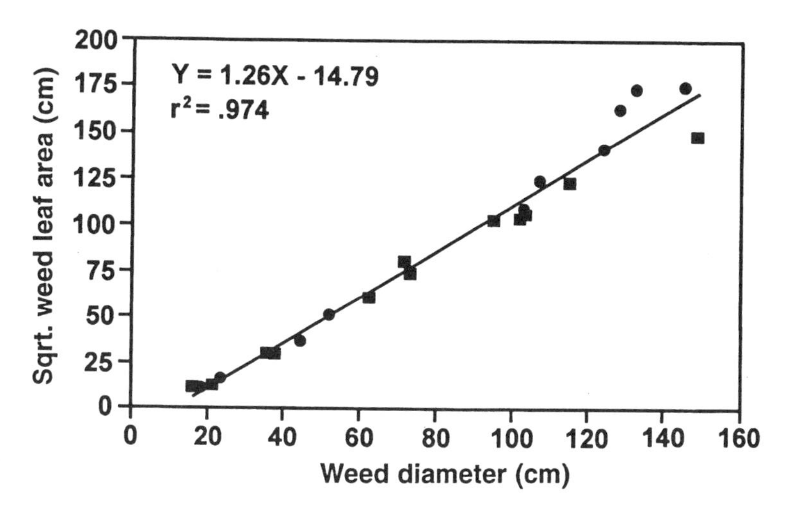

Figure 7. Figure 7. Relationship between weed leaf area and weed canopy diameter for common cocklebur grown at Clayton, NC in 1984 and 1985. Data for weeds grown in competition with soybean are represented by ■, and data for weeds grown in a cleared-out section of row by ●. (Data from Gunsolus, 1986.)

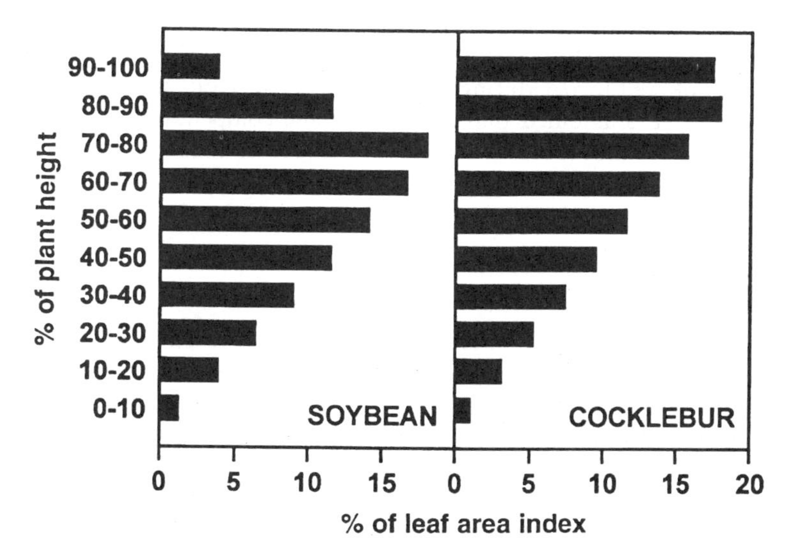

Figure 8. Leaf area distributions by height used in SOYWEED/LTCOMP.

distributed according to a weighted average of the densities calculated from triangular and uniform density functions (Figure 8).

Light interception by the mixed canopy (I_{jT}) in each layer is calculated by assuming independence in the probability that a light beam will strike either crop or weed:

$$I_{jT} = I_{jc} + I_{jw} - I_{jc} I_{jw} \quad (27)$$

The potential interception by crop (I_{jc}) or weed(I_{jw}) in layer j is computed assuming an exponential extinction of light through the canopy:

$$I_{jT} = 1 - e^{(-k_i LAI_{ji})} \qquad \text{for } I = c, w \quad (28)$$

$$I_{ji} = \left[1 - \sum_{m=1}^{j-1} I_{mT}\right] [1 - e^{(-k_i LAI_{ji})}] \qquad \text{for } j = 2, 3. \quad (29)$$

Where $k_{i'}$ is the crop or weed extinction coefficient. Total interception in each layer is apportioned to crop or weed based on their contribution to total leaf area weighted by the extinction coefficient:

$$f_{ji} = k_i LAI_{ji} / (k_c LAI_{jc} + k_w LAI_{jw}) \quad (30)$$

where f_{ji} represents fraction of total interception in layer j apportioned to species i.

Calculating interception by components of a mixed canopy with this procedure assumes weed and crop leaf area are thoroughly intermingled horizontally within the area of influence. However, weeds emerging with the crop which become competitive, usually emerge in skips in the row or a few inches to the side of the row (Murphy and Gossett, 1981). Thus, weed and crop foliage are likely not well-mixed during early growth. To avoid underestimation of early growth, it is assumed that no interspecific competition for light occurs until total LAI within the area of influence is greater than 1.0.

For those areas of the field outside a weed area of influence, crop light interception is also calculated assuming an exponential extinction of light through the canopy:

$$I_c = 1 - e^{(-k_c LAI_c)} \quad (31)$$

A field average for crop light interception is computed based on the proportions of the field that are within and without weed areas of influence. Relative gross photosynthesis is determined from average crop light interception using the equations of SOYGRO V5.41 (Wilkerson et al., 1983; Jones et al., 1988). Once this field average value for gross photosynthesis has been determined, the average increases in soybean leaf, stem, root and pod dry weights across the field are determined according to the SOYGRO equations. This assumes that weed competition has no major effect on crop dry matter partitioning. Calculation of soybean growth then proceeds exactly as it does in the weed-free case.

Once weed light interception has been calculated, weed growth is computed using the same equations as for weeds grown alone. Weed area of influence is increased based on the increase in weed leaf area. New crop leaf weight and area are assigned either to plants within a weed area of influence or to plants not affected by weeds on a proportional basis. Leaf senescence is also subtracted on a proportional basis. Increase in soybean leaf area within a weed area of influence is computed by adding leaf area grown during the day, subtracting leaf area senesced, then adding the leaf area added through expansion of the area of influence.

Competition for water. Root growth and water uptake and movement are modeled as in SOYGRO V5.41 (Jones et al., 1988). Weed and crop roots are grown independently. That is, the presence of the other species does not influence rate of root growth or vertical root distribution except indirectly through the influence on available soil water. The demand for water is determined by calculating a total demand for the field from total leaf area present. This demand is apportioned between weeds and crop according to their contribution to total leaf area.

Simulation results. SOYWEED/LTCOMP has been compared to two years of growth analysis data for widely-spaced common cocklebur populations (Gonsolus, 1986; Wiles and Wilkerson, 1991). Simulated results match field data fairly well. Sensitivity analyses showed that simulated crop responses to changes in weed emergence date and time before weed removal were consistent with observations reported in the literature (Wiles and Wilkerson, 1991). The importance of modeling weed/crop competition within an area of influence for low weed populations can be seen by comparing simulated weed growth and crop yield reduction with and without an area of influence (Figure 9). For the conditions pertaining to the field experiment (a very competitive weed species, crop and weed nearly equal in height 1.75 m, weed leaf area concentrated more top-heavily than crop leaf area, and weeds 3.0 m apart), assuming that weed leaf area was distributed homogeneously across the field rather than confined to an area of influence resulted in a 28% underestimation of crop yield. If weed height is reduced by one-half, then having weed competition limited to an area of influence is not nearly so important (Figure 10). Reducing the weed extinction coefficient from 0.77 to 0.61 to be identical to that of the crop also reduces the importance of horizontal leaf distribution Figure 11.

In all cases, spreading the weed leaf area over the whole field, rather than confining it to an area of influence overestimates weed light interception. This is due to the nature of the exponential equations which are used in this and other competition models. For example, suppose each weed is assumed to have 1.0 m^2 of leaf area above the crop canopy; suppose there is one weed in every 10 m^2 of the field, then the proportion of available light which will be intercepted by the weed if the weed leaf area is spread across the whole 10 m^2 is 0.074 for an extinction coefficient of 0.77 ($LAI = 0.1$). If the leaf area is confined to an area of influence of size 1.0 m^2 ($LAI = 1.0$ within the area of influence and $LAI = 0$ elsewhere), then the proportion of light which is intercepted is calculated to be 0.054 (fifty-four percent of the light that hits the area of influence

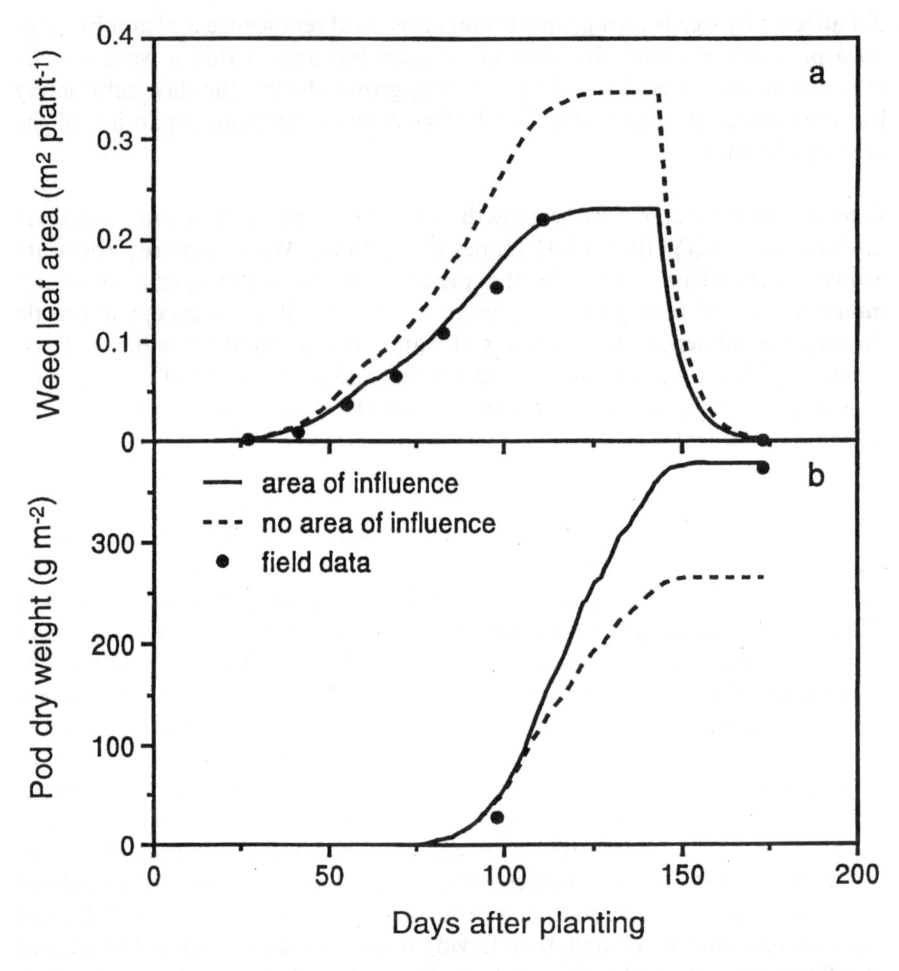

Figure 9. Comparison of the area of influence approach to distribution of weed leaf area to an approach which distributes weed leaf area homogeneously across the field. Simulations assume the conditions for field experiments performed in 1985 and presented by Gunsolus, 1986.

is intercepted by the weed, but no light is intercepted by the weed in the remaining 9.0 m^2).

Confining weed leaf area to an area of influence is even more important in a real field situation. Several studies have demonstrated that weed populations are likely to be clumped (Marshall, 1988; Thornton et al., 1990; Wiles et al., 1992; Mortensen et al., 1993), rather than distributed randomly or uniformly across a field. Wiles et al. (1992) and Mortensen et al. (1993) demonstrated that weed spatial pattern can often be represented by a negative binomial distribution, such as that shown in Figure 12. One way of more accurately simulating weed competition when weeds are clumped is to fit a negative

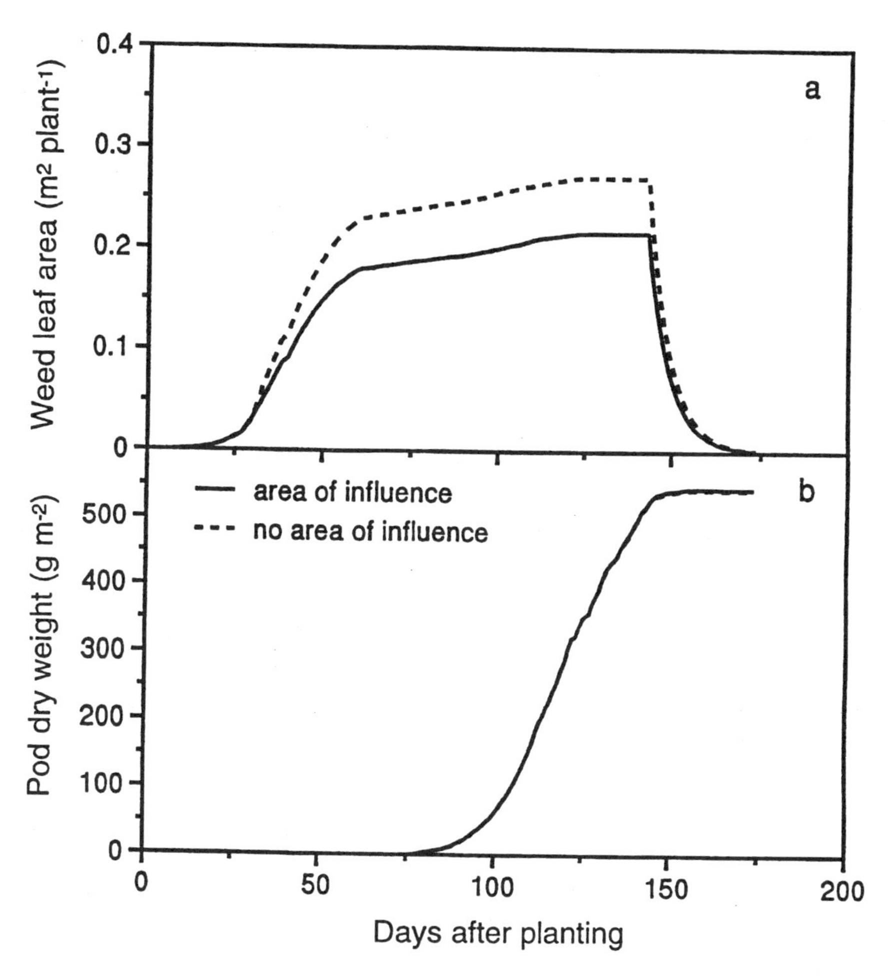

Figure 10. Comparison of the area of influence approach to distribution of weed leaf area to an approach which distributes weed leaf area homogeneously across the field. Simulations assume the conditions for field experiments performed in 1985 and presented by Gunsolus, 1986, except that the rate of increase for weed height has been halved.

binomial distribution to the data, and break the field into areas having different densities. For example, for a field with a distribution such as in Figure 12, with mean density of 1.0 weed per quadrat, if the field is divided into quadrats consisting of 10 m of row, more than 55% of the quadrats should contain no weeds, about 22% should contain one weed, and the remaining quadrats should contain from three to 17 weeds. Yield loss for the whole field can be estimated by simulating from zero to 17 weeds per quadrat, then taking a weighted average of the simulated yield losses. In the example presented in Figure 12, estimated yield loss will be 6.8% if weeds are assumed to be clumped, 7.0% if

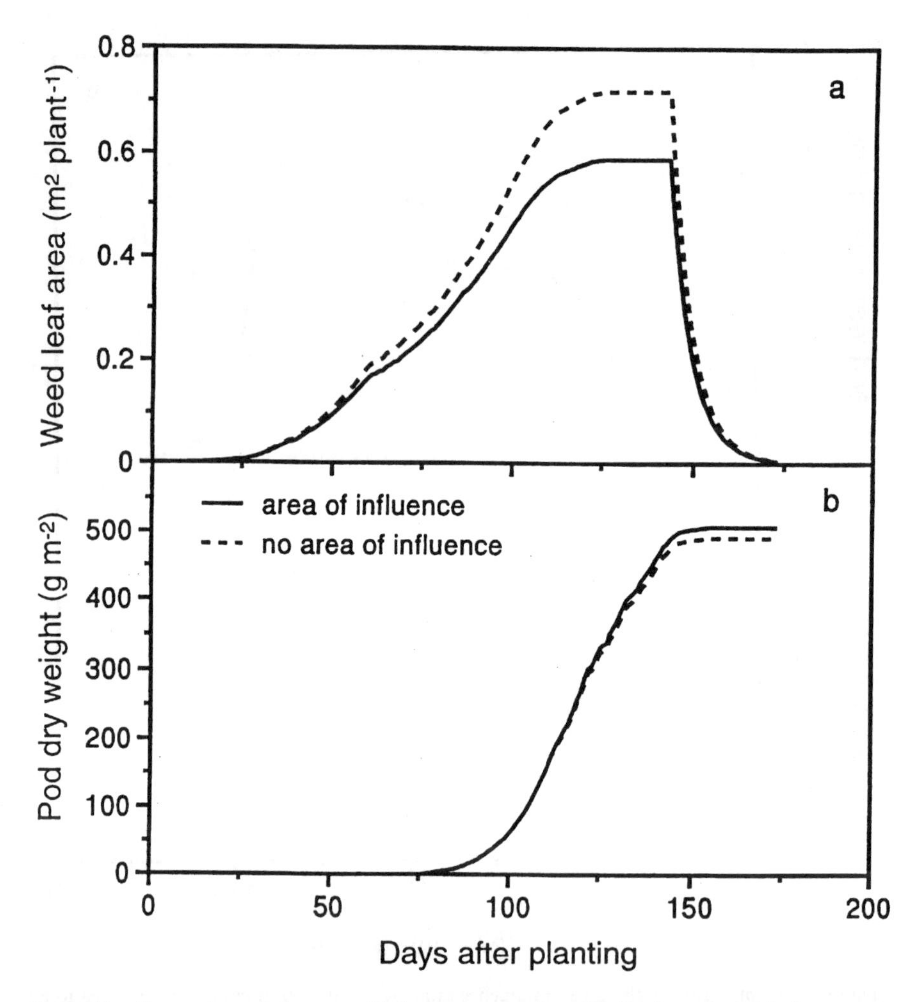

Figure 11. Comparison of the area of influence approach to distribution of weed leaf area to an approach which distributes weed leaf area homogeneously across the field. Simulations assume the condition for field experiments performed in 1985 and presented by Gunsolus, 1986, except that the extinction coefficient for weeds has been reduced to that used for the crop.

weeds are distributed uniformly but compete only within an area of influence, and 18% if weeds compete equally across the whole field.

SOYWEED/LCOMP is currently designed for simulation of the crop and one weed species at a time. With the area of influence approach to modeling competition for light, simulating multi-species competition for light becomes very complicated. One approach to simulating multi-species competition utilizing the area of influence has been presented by Wiles et al. (1991). In a program called WEEDING which was developed as an instructional game, multi-species competition is simulated by translating all weed species into equivalent numbers

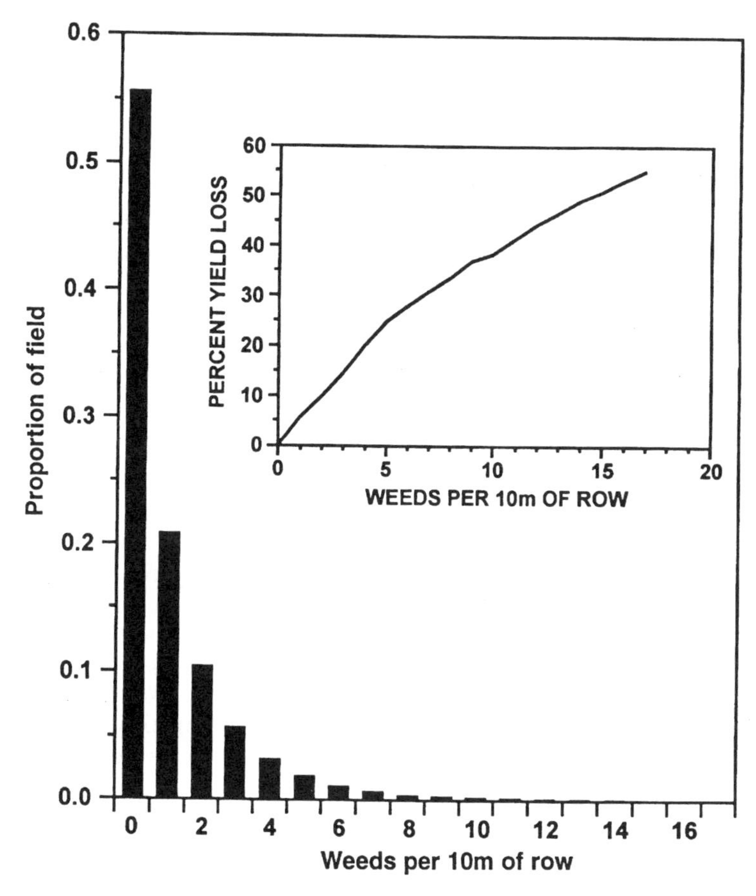

Figure 12. Example of a negative binomial distribution with mean density of 1.0 and aggregation coefficient of 0.6, and the simulated yield loss for quadrats containing different weed population densities.

of common cocklebur. The program uses a ranking of weed competitiveness developed by Coble (1986) and utilized in the decision model HERB (Wilkerson et al., 1990). Each weed species has been assigned a relative competitive index on a scale of 1 to 10. The density of each weed species is translated into an equivalent density of common cocklebur based on the relative competitive indices. Growth and competition of this cocklebur density equivalent are then simulated. This approach has an advantage over algorithms currently used in decision models in that estimated yield loss will vary with field-specific conditions. However, it assumes that all weed species will respond like common cocklebur to changing conditions. A better approach to handling multi-species populations might be to simulate crop/weed competition for each species individually. Damage from the multi species complex could then be estimated by using the simulation results to adjust each species competitive index for use in the decision model calculations. Dividing weed species into a limited number of categories according to growth characteristics (Graf et al., 1990) seems to be a practical approach to limiting the number of weed models which need to be developed.

In modifying SOYGRO V5.41 to incorporate weed competition, all sensitivity to row spacing and planting geometries are removed. This is because crop light interception is now solely a function of crop LAI. In reality, row spacing does matter for much the same reasons that confining weed leaf area to an area of influence matters: by spreading out crop leaf area across the whole ground area, the model will overestimate light interception in widely-spaced rows prior to canopy closure. If model parameters are adjusted to correctly simulate this situation, then the model will underestimate growth in narrow-row plantings. Work is currently underway on a simple light interception model which will take into account daylength and planting geometry, as well as leaf area of the competing species.

Simulation of multiple pest damage effects using CERES-Rice

The following examples to illustrate the potential of pest-crop modeling are based on the one-way approach (see section on Quantifying pest dynamics for crop-pest modeling), in which observed or assumed pest levels are 'read' from input files and used in the simulation. In the first example, yields from plots subjected to various multiple pest infestation treatments in field experiments conducted at the IRRI farm in the Philippines during wet seasons 1990 and 1991 and during dry season 1991 were estimated by the multiple pest-coupled CERES-Rice model (Pinnschmidt and Teng, 1994; Pinnschmidt et al., 1994, 1995) and plotted versus observed yields. The pests and damage types considered in the simulation were: leaf blast (LB), collar blast (CB), panicle blast (PB), sheath blight on the leaf (SHB), sheath blight on the sheath (SHBs), bacterial leaf blight (BLB), other leaf diseases (OLD), defoliation (DEF), stem borer dead hearts and white heads (DHWH), and rat damage (R), using the coupling points specified in Table 9 to simulate their damage effects. The precision of yield expectations was considerably improved when pest effects were considered in the simulation (Figure 13). In the example, any management decision that is based on a pest-free yield expectation might be totally wrong while it will be improved when the yield expectation is adjusted for pest effects.

Pest scenarios that varied with respect to onset times t_0 and maximum pest levels X_{max} were assumed for the single pests leaf blast which was assumed to follow an unimodal progress curve, and for defoliation, sheath blight, and weed infestation which were assumed to follow sigmoidal curves (Figure 1). Yield consequences of these scenarios were simulated. The results indicate nonlinear increases in yield losses with increasing X_{max} and decreasing t_0 and suggest that late pest infestations at low severity levels might be tolerable while early

Table 9. Simulated yield loss (in %) due to combined hypothetical infestations of leaf blast, sheath blight, defoliators, and weeds in rice. Simulated by the multiple pest-coupled CERES-Rice. From: Pinnschmidt et al., 1995.[a,b,c]

X_{max} weeds (m^2/m^2)	X_{max} defoliation (%)	X_{max} leaf blast (%)			
		0		10	
		X_{max} sheaf blight (%)			
		0	30	0	30
0	0	0	18.7	30.6	30
	40	11.6	27.1	43.3	53.4
2.4	0	41.6	51.2	62.0	69.1
	40	45.8	54.3	65.7	71.7

[a] X_{max} = maximum pest level respectively asymptote of pest progress curve.
[b] Other inputs: leaf blast: $t_0 = 20$ days after seeding, unimodal progress curve sheath blight: $t_0 = 50$ days after seeding, sigmoidal progress curve defoliation: $t_0 = 40$ days after seeding, sigmoidal progress curve weeds: $t_0 = 10$ days after seeding, sigmoidal progress curve (t_0 = time of pest onset in days after seeding).
[c] For physiological pest damage coupling points used, see Table 6.

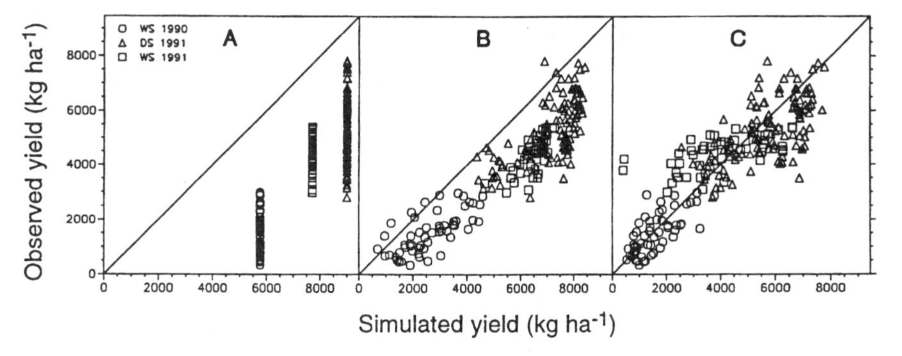

Figure 13. Observed yields versus yields estimated by the multiple pest-coupled version of CERES-Rice. A: pest damage effects not simulated; B: pest damage effects simulated, estimation of damage coefficients based on educated guesses (see Table 6); C: like B, but after optimization of damage coefficients (Pinnschmidt and Luo, 1993, unpublished). *WS* = wet season, *DS* = dry season. From: Pinnschmidt and Teng, 1994.

ones always bear a risk of high yield losses (Figure 14). Iso-loss curves such as those show in Figure 14 can be used to determine what kind of pest scenario needs to be controlled. This will be of direct help for decision making in pest management.

Economists have argued that multiple pest situation might have yield consequences different from just separately summing-up the yield effects of the individual pest species. This would alter decision criteria in pest management, such as control thresholds, depending on the multiple pest scenario (Blackshaw, 1986; Palis et al., 1990). Some scenarios of combined infestations of sheath blight and weeds were therefore assumed and simulated: t_0 was 10 days after

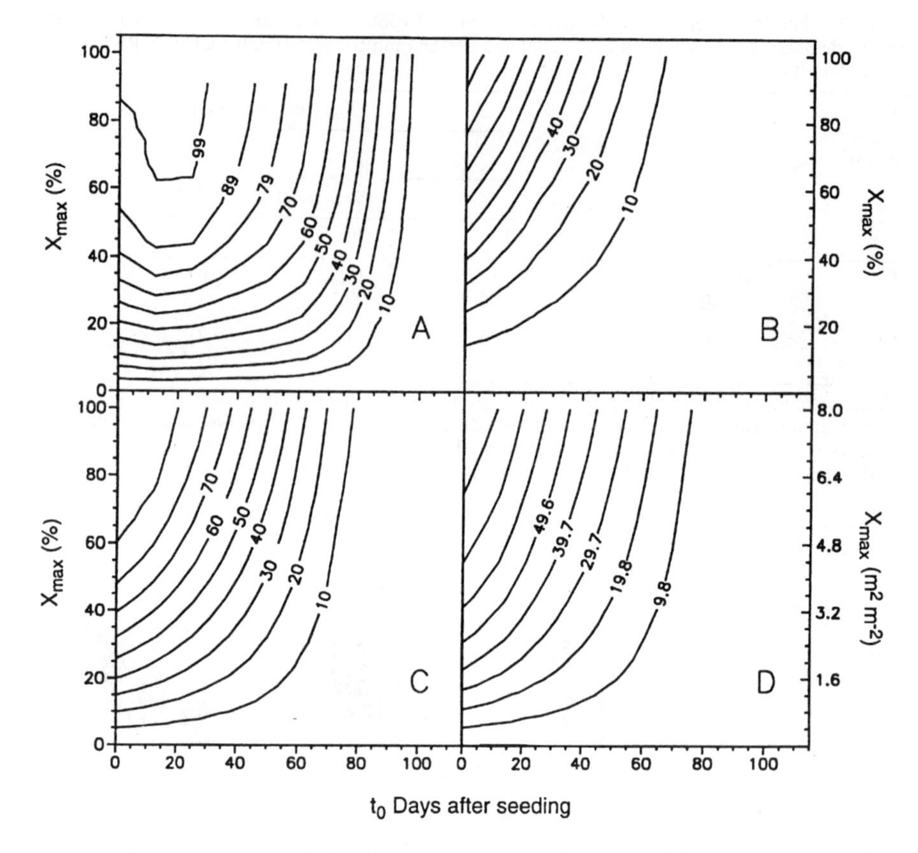

Figure 14. Iso-loss curves depicting simulated yield losses (in %) resulting from pest scenarios for leaf blast (A), defoliation (B), sheath blight (C), and weeds (D). Variable t_0 = time of pest onset, x_{max} = maximum pest level. See Table 6 for coupling points used. Unimodal progress curves were assumed for leaf blast, sigmoidal progress curves for the other pests (see Figure 1). Simulated by CERES-Rice. From: Pinnschmidt et al., 1994a.

seeding for weeds and 650 days after seeding for sheath blight; X_{max} of the progress curve for each pest was varied. The iso-loss curves resulting from these simulations are almost linear and indicated additive multiple pest yield effects (Figure 15). However, other simulation studies have indicated less-than-additive effects of multiple pests (Boote et al., 1993; Pinnschmidt and Teng, 1994; Pinnschmidt et al., 1995) such as shown in Table 9 for multiple pest scenarios of leaf blast, sheath blight, defoliation, and weeds.

As an example of how to use the approach to explore the effects of pest control measures in order to optimize control strategies, the feeding effects of a hypothetical population of defoliators were simulated, assuming applications of a 100% efficient control measure at different times after the onset of pest feeding. It was assumed that the crop would remain pest-free after control

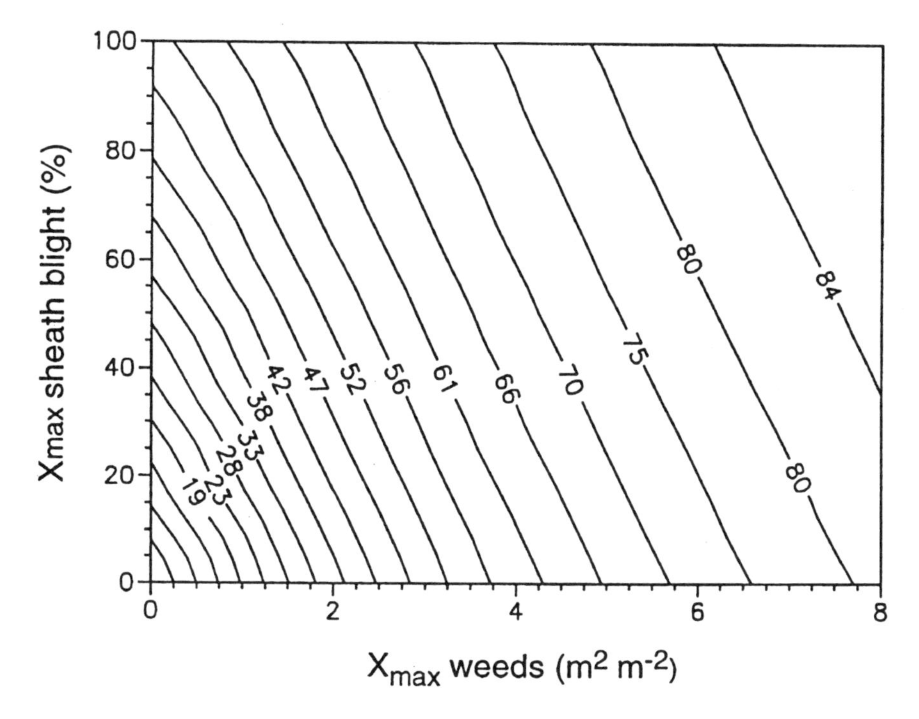

Figure 15. Iso-loss curves for simulated % yield losses resulting from combined infestation scenarios of sheath blight and weeds. See Table 6 for coupling points used. Sigmoidal progress curves were assumed for both pest (see Figure 1). t_0 for weed infestation = 10 days after seeding, t_0 for sheath blight infestation = 50 days after seeding. Simulated by CERES-Rice. From; Pinnschmidt and Teng, 1994.

measures were applied. These control scenarios were simulated for two different onset times (t_0) of leaf feeding, each with two different maximum potential feeding rates (X_{max}) of the pest population. The results indicated that the yield loss will increase sigmoidally with delayed control, but the relative degree of benefit obtained from a control (= the reduced yield loss) will primarily depend on the onset time of the pest infestation and on the feeding potential of the pest population (Figure 16).

One of the most interesting aspects of a pest-coupled crop models to forecast not only the probable biological effects, but also the economic benefits of control measures. The example presented here are from a two-way approach where pest dynamics were simulated by simple population models interlinked with CERES-Rice (Pinnschmidt et al., 1990b). Figure 17 illustrates the simulated effects of the timing (A), and of the number (B) of stem borer control measures on the development of stem borer larvae (upper portion of the figure) and corresponding biomass accumulation and yield loss (lower portion of the figure). Pest development was driven by healthy tillers per area and the damaging effects of stem borer larvae consisted in killing of growing points which

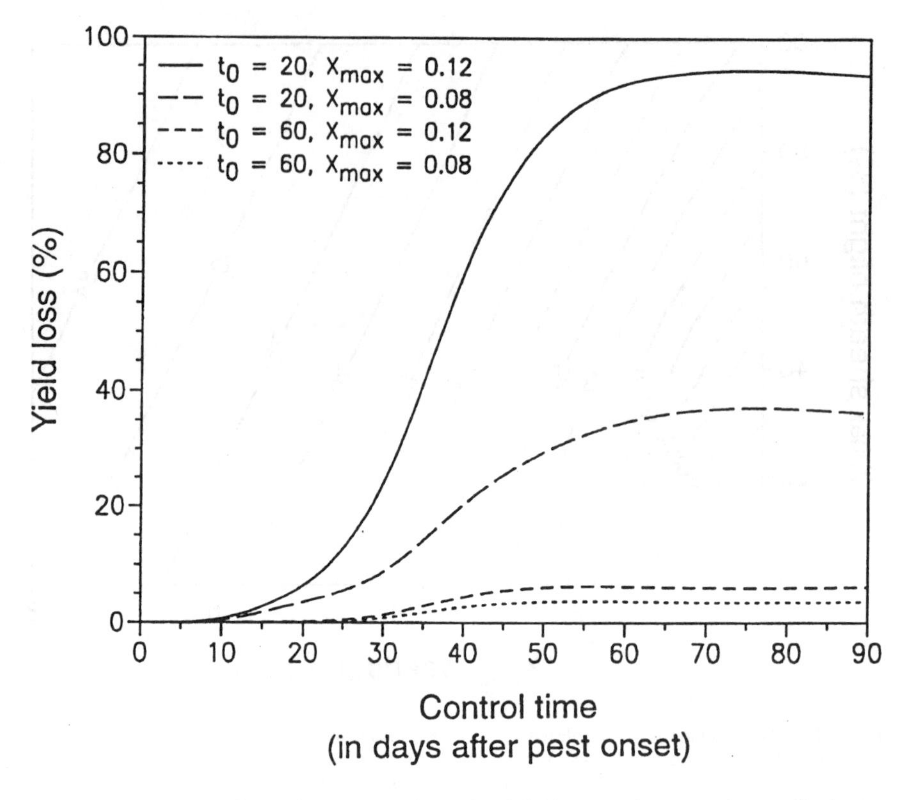

Figure 16. Simulated effect of the onset time of defoliation (t_0, in days after seeding) of a hypothetical population of leaf feeders, maximum potential defoliation rate (xmax, in m^2 leaf area/m^2 field area/day), and time of control on yield loss. Simulated by CERES-Rice. A unimodal progress curve was assumed for the potential defoliation rate (see Figure 1). The crop was assumed to remain 100% free from new defoliation after application of control measures. From: Pinnschmidt and Teng, 1994.

reduced the number of effective tillers and the leaf area index (LAI). Arrows in the upper portion of Figure 17 indicate the time when an insecticide with 80% 'knock down' effect and no residual effect was applied. An early control resulted in less yield loss, compared to a late control. The yield loss was substantial in any case, because the stem borer population was always able to build-up again after the insecticide application. If insecticide was applied several times subsequently, each control measure increased the biomass production and decreased the yield loss substantially, except for the last application. Assuming an obtainable yield of around 7 t ha^{-1} and a pessimistic economic scenario, that is, a market price of US\$0.13 kg^{-1} brown rice and US\$30.77 costs per control measure, the highest benefit/cost ratio (ca. 4.0) is obtained for the strategy with two control measures. Applying only one control at 75 days after seeding gives a better benefit/cost ratio (ca. 3.6) than three

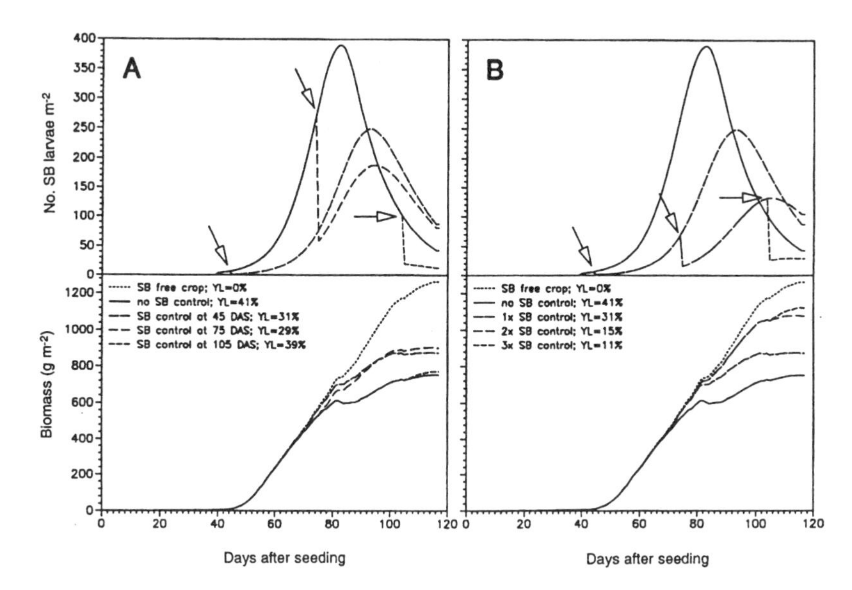

Figure 17. Simulated effects of the time (*A*) and the number (*B*) of stem borer (*SB*) control measures on a hypothetical population of stem borer larvae, crop biomass accumulation, and yield loss (*YL*). The time of control measures (*DAS* = days after seeding) is indicated by arrows. Control measures were assumed to have 80% 'knock down' efficiency and no residual effects. Simulated by CERES-Rice interlinked with a simple stem borer population model. For more details, see text.

subsequent control measures (ca. 3.0). However, with respect to the marginal net return, the strategy with three subsequent control applications performs best, resulting in a net return of around US$188.

Future development

Due to their physiologically based structure and dynamic nature, pest-coupled crop models allow mimicry of yield losses in a real-time pest-crop system. They have a great potential not only for adjusting crop simulations for pest stress and thus for arriving at more realistic yield expectations, but also for exploring yield consequences of hypothetical pest infestation scenarios. The approach is strictly mechanistic and therefore has the potential to be applied across pest situations and cropping conditions. Extensive simulation studies for single and multiple pests, varying patterns of pest progress, varying crop age, environment, and cropping practices will improve our understanding of pest-loss relations at the field level. Economic values such as yield goal, price per unit of yield, control costs, benefit-cost ratio, and marginal net return must be taken into consideration in the future as well as stochastic elements. Thresholds can be obtained that indicate when actions to control specific pests are justifiable. Risk analyses will certainly play an important role in the future. Least-loss strategies can thus be developed and pesticide application schemes be optimized. Pest-coupled crop models can thus help to improve decision making in IPM, as emphasized by Teng (1988).

Another way in which pest-crop models could be used is to present information from extensive simulation studies in a 'condensed' form. For example, iso-loss curves, least-loss 'look-up' tables, and other means can be produced based on simulating outputs. Nonlinear regression models could be produced based on simulation outputs. Nonlinear regression models could be fit to simulation outputs such as presented in the section entitled Simulation of multiple pest damage effects using CERES-Rice where yield loss, net return or the like will be dependent variables and pest onset time (t_0), maximum pest level (X_{max}), time of control, and so on will be independent variables. For assisting in practical decision making, these regression models could be solved even on a pocket calculator without the potential user having to deal with complicated simulation models any longer.

A few presuppositions however will have to be met before pest-coupled crop models can be widely applied. The damage mechanisms of pests must be known and data must be generated to quantify the relationship between pest level and corresponding effect on crop physiological variables and processes. Interactions between pest damage effects and environment have to be studied. For example, if high temperatures combined with excess solar radiation cause photosynthesis to decline or if drought stress induces leaf senescence, how will pest damage interact with these? Will the interactions occur in an additive, less-than-, or greater-than-additive way? Likewise, interactions among damage effects of multiple pests might need more attention. The structure of a crop model must be sophisticated enough to enable the model to mimic the relevant physiological state and rate variables in the crop model that are compatible with the pest damage effects to be simulated. If a crop model lacks essential features, it will need to be revised. Furthermore, the role of active compensation for pest damage in plants requires elaboration and the crop model structure has to handle this phenomenon appropriately. Pest-coupled crop models, like other models, need to be validated on field data. To ensure transportability of the approach, the field data should also represent a broad range of conditions.

All the above will have to be addressed in future research: in 'micro-level' trials aiming at a qualitative and quantitative understanding of the nature of pest damage effects at the physiological level and in omacro-level' trials that aim at measuring the consequences of pest infestation at the crop level in order to calibrate and validate pest effect simulations (Pinnschmidt and Teng, 1994; Pinnschimdt et al., 1994). Both types of experimentation go hand-in-hand and have to be accompanied by model development. Micro-level trials have a high explanatory value and provide evidence. They yield damage coefficients and functional damage relations to be used for physiologically based pest effect simulation in pest-coupled crop models. However, their results have no practical applicability until they are incorporated into crop models and validated in macro-level trials. On the other hand, macro-level trials rarely yield a thorough understanding of pest effects on crops unless they are supported by micro-level trials. Nevertheless, their data might eventually be used to derive parameter estimates for physiological pest effect modeling by means of optimization techniques (Pinnschmidt et al., 1994) which might lead to projections of pest-induced yield losses similar to those obtained from using parameter estimates of micro-level trials (Pinnschmidt et al., 1995).

Much has been accomplished through the IBSNAT Project in understanding pest effects on crops, and in developing techniques to couple the effects to crop models for simulating yield loss. The actual use of the coupled models and their outputs for decision support, however, is still lagging behind the progress made in modeling. It is this aspect of modeling that deserves increase attention by researchers worldwide as it is likely to have most benefit for improving pest management.

References

Aarts H F M, de Visser C L M (1985) A management information system for weed control in winter wheat. Pages 679–686 in proc British Crop Protection Conference-Weeds, Vol 2, Brighton Metropole, England 18–21 Nov 1985, BCPC Publications, Croydon, England.

Alocilja E C, Ritchie J T (1988) Upland rice simulation and its use in multicriteria optimization. Research series #1, IBSNAT Project, University of Hawaii, Honolulu, Hawaii, USA 95 pages.

Barrentine W L (1974) Common cocklebur competition in soybeans. Weed Science 22:600–603.

Barrentine W L, Oliver L R (1977) Competition threshold levels and control of cocklebur in soybeans. Tech Bull 83. Mississippi Agriculture and Forestry Experiment Station, USA.

Bastions L (1991) The ratio between virtual and visual lesion size as a measure to describe reduction in leaf photosynthesis of rice due to leaf blast. Phytopathology 81(6):611–615.

Batchelor W D, Jones J W, Boote K J, Pinnschmidt H O (1992) Assessing pest and disease damage with DST. Agrotechnology Transfer 16:1–8. University of Hawaii, Honolulu, Hawaii, USA.

Batchelor W D, Jones J W, Boote K J, Pinnschmidt H O (1993) Extending the use of crop models to study pest damage. Transaction of the ASAE (36):551–558. St. Joseph, Michigan, USA.

Baysinger J A, Sims B D (1991) Giant ragweed (*Ambrosia trifida*) interference in soybeans (Glycine max). Weed Science 39:358–362.

Benigno E A, Shephard B M, Rubia E G, Arida G S, Penning de Vries F W T, Bandong J P (1988) Simulation of rice leaffolder population dynamics in lowland rice. IRRI Research Paper Series No. 135, 8 pages. International Rice Research Institute, Los Baæos, Philippines.

Blackshaw R P (1986) Resolving economic decisions for the simultaneous control of two pests, diseases or weeds. Crop Protection 2:93–99.

Bloomberg J R, Kirkpatrick B L, Wax L M (1982) Competition of common cocklebur (*Xanthium pennsylvanicum*) with soybean (*Glycine max*). Weed Science 30:507–513.

Boote K J, Jones J W, Mishoe J W, Berger R D (1983) Coupling pests to crop growth simulators to predict yield reductions. Phytopathology 73:1581–1587.

Boote K J, Jones J W, Mishoe J W, Wilkerson G G (1986) Modeling growth and yield of groundnut. In proceedings international symposium Agrometeorology of groundnut, ICRISAT Sahelian Center, Niamey, Niger. 21–25 August 1985.

Boote K J, Batchelor W D, Jones J W, Pinnschmidt H O, Bourgeois G (1993) Pest damage relations at the field level. Pages 277–296 in Penning de Vries F W T, Teng P S, Metselaar K (eds.)Systems approaches for agricultural development. Kluwer Academic Publishers, Dordrecht, The Netherlands.

Bridges D C, Brecke B J, Barbour J C (1992) Wild poinsettia (*Euphorbia heterophylla*) interference with peanut (*Arachis hypogea*). Weed Science 40:37–42.

Campbell C L, Madden L V (1990) Introduction to plant disease epidemiology. John Wiley and Sons, New York, USA 532 pp.

Coble H D (1986) Development and implementation of economic thresholds for soybean. Pages 295–307 in Frisbie R E, Adkisson P L (eds) CIPM: Integrated pest management on major agricultural systems. Texas A&M University, College Station, Texas, USA.

Cousens R D (1992) Weed competition and interference in cropping systems. Pages 113–117 in proceedings of the First International Weed Control Congress, Melbourne, Australia. Volume 1, Weed Science Society of Victoria.

Dent J B (1978) Crop disease management: a system research approach. Pages 30–1 to 30–10 in R C Close (Ed.) Epidemiology and crop loss assessment, Lincoln College, University of Canterbury, New Zealand.

Fellows G M, Roeth F W (1992) Shattercane (*Sorghum bicolor*) interference in soybean (*Glycine max*). Weed Science 40:68–73.

Fieberg W W, Shilling D G, Knauft D A (1991) Peanut genotype response to interference from common cocklebur. Crop Science 31:1289–1292.

Frazier B D, Gilbert N (1976) Cocinellids and aphids: A quantitative study of the impact of adult lady birds (*Coleoptera: Coccinellidae*) preying on field populations of pea aphids (*Homoptera: Aphididae*). Journal of the Entomological Society of British Columbia, 77:33–56.

Gaunt R E (1990) Empirical disease-yield loss models. Pages 185–192 in Crop loss assessment in rice. IRRI, Los Baæos, Laguna, Philippines.

Geddes R D, Scott H D, Oliver L R (1979) Growth and water use by common cocklebur (*Xanthium pennsylvanicum*) and soybeans (*Glycine max*) under field conditions. Weed Science 27:206–212.

Graf B, Lamb R, Heong K L, Fabellar L (1991) A simulation model for the population dynamics of rice leaffolders (*Lepidoptera*: *Pyralidae*) and their interactions with rice. Journal of Applied Ecology 29:558–570.

Graf B, Gutierrez A P, Rakotobe O, Zahner P, Delucchi V (1990) A simulation model for the dynamics of rice growth and development: Part II – The competition with weeds for nitrogen and light. Agricultural Systems 32:367–392.

Gunsolus J L (1986) Reciprocal interference effects between weeds and soybeans (*Glycine max*) measured by area of influence methodology. Ph.D. dissertation, North Carolina State University, Raleigh, North Carolina, USA. (Diss. Abstr. DA 8608058).

Gutierrez A P, Falcon L A, Loew W B, Leipzig P A, Van den Bosch R (1975) An analysis of cotton production in California: A model of Acala cotton and the effect of defoliators on its yield. Envionmental Entomology 4:125–136.

Gutierrez A P, Butler G D, Wang Y, Westphal D (1977) The interaction of pink bollworm (*Lepidoptera*: *Gelichidae*) cotton and weather: a detailed model. Canadian Entomology 109:1457–1468.

Gutierrez A P, De Vay J E, Pullman G S, Friebertshaeuser G E (1983) A model of *Verticillium* wilt in relation to cotton growth and development. Phytopathology 73:80–95.

Jones J W (1993) Decision support systems for agricultural development. Pages 459–471 in Penning de Vries F W T, Teng P S, and Metselaar K (Eds.) Systems Approaches for Agricultural Development, Kluwer Academic Publishers, Dordrecht, The Netherlands.

Jones J W, Boote K J, Jagtap S S, Hoogenboom G, Wilkerson G G (1988) SOYGRO V5.41: Soybean crop growth simulation model user's guide. Agricultural Engineering Department and Agronomy Department, University of Florida, Gainesville, USA. Florida Experiment Station Journal No. 8304.

Keisling T C, Oliver L R, Crowley R H, Baldwin F L (1984) Potential use of response surface analyses for weed management in soybeans (*Glycine max*). Weed Science 32:552–557.

Kiniry J R, Williams J R, Gassman P W, Debaeke P (1992) A general, process-oriented model for two competing plant species. Trans. ASAE 35:801–810.

Kropff M J (1988) Modeling the effect of weeds on crop production. Weed Research 28:465–471.

Kropff M J, Lotz L A P (1992) Optimization of weed management systems: The role of ecological models of interplant competition. Weed Technology 6:462–470.

Kropff M J, Spitters C J T, Schnieders B J, Joenje W, de Groot W (1992) An eco-physiological model for interspecific competition, applied to the influence of *Chenopodium album L.* on sugar beet. II. Model evaluation. Weed Research 32:451–463.

Legere A, Schreiber M M (1989) Competition and canopy architecture as affected by soybean (*Glycine max*) row width and density of redroot pigweed (*Amaranthus retroflexus*). Weed Science 37:84–92.

Luo Y, Teng P S, Fabellar N, TeBeest D O (1997) A rice-leaf blast combined model for simulation of epidemics and yield loss. Agricultural Systems 53: 27–39.

Marshall E J P (1988) Field-scale estimates of grass weed population in arable land. Weed Research 28:191–198.
Marwat K B (1988) Interference of common cocklebur (*Xanthium strumarium L.*) and velvetleaf (*Abutilon theophrasti medic.*) with soybean strands at different densities and planting patterns. Ph.D. dissertation. Univ of Illinois, Urbana-Champaign, Illinois, USA.
McWhorter C G, Hartwig E E (1972) Competition of johnsongrass and cocklebur with six soybean varieties. Weed Science 20.56–59.
Monks D W, Oliver L R (1988) Interactions between soybean (*Glycine max*) cultivars and selected weeds. Weed Science 36:770–774.
Mortensen D A, Coble H D (1989) The influence of soil water content on common cocklebur (*Xanthium strumarium*) interference in soybeans (*Glycine max*). Weed Science 37:76–83.
Mortensen D A, Coble H D (1991) Two approaches to weed control decision-aid software. Weed Technology 5:445–452.
Mortensen D A, Johnson G A, Young L J (1993) Weed distribution in agricultural fields. Pages 113–124 in Robert P, Rust R H (Eds.) Soil specific crop management. American Society of Agronomy, Inc., Madison, Wisconsin, USA.
Murphy T R, Gossett B J (1981) Influence of shading by soybeans (*Glycine max*) on weed suppression. Weed Science 29:610–615.
Norman J M (1979) Modeling the complete crop canopy. Pages 249–277 in Barfield B J, Gerber J F (Eds.) Modification of the aerial environment of crops. ASAE Monograph No. 2, St. Joseph, Michigan, USA.
Oliver L R (1979) Influence of soybean (*Glycine max*) planting date on velvetleaf (*Abutilon theophrasti*) competition. Weed Science 27:183–188.
Palis F, Pingali P L, Litsinger J A (1990) A multiple-pest economic threshold for rice (a case study in the Philippines) Pages 229–242 in Crop loss assessment in rice. IRRI, Los Baños, Laguna, Philippines.
Pereira A R, Shaw R H (1980) A numerical experiment on the mean wind structure inside canopies of vegetation. Agricultural Meteorology 22:308–318.
Pinnschmidt H O (1997) Simulation of host pathogen dynamics, crop loss, and effects of fungicide applications. Chapter 33 in Francl L J, Neher D A (Eds.) Exercises in plant disease epidemiology. APS Press, St. Paul, Minnesota, USA.
Pinnschmidt H O, Batchelor W D, Teng P S (1995) Simulation of multiple species pest damage on rice. Agricultural Systems 48:193–222.
Pinnschmidt H O, Luo Y, Teng P S (1994) Methodology for quantifying rice yield effects of blast. Pages 318–408 in Ziegler R S, Leong S A, Teng P S (Eds.) Rice blast disease. CAB International, Wallingford Oxon OX10 8DE, UK
Pinnschmidt H O, Teng P S, Yuen J E (1990a) Pest effects on crop growth and yield. Pages 26–29 in Teng P S, Yuen J E (Eds.) Proceedings of the workshop on modeling pest-crop interactions, University of Hawaii at Manoa, Honolulu, Hawaii, January 7–10, 1990. IBSNAT Project Univ of Hawaii, Honolulu, USA.
Pinnschmidt H O, Teng P S, Yuen J E, Djurle A (1990b) Coupling pest effects to the IBSNAT CERES crop model for rice. Paper presented at the APS meeting in Grand Rapids, Michigan, August 1990. Phytopathology 80:997.
Pinnschmidt H O, Teng P S (1994) Advances in modelling multiple insect-disease-weed effects on rice and implications for research. Pages 381–408 in Teng P S, Heong K L, Moody K (Eds.) Rice pest science and management. April 1992, IRRI, Los Baæos, Laguna, Philippines.
Qasem J R (1992) Pigweed (Amaranthus spp.) interference in transplanted tomato (*Lycopersicon esculentum*). Journal of Horticultural Science 67:421–427.
Rabbinge R, Carter N (1993) Application of simulation methods in the epidemiology of pests and diseases: an introductory review. SROP WPRS Bulletin 6(2):17–30.
Rabbinge R, Ward S A, van Laar H H (Eds.) (1989) Simulation and systems management in crop protection. CABO Simulation Monographs 32. PUDOC, Wageningen, The Netherlands, 420 pp.
Rimmington G M (1984) A model of the effect of interspecies competition for light on dry-matter production. Australian Journal Plant Physiology 11:277–286.
Rouse D I (1988) Use of crop growth-models to predict the effects of disease. Annual Review of Phytopathology 26:183–201.
Shepard B M, Ferrer E R (1990) Sampling insects and diseases in rice. Pages 107–130 in Crop loss assessment in rice. IRRI, Los Baæos, Laguna, Philippines.

Spitters C J T (1989) Weeds: population dynamics, germination and competition. Pages 182–216 in Rabbinge R, Ward S A, and van Laar H H (eds). Simulation and systems management in crop protection. PUDOC, Wageningen, The Netherlands.

Spitters C J T, Aerts R (1983) Simulation of competition for light and water in crop-weed associations. Aspects of Applied Biology 4:467–481.

Stoller E W, Harrison S K, Wax L M, Regnier E E, Nafziger E D (1987) Weed interference in soybeans (*Glycine max*). Review of Weed Science 3:155–181.

Teng P S (1985) A comparison of simulation approaches to epidemic modeling. Annual Review of Phytopathology 23:351–379.

Teng P S (1987) Crop loss assessment and pest management. American Phytopathological Society, St. Paul, Minnesota, USA. 270 pp.

Teng P S (1988) Pests and pest-loss models. Agrotechnology Transfer 8:1, 5–10, IBSNAT Project, University of Hawaii, Honolulu, Hawaii, USA.

Teng P S (1994a) Integrated pest management in rice. Experimental Agriculture 30:115–137.

Teng P S (1994b) The epidemiological basis for blast management. Pages 408–433 in Zeigler R S, Leung S, Teng P S (Eds.) Rice blast disease. CAB International, Wallingford Oxon OX10 8DE, UK.

Teng P S and Rouse D I (1984) Understanding computers- applications in plant pathology. Plant Disease 68:539–543.

Teng P S and Savary S (1992) Implementing the systems approach in pest management. Agricultural Systems 40: 237–264.

Thornton P K, Fawcett R H, Dent J B and Perkins T J (1990) Spatial weed distribution and economic thresholds for weed control. Crop Protection 9:337–342.

Waggoner P E (1986) Progress curves of foliar diseases: their interpretation and use. Pages 3–37 in Leonard K J and Fry W E (eds) Plant disease epidemiology-population dynamics and management, Volume 1. MacMillan Publishing Co, New York, USA.

Walker P T (1990) Quantifying insect populations and crop damage. Pages 55–66 in Crop loss assessment in rice. IRRI, Los Baæos, Laguna, Philippines.

Wax L M and Pendleton J W (1968) Effect of row spacing on weed control in soybeans. Weed Science 16:462–465.

Wiles L J and Wilkerson G G (1991) Modeling competition for light between soybean and broadleaf weeds. Agricultural Systems 35:37–51.

Wiles L J, Oliver G W, York A C, Gold H J, and Wilkerson G G (1992) Spatial distribution of broadleaf weeds in North Carolina soybean (*Glycine max*) fields. Weed Science 40:554–557.

Wilkerson G G, Jones J W, Coble H D, and Gunsolus J L (1990) SOYWEED: A simulation model of soybean and common cocklebur growth and competition. Agronomy Journal 82:1003–1010.

Wilkerson G G, Jones J W, Boote K J, Ingram K T, and Mishoe J W (1983) Modeling soybean growth for crop management. Transactions of the American Society of Agricultural Engineers 26:63–73.

Wilkerson G G, Modena S A, and Coble H D (1991) HERB: Decision model for postemergence weed control in soybean. Agronomy Journal 83:413–417.

Williams J R, Jones C A, and Dyke P T (1984) A modeling approach to determining the relationship between erosion and soil productivity. Transactions of the American Society of Agricultural Engineers 27:129–144.

Zadoks J C (1985) On the conceptual basis of crop loss assessment: The threshold theory. Annual Review of Phytopathology 23:455–473.

Zimdahl R L (1980) Weed-crop competition: A review. International Plant Protection Center, Corvallis, Oregon, USA. 196 pp.

The use of crop models for international climate change impact assessment*

C. ROSENZWEIG[1] and A. IGLESIAS[2]

[1] *Goddard Institute for Space Studies and Columbia University, 2880 Broadway, New York, New York 10025, USA*

[2] *Centro de Investigación Forestal – Instituto Nacional de Investigaciones Agrarias (INIA), Madrid, Spain*

Key words: climate change, world food supply, crop model, adaptation

Abstract

The methodology for an assessment of potential impacts of climate change on world crop production, including quantitative estimates of yield and water-use changes for major crops, is described. Agricultural scientists in 18 countries estimated potential changes in crop growth and water use using compatible crop models and consistent climate change scenarios. The crops modeled were wheat, rice, maize and soybean. Site-specific estimates of yield changes for the major crops modeled were aggregated to national levels for use in a world food trade model, the Basic Linked System. The study assessed the implications of climate change for world crop yields for arbitrary and GCM equilibrium and transient climate change scenarios. The climate change scenarios were tested with and without the direct physiological effects of CO_2 on crop growth and water use, as reported in experimental literature. Climate change impacts on crop yields incorporating farm-level adaptation were also simulated, based on different assumptions about shifts in crop planting dates, changes in crop variety, and level of irrigation.

Introduction

Scientists predict significant global warming in the coming decades arising from increasing atmospheric carbon dioxide and other trace gases (IPCC, 1990a; 1992; 1996a). Substantial changes in hydrological regimes are also forecast to occur. Understanding the potential effects of these changes on agriculture is an important task, because agriculture provides food for the world's population, now estimated at 5.8 billion and projected to rise to over 9 billion in the coming century (UN, 1996). Despite technological advances such as improved crop varieties and irrigation systems, weather and climate are still key factors in agricultural productivity. For example, weak monsoon rains in 1987 caused large shortfalls in crop production in India, Bangladesh, and Pakistan, contributing to reversion to wheat importation by these countries

*This chapter is based on Rosenzweig and Iglesias (1994), The use of crop models for international climate change impact assessment: study design, methodology, and conclusions, in Rosenzweig and Iglesias (Eds.), Implications of Climate Change for International Agriculture: Crop Modeling Study. US Environmental Protection Agency, EPA 230-B-94-003. Washington, DC.

G.Y. Tsuji et al. *(eds): Understanding Options for Agricultural Production, 267–292*

(World Food Institute, 1988). Despite adequate supplies elsewhere, the 1980s also saw the continuing deterioration of food production in Africa, caused in part by persistent drought and low production potential. This resulted in international relief efforts to prevent widespread famine. These examples emphasize the close links between agriculture and climate, the international nature of food trade and food security, and the need to consider the impacts of climate change in a global context.

Early research focused on regional and national assessments of the potential effects of climate change on agriculture (IPCC, 1990b). The methodology for regional and national climate impact studies was thus developed and tested. However, the studies, for the most part, treated each region or nation in isolation, without relation to changes in production in other places. Simultaneous changes in production in all major food-producing regions may lead to altered world supply and demand (and prices), and hence competitiveness in any given region. Understanding potential global changes should help in the meaningful interpretation of regional climate change impact studies.

This chapter describes the crop modeling activities of an international study on potential agricultural impacts of climate change. The study involved the collaborative efforts of agricultural scientists in 18 countries. Modeling tools included dynamic process crop models, climate change scenarios, and a world food trade model (Figure 1). The climate change scenarios consisted of climate sensitivity tests and scenarios based on global climate models (GCMs, also known as general circulation models). Common methodology was developed for the simulation of climate change impacts on major agricultural crops and for the analysis of the results. The International Benchmark Sites Network for Agrotechnology Transfer (IBSNAT, 1989) compatible crop growth models were utilized at over 100 sites (Figure 2). The crop models simulate the direct physiological effects of increased atmospheric CO_2 on crop growth and water use. They allow for the simulation of both rainfed and irrigated agricultural systems and other potential farmer adaptations to climate change. Results of the dynamic crop growth simulations were then used to estimate global changes in crop yields for use in a world food trade model, the Basic Linked System (Fischer et al., 1988).

In this chapter, we describe the methodology and results of the crop modeling portion of the international study. Results of the entire study, including the economic and food security analysis based on the Basic Linked System simulations, are described in Rosenzweig and Parry (1994).

Background and previous studies

Research on the effects of climate on agriculture has been extensive for many years. Much of the previous research on climate impact assessment sought to isolate the effects of climate on agricultural activity, whereas lately there has

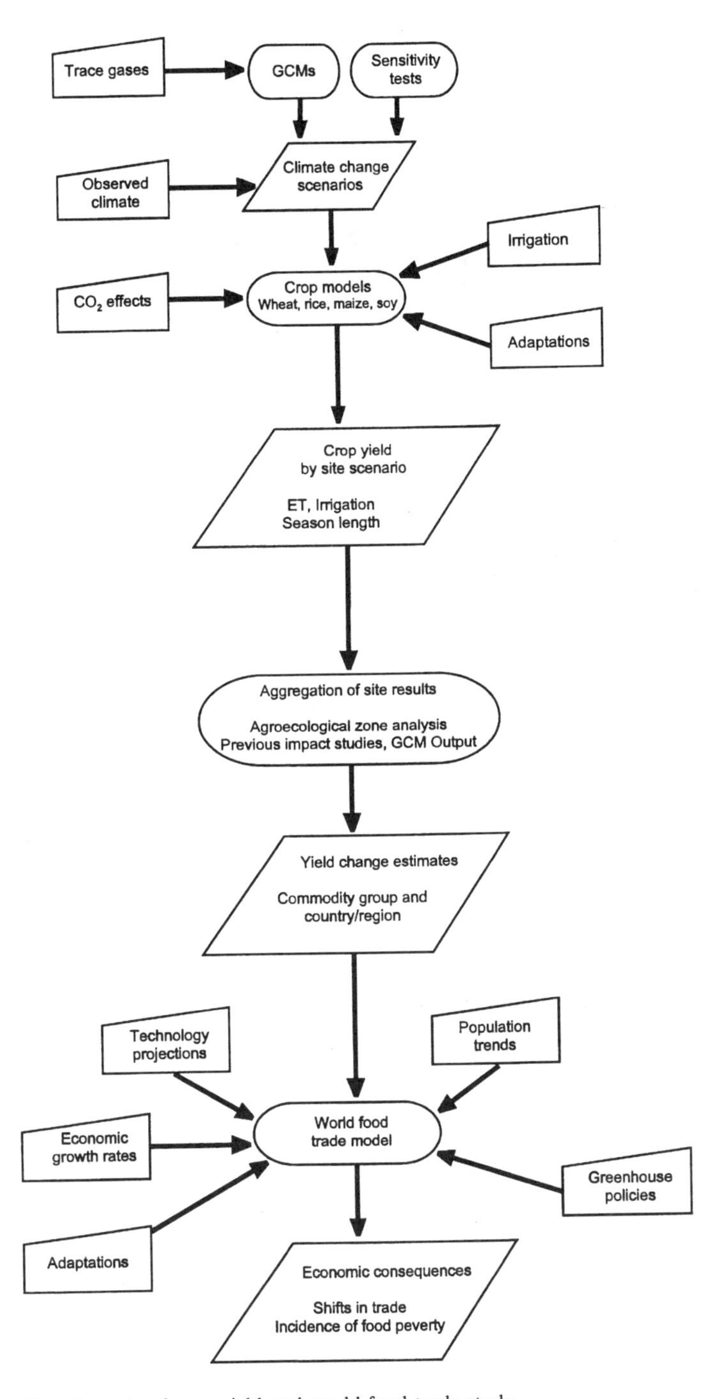

Figure 1. Key elements of crop yield and world food trade study.

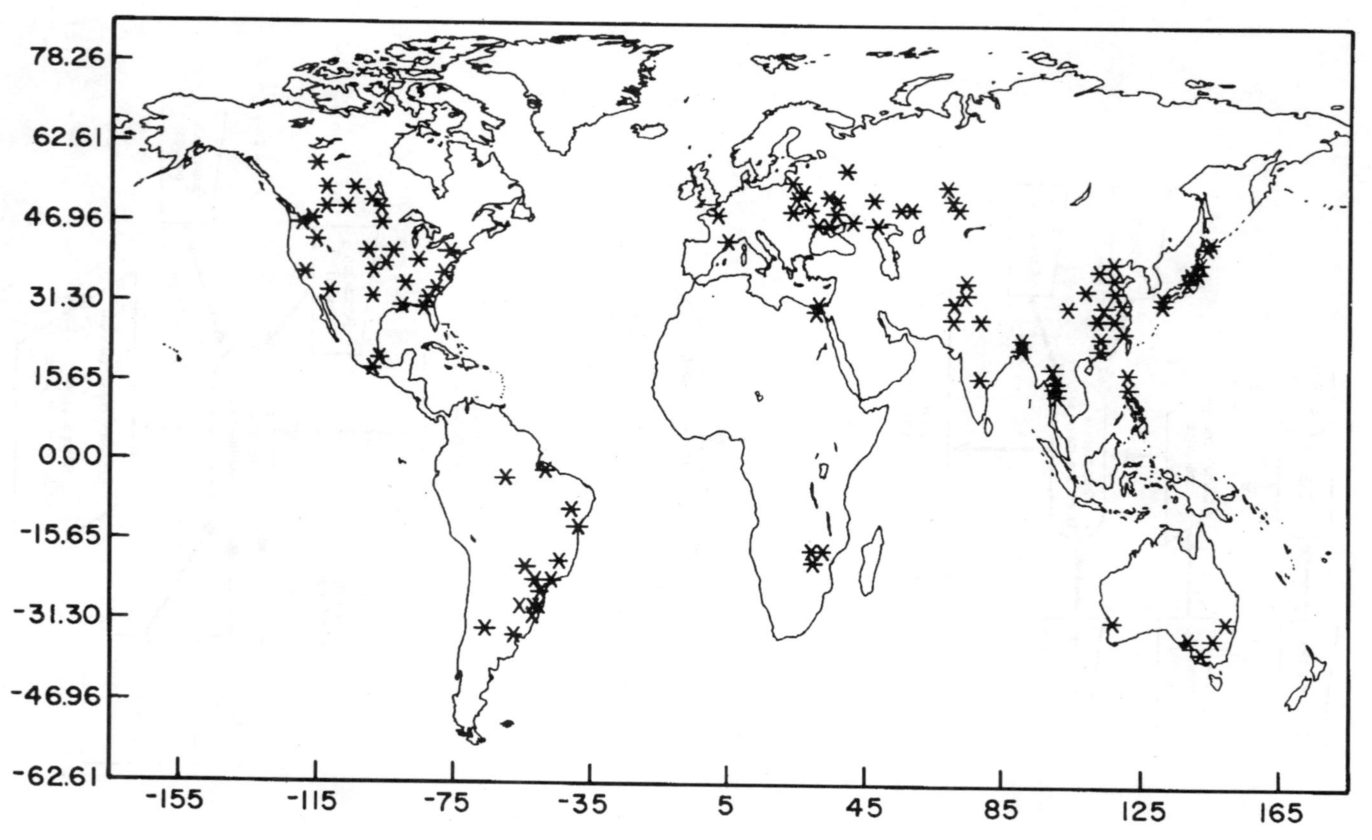

Figure 2. Crop model sites.

been a growing emphasis on understanding the interactions of climatic, environmental, and social factors in a wider context (Parry et al., 1988). An example of the earlier approach is the US Department of Transportation's study on the possible effects of atmospheric ozone depletion which (in part) used regression models to determine statistical relationships between climate and agriculture, a method which does not explain the processes underlying the relationships. The National Defense University also studied long-term effects of climate change on crop yields and agricultural production with a relatively simple cause-and-effect approach (NDU, 1980).

More integrated approaches to climate change impact assessments have been described by Callaway et al. (1982), Nix (1985), and Warrick et al. (1986). These studies and reviews advocate comprehensive research methods that integrate economic and political factors as well as biophysical ones. Some integration has been achieved in national impact studies completed in the United States (Adams et al., 1990; Smith and Tirpak, 1989), Canada (Smit, 1989), Australia (Pearman, 1988), the UK (UK Department of the Environment, 1991), and New Zealand (Martin et al., 1990), and in regional studies including high-latitude and semi-arid agricultural areas (Parry et al., 1988) and the U.S. Midwest (Rosenberg and Crosson, 1991).

Regional, national, and global studies have been summarized in the Intergovernmental Panel on Climate Change (IPCC) Working Group II Reports (IPCC, 1990b; 1996b). On the global scale, Kane et al. (1991) have analyzed the sensitivity of world agriculture to potential climate changes and found that the overall effect of moderate climate change on world and domestic economies may be small, because reduced production in some areas is balanced by gains in others. Leemans and Solomon (1993) found that climate change as depicted by one GCM for doubled levels of atmospheric CO_2 would affect the yield and distribution of world crops, leading to production increases at high latitudes and production decreases in low latitudes.

Climate change scenarios

Because of the uncertainties surrounding prediction of climate change, it is common to employ climate scenarios (Wigley, 1987; Lamb, 1987) to estimate the impacts of climate change on a system (in this case agricultural production, both potential and actual, and world food trade). Climate scenarios are sets of climatic perturbations which are used with impact models to test the sensitivity of the system to the projected changes. The design of an impact study should include more than one scenario, so that a range of possible effects may be defined. Realism is increased when the climate change scenarios are internally consistent, i.e., the climate variables within the scenario should vary in a physically realistic way (Wigley, 1987). For the international crop modeling study, a set of doubled CO_2 GCM climate change scenarios and climate sensitivity tests was selected for use by all participants.

Table 1. GCM climate change scenarios.

GCM	Year*	Resolution**	Change in average global	
			Temperature	Precipitation
GISS[1]	1982	7.83° × 10°	+4.2°C	+11%
GFDL[2]	1988	4.4° × 7.5°	+4.0°C	+8%
UKMO3	1986	5.0° × 7.5°	+5.2°C	+15%

*When calculated.
**Latitude × longitude.
[1]Hansen et al,. 1983.
[2]Manabe and Wetherald, 1987.
[3]Wilson and Mitchell, 1987.

Sensitivity tests

One approach to analyze the possible impacts of different climate on crop yield is to specify incremental changes to temperature and precipitation and to apply these changes uniformly to the baseline climate. Arbitrary climate sensitivity tests were conducted to test crop model responses to a range of temperature (+2°C and +4°C) and precipitation changes (±20%). (While GCMs project precipitation increases on the global scale associated with warmer temperatures, precipitation decreases are projected in some locations.) Sensitivity studies allow the consideration of the question: 'What type, magnitude, and rate of climate change would seriously perturb the agricultural system in question?'

GCM equilibrium and transient scenarios

Scenarios are often devised by changing an original set of climatological data by prescribed anomalies. These anomalies may be derived from historical climate or from global climate models. GCMs provide the most advanced tool for predicting the potential climatic consequences of increasing radiatively active trace gases in a consistent manner. However, climate models have not yet been validated to project changes in climate variability, such as changes in the frequencies of drought and storms, even though these could affect crop yields significantly.

Mean annual changes in climate variables from doubled CO_2 simulations of three GCMs from the Goddard Institute for Space Studies (GISS, Hansen et al., 1983), the Geophysical Fluid Dynamics Laboratory (GFDL, Manabe and Wetherald, 1987), and the United Kingdom Meteorological Office (UKMO, Wilson and Mitchell, 1987), were applied to observed (baseline) daily climate to create climate change scenarios for each site (Table 1). GCMs were used to create climate change scenarios because they produce climate variables that are internally consistent; thus they allow for comparisons between or among regions.

The method used the differences between $1 \times CO_2$ and $2 \times CO_2$ monthly GCM temperatures, and the ratio between $2 \times CO_2$ and $1 \times CO_2$ monthly precipitation and solar radiation amounts ($1 \times CO_2$ refers to modeled current climate conditions and $2 \times CO_2$ refers to the modeled climate that would occur with an equivalent radiative forcing of doubled CO_2 in the atmosphere). Temperature ranges of the three GCMs are near the high end of the IPCC range of predicted warming for doubled CO_2 (IPCC, 1990a). In general, the GCMs predict increases in global precipitation associated with warming because warmer air can hold more water vapor.

Most of our knowledge concerning the climate response to greenhouse-gas forcing has been obtained from equilibrium response GCM experiments. These are experiments which consider the steady-state response of the model's climate to step-function changes in atmospheric CO_2. Recent evidence from GCM experiments incorporating time-dependent greenhouse-gas forcing suggest that there may be important differences between the equilibrium and transient responses, based in part on differential warming of land and ocean areas (Hansen et al., 1988; IPCC, 1992, 1996a).

This study also considered a set of transient climate scenarios (as opposed to the atmospheric equilibrium scenarios), derived from the GISS transient climatic simulations (Hansen et al., 1988) for the 2010s, 2030s, and 2050s, and assuming CO_2 concentrations of 405, 460, and 530 ppm, respectively. Transient scenarios for each site were developed using the same procedure as that used for the equilibrium scenarios.

Limitations

Current global climate models which have been used for CO_2 studies employ grids on the order of 4° latitude by 5° longitude, or greater. At this resolution, many smaller-scale elements of climate are not properly represented, such as warm and cold fronts and hurricanes, as well as the diversities of ecosystems and land use. Accurate modeling of hydrological processes is particularly crucial for determining climate change impacts on agriculture, but GCM simulation of infiltration, runoff, and evaporation, and other hydrological processes is highly simplified. Precipitation, in particular, is sometimes poorly simulated in GCMs.

There is also uncertainty in the prediction of the rate and magnitude of climate change. Ocean heat transport is a key but not well understood process that affects how fast the climate may warm. The doubled CO_2 climate change scenarios used in this study have assumed an abrupt doubling of the CO_2 concentration in the atmosphere and then have allowed the simulated climate to come to a new equilibrium. This step change in the atmosphere is unrealistic, since trace gases are increasing gradually. The transient scenario does simulate the response of gradually increasing radiatively active gases and is more realistic in this respect.

Table 2. Current world crop yield, area, production, and percent world production aggregated for countries participating in study (FAO, 1988).

Crop	Yield t ha^{-1}	Area ha × 1000	Production t × 1000	Study countries %
Wheat	2.1	230,839	481,811	73
Rice	3.0	143,603	431,585	48
Maize	3.5	127,393	449,364	71
Soybeans	1.8	51,357	91,887	76

Crop models

The crop modeling study estimated how climate change and increasing levels of carbon dioxide may alter yields and water use of world crops in both major production areas and vulnerable regions. The crops modeled were wheat, rice, maize, and soybeans. Table 2 shows the percentages of world production modeled in this study for these crops. Even though only two countries (Brazil and USA) simulated soybean production, their combined output accounts for 76% of world total. Less of the total world rice production was simulated than total production of the other crops. This is because India, Indonesia, and Vietnam have significant rice production not included in the study. Together, these crops account for more than 85% of the world traded grains and legumes, although only approximately 4% of rice produced is traded compared with about 20% of wheat. Rice is included in the study because of its importance to the food security of Asia.

The IBSNAT models employ simplified functions to predict the growth of crops as influenced by the major factors that affect yield, i.e., genetics, climate (daily solar radiation, maximum and minimum temperatures, and precipitation), soils, and management (IBSNAT, 1989). The models used were CERES-Wheat (Ritchie and Otter, 1985; Godwin et al., 1989), CERES-Maize (Jones and Kiniry, 1986; Ritchie et al., 1989), CERES-Rice (both paddy and upland) (Singh et al. 1993), and SOYGRO (soybean) (Jones et al., 1989).

The IBSNAT models were selected for use in this study because they have been validated over a wide range of environments (e.g., Otter-Nβcke et al., 1986) and are not specific to any particular location or soil type. They are thus suitable for use in international studies in which crop growing conditions differ greatly. The validation of the CERES and SOYGRO models over different environments also serves to enhance predictive capability concerning the climate change scenarios, in cases when predicted climates are similar to existing climates in other regions. Furthermore, because management practices, such as cultivar, planting date, plant population, row spacing, and sowing depth, may be varied in the models, they permit experiments that simulate management adjustments by farmers to climate change.

Modeled processes include phenological development, i.e., duration of growth stages, growth of vegetative and reproductive plant parts, extension growth of leaves and stems, senescence (aging) of leaves, biomass production and parti-

Table 3. Photosynthetic ratios and stomatal resistances used to simulate direct physiological CO_2 effects in the IBSNAT models (555 ppm CO_2/330 ppm CO_2).

	Photosynthesis ratio*	Stomatal resistance**, s m^{-1}
Soybean	1.21	49.7/34.4
Wheat	1.17	49.7/34.4
Rice	1.17	49.7/34.4
Maize	1.06	87.4/55.8

*Based on experimental work reviewed by Cure (1985).
**Based on experimental work by Rogers et al. (1983).

tioning among plant parts, and root system dynamics. The CERES and SOYGRO models also have the capability to simulate the effects of nitrogen deficiency and soil-water deficit on photosynthesis and pathways of carbohydrate movement in the plant.

Physiological CO_2 effects

Most plants growing in atmospheric CO_2 higher than ambient levels exhibit increased rates of net photosynthesis (i.e., total photosynthesis minus respiration). High CO_2 also reduces the stomatal openings of some crop plants. By so doing, CO_2 reduces transpiration per unit leaf area while enhancing photosynthesis. Thus it often improves water-use efficiency (the ratio of crop biomass accumulation or yield and the amount of water used in evapotranspiration). Experimental effects of CO_2 on crops have been reviewed by Acock and Allen (1985) and Cure (1985). In a compilation of greenhouse and other experimental studies, Kimball (1983) estimated a mean crop yield increase of $33 \pm 6\%$ for a doubling of CO_2 concentration from 300 to 600 ppm for a range of important agricultural crops.

The IBSNAT models have been modified to simulate the changes in photosynthesis and evapotranspiration caused by higher levels of CO_2. These modifications, based on methods derived from Peart et al. (1989), were used in the crop yield – climate change scenario modeling to study the relative magnitudes of the direct physiological and the climatic effects of increased CO_2. Ratios were calculated between measured daily photosynthesis and evapotranspiration rates for a canopy exposed to a range of high CO_2 values, based on published experimental results (Allen et al., 1987; Cure and Acock, 1986; and Kimball, 1983). Instantaneous midday values were then modified to give daily integrated increases, allowing for lower light intensities in morning and evening. In the crop models, the photosynthesis ratios (Table 3) were applied to the maximum amount of daily carbohydrate production which is based on incoming solar radiation.

To account for the effect of elevated carbon dioxide on stomatal closure and increased leaf area index, and hence on potential transpiration, the evapotranspiration formulation of the IBSNAT models was changed to include the ratio of transpiration under elevated CO_2 conditions to that under ambient conditions. To derive the ratio, Peart et al. (1989) applied the Penman-Monteith equation as written in France and Thornley (1984) to the same canopy and environment, except for differing CO_2 concentrations. The leaf resistances were calculated as a function of the differing CO_2 concentrations using equations developed by Rogers et al. (1983) based on experimental data for maize and soybean (used for all C_3 crops) (Table 3). The ratio procedure results in a lower transpiration rate for higher CO_2 levels on a daily per unit leaf area basis. Seasonal evapotranspiration, however, may not change proportionately, and may even increase, because of the greater leaf area grown under elevated CO_2 conditions.

The simulation of direct CO_2 effects for soybeans, wheat, and maize under current climate conditions have been compared with experimental results (Peart et al., 1989; Jones and Allen, personal communication; Rosenzweig, 1990). The wheat and soybean results compare well with experimental results, but maize simulations tended to overestimate yield increases due to high CO_2 at sites with low annual precipitation.

For the crop modeling study, CO_2 concentrations were estimated to be 555 ppm in 2060 (based on Hansen et al., 1988). Because greenhouse gases other than CO_2, such as methane (CH_4), nitrous oxide (N2O), and the chlorofluorocarbons (CFCs), are also increasing, an 'effective CO_2 doubling' has been defined as the combined radiative forcing of all greenhouse gases having the same forcing as doubled CO_2 (usually defined as 600 ppm). The effective CO_2 doubling will occur around the year 2030, if current emission trends continue. The climate change caused by an effective doubling of CO_2 may be delayed by 30 to 40 years or longer.

Limitations

The IBSNAT models contain many simple, empirically-derived relationships that do not completely mimic actual plant processes. These relationships may or may not hold under differing climatic conditions, particularly the higher temperatures predicted for global warming. For example, most of the data used to derive the relationships in the crop models were obtained with temperatures below 35 C whereas the projected temperatures for doubled CO_2 are often 35 or even 40 C during the growing period. Other simplifications of the crop models are that weeds, diseases, and insect pests are controlled; there are no problem soil conditions such as high salinity or acidity; and there are no catastrophic weather events such as heavy storms. The crop models simulate the current range of agricultural technologies available around the world; they do not include potential improvements in such technology, but may be used to test the effects of some potential improvements, such as improved varieties and irrigation schedules.

Calibration and validation

The IBSNAT crop models have been created with the express purpose of broad applicability across a wide range of environments. Individual investigators in the 18 countries participating in the climate change impact study calibrated and validated the IBSNAT crop models using local experimental data, where possible. Where such procedures were not possible, previous calibrations of cultivars based on IBSNAT minimum data set methods were relied on, as well as previous validations.

Genetic coefficients for different crop genotypes were estimated from data gathered at local agricultural experimental stations. Each parameter was calibrated directly from observed results (phase duration, biometric ratio, and growth rates) in order to obtain rough parameter values. These values were then used in model runs and adjusted in order to obtain a best fit to observed data. Overall, the results of validation were acceptable for the experiments conducted.

Crop modeling procedures

The participating agricultural scientists carried out a set of crop modeling simulation experiments for baseline climate, GCM doubled CO_2 and transient climate change scenarios with and without the physiological effects of CO_2 and sensitivity tests. This involved the following tasks:

(1) Definition of the geographical boundaries of the major production regions of the country, and estimation of the current production of major crops in those regions.
(2) Provision of observed climate data for representative sites within these regions for the baseline period (1951–1980), or for as many years of daily data as were available, and specification of the soil, crop, and management inputs necessary to run the crop models at the selected sites.
(3) Validation of the crop models with experimental data from field trials.
(4) Simulation experiments with the crop models utilizing baseline data and climate change scenarios, with and without the direct effects of CO_2 on crop growth, with irrigated production, sensitivity tests, and adaptation responses. Adaptation response simulations included shifts in planting date and crop varieties. Modeled yield changes and other results, such as changes in growing season length arising from climate change, were then reported.
(5) Identification and evaluation of alterations in agricultural practices that would lessen any adverse consequences of climate change.

The following sections describe the estimation procedures used to aggregate site results to national level yield changes, the design of the adaptation analysis, and the regional and global results of the crop model simulations.

Estimation of national yield changes

The crop modeling results from the procedures described above were used to estimate potential changes in national crop yields for 117 countries. This was done first for the crops and countries in the crop modeling study. These results were then extended to other crops based on agronomic characteristics, and to other countries and regions based on similarities in agroecological zones, previous climate change impact studies, and on comparison of GCM climate change scenarios for a full complement of global estimates of potential impacts of climate change on crop yields.

Crop model results from over 100 sites in the 18 countries were aggregated by weighting current regional production to produce national yield change estimates. This is an intermediate step that permits the extrapolation of results from crop modeling experiments at individual sites to national yield changes. The agricultural scientists in each country selected sites representative of major agricultural regions, described the agricultural practices of the regions, and provided regional and national production data for estimation of regional contributions to the national yield changes. All the crop modeling aggregation results were either calculated by the agricultural scientists themselves or developed jointly with them.

In the most complete national studies, enough sites were modeled to represent all the major agroecological regions. In other cases, modelers were asked to analyze the sites modeled and the regions in their country in order to extend the results as appropriately as possible. The regional yield estimates represent the current mix of rainfed and irrigated acreage, the current crop varieties, nitrogen management and soils, as provided by the country participants. In most cases only one crop variety and one soil were modeled at each site.

A database was created containing current regional and national production data for the 18 countries, production data being provided primarily by the participants and also by FAO (1988). For the USA, the data source was the US Department of Agriculture (USDA) Crop Production Statistical Division; for the former Soviet Union the data source was the USDA International Service.

Results were aggregated similarly for the 11 countries where wheat was modeled (Table 4). There are large differences among national results; for example, Brazil, Egypt, Pakistan, and Uruguay show significant decreases in wheat yields even with the direct effects of CO_2, while simulated wheat yields in Canada, and the USSR primarily increase under the climate change scenarios with the direct effects of CO_2. The other crops were aggregated using the same methodology.

The crop yield estimates incorporate some major improvements: (1) consistent crop simulation methodology and climate change scenarios; (2) weighting of model site results by contribution to regional and national, and rainfed and irrigated production; and (3) a quantitative foundation for estimation of physio-

Table 4. Current production and changes in simulated wheat yields under GCM 2 × CO_2 climate change scenarios, with and without the direct effects of CO_2[1].

Country	Current production				Change in simulated yields					
	Yield t ha^{-1}	Area Mha	Production Mt	Total %	GISS[2]	GFDL[2]	UKMO[2]	GISS[3]	GFDL[3]	UKMO[3]
Australia	1.38	11,546	15,574	3.2	−18	−16	−14	8	11	9
Brazil	1.31	2,788	3,625	0.8	−51	−38	−53	−33	−17	−34
Canada	1.88	11,365	21,412	4.4	−12	−10	−38	27	27	−7
China	2.53	29,092	73,527	15.3	−5	−12	−17	16	8	0
Egypt	3.79	572	2,166	0.4	−36	−28	−54	−31	−26	−51
France	5.93	4,636	27,485	5.7	−12	−28	−23	4	−15	−9
India	1.74	22,876	39,703	8.2	−32	−38	−56	3	−9	−33
Japan	3.25	237	772	0.2	−18	−21	−40	−1	−5	−27
Pakistan	1.73	7,478	12,918	2.7	−57	−29	−73	−19	31	−55
Uruguay	2.15	91	195	0.0	−41	−48	−50	−23	−31	−35
Former USSR										
winter	2.46	18,988	46,959	9.7	−3	−17	−22	29	9	0
spring	1.14	36,647	41,959	8.7	−12	−25	−48	21	3	−25
USA	2.72	26,595	64,390	13.4	−21	−23	−33	−2	−2	−14
WORLD[4]	2.09	231	482	72.7	−16	−22	−33	11	4	−13

[1] Results for each country represent the site results weighted according to regional production. The world estimates represent the country results weighted by national production.
[2] GCM 2 × CO_2 climate change scenario alone.
[3] GCM 2 × CO_2 climate change scenario with direct CO_2 effects.
[4] World area and production × 1,000,000.

logical CO_2 effects on crop yields. Another set of estimates incorporating the effects of farmer adaptation to climate change was also produced. All results were forwarded to the world food trade model in terms of percent change from current yields, rather than absolute values. Analysis of relative changes in yields are more appropriate, given the many uncertainties involved in analysis of climate change impacts. It is important to note, however, that percent change in yield depends on absolute values of base yields which are different for the three crops. For instance, soybean base yields are low, so that a large percentage change does not represent a large absolute change in yield.

Results

Sensitivity tests

While the arbitrary sensitivity tests are dissociated from the processes that influence climate, they simulate a controlled experiment and provide better understanding of the factors affecting crop model responses. They can also help to identify climatic thresholds of critical impacts. Climate sensitivity tests were carried out in 13 countries for combinations of 0, 2, and 4°C temperature increases coupled with precipitation changes of 0%, +20%, and −20%. Changes were considered relative to the baseline yield at 330 ppm CO_2.

Without the direct effects of CO_2, crops averaged over all sites showed an increasingly negative response to increased temperatures, with percent decreases in yields approximately doubling from the +2 to +4°C cases. When direct CO_2 effects are included, wheat, soybean, and rice yields increase about 15% with a 2°C temperature rise, but decrease at +4°C, indicating a possible threshold of compensation of direct CO_2 effects for temperature increases between 2 and 4°C as simulated in the IBSNAT crop models (Figure 3).

GCM scenarios

Climate change scenarios without the direct physiological effects of CO_2 caused decreases in simulated crop yields in many cases, while the direct effects of CO_2 mitigated the negative effects primarily in mid and high latitudes (Table 4). Potential changes in crop yields varied for the GISS, GFDL, and UKMO climate change scenarios (Figure 4). However, latitudinal differences were apparent in all scenarios. Changes in crop yields in the higher latitudes were less severe than crop yield changes in lower latitudes. In some cases, yield changes in mid and high latitudes were positive.

The GISS and GFDL climate change scenarios produced a range of yield changes from +30 to −30%, although there were regional differences. The GISS scenario is, in general, more detrimental to crop yields in Asia and South America, and the GFDL scenario is more harmful in North America and the former USSR. The UKMO climate change scenario, which has the greatest

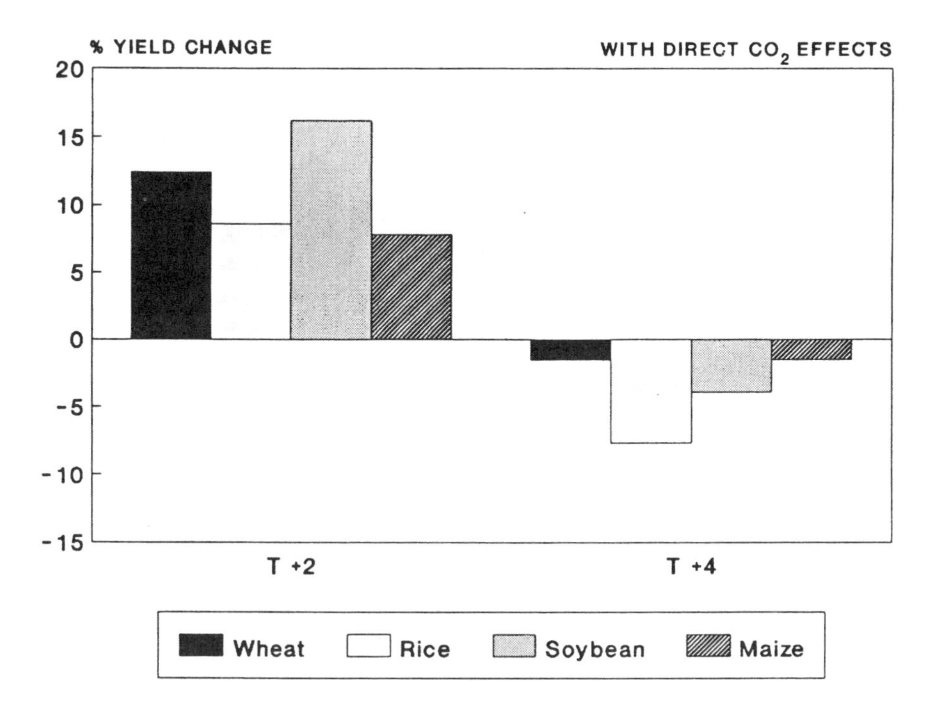

Figure 3. Aggregated IBSNAT crop model yield changes for +2°C and +4°C temperature increase. Country results are weighted by contribution of national production to world production. Direct effects of CO_2 on crop growth and water use are taken into account.

warming (5.2°C global surface air temperature increase), generally causes the largest yield declines (up to 50%).

The magnitudes of the estimated yield changes vary by crop. Maize production is most negatively affected, probably because of its lower response to the physiological effects of CO_2 on crop growth, while soybean is least affected because it responds significantly to increased CO_2, at least in the climate change scenarios with lower estimated mean global surface air temperature warming. Simulated yield losses are caused by a combination of factors, depending on location and nature of the climate change scenario. Primary causes of detrimental impacts on yield are:

(1) Shortening of the growing period (especially grain filling stage) of the crop. This occurred at some sites in all countries.
(2) Decrease of water availability caused by increased evapotranspiration and loss of soil moisture and in some cases a decrease in precipitation in the climate change scenarios. This occurred in Argentina, Brazil, Canada, France, Japan, Mexico, and USA.
(3) Poor vernalization. Many temperate crops require a period of low temperature in winter to initiate or accelerate the flowering process. Low vernal-

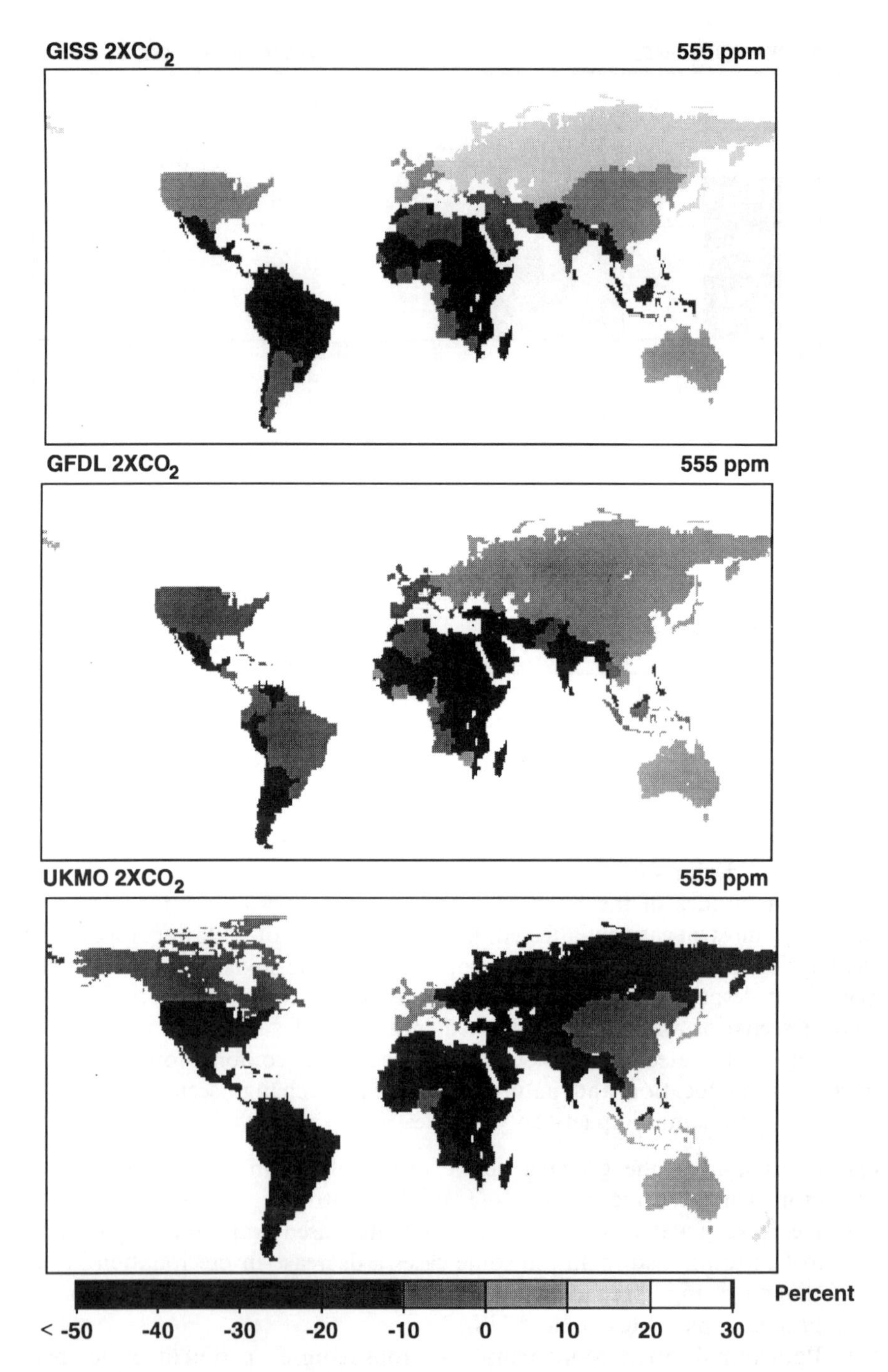

Figure 4. Estimated change in average grain yield (wheat, rice, coarse grains, and protein feed) for the GISS, GFDL, and UKMO climate change scenarios with direct CO_2 effects.

ization results in low flower bud initiation and ultimately reduced yields. This caused decreases in winter wheat yields in Canada and the former USSR.

Simulated yield increases in the mid- and high-latitudes were caused primarily by:

(1) The positive physiological effects of CO_2. At sites with cooler initial temperature regimes, increased photosynthesis more than compensated for the shortening of the growing period caused by warming.
(2) The lengthened growing season and the amelioration of cold temperature effects on growth. At some sites near the boundaries of current agricultural production, increased temperatures extended the frost-free growing season and provided regimes more conducive to increased crop productivity.

Transient scenarios

When the crop models were run with transient climate changes projected for the 2010s, 2030s, and 2050s from the GISS transient run A, national yield responses displayed differing trajectories of change leading to the doubled CO_2 equilibrium results, estimated to occur in 2060 (Figure 5). Crop yields in many countries were projected to respond non-linearly, displaying different levels of response in different decades. Aggregated national wheat yields exhibited the widest range of effects and were the most non-linear of the three crops. Simulated wheat yield responses in India, Egypt, and Japan changed sign at least once during the period from the 2010s to the 2050s. Transient soybean yield changes in the US and Brazil were the most positive of the three crops, while maize yield changes tended to be the most negative throughout the transient simulation period.

Adaptation study

Farmers will react dynamically to changing environmental conditions. Country participants tested the efficacy of several types of adaptations in crop simulation experiments. The adaptation strategies tested involved changes in current management practices such as planting date, fertilizer, and irrigation; and changes in crop variety either to existing varieties or hypothetical new varieties. Several participants considered the expansion of crops (for example, winter wheat in the former Soviet Union and Canada and rice in China) to areas that are temperature limited under the current conditions. The primary adaptation strategies tested are described below.

Planting date

The most likely response of farmers to warmer temperature would be to plant earlier to utilize the cooler early season and to avoid high temperatures during

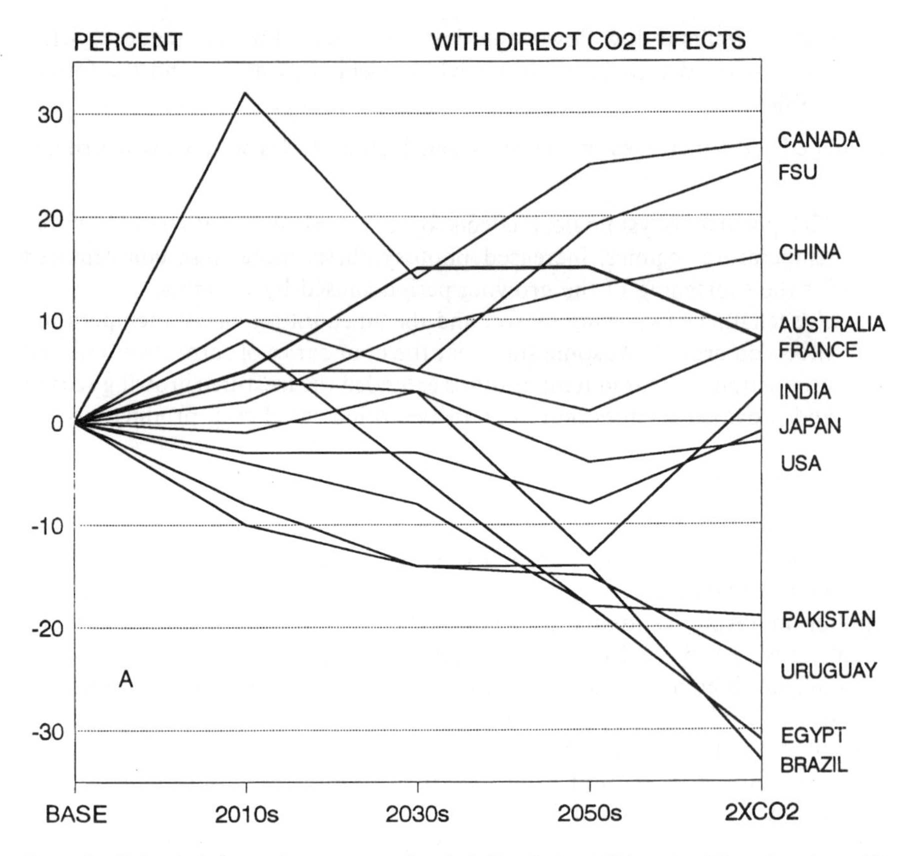

Figure 5. Estimated change in average national yields of wheat (A), maize (B), and soybean (C) for the GISS transient run A and doubled CO_2 climate change scenarios. Direct effects of CO_2 on crop growth and water use are taken into account.

the grain-filling period. All the crop model study participants suggested earlier planting as a strategy for adapting to global warming. A relatively small change in planting date, perhaps up to four weeks, should be easily supportable, but longer shifts in sowing date may adversely affect soil moisture and solar radiation at planting. In some country studies (Argentina, Philippines, and China) the sensitivity tests on planting date suggested large changes in seasonal agricultural production, implying major changes in the agricultural systems. In Argentina, planting date shifts of up to 4 months earlier or one month later were suggested by adaptation simulations.

Irrigation

The crop model simulations demonstrated that climate change may bring significant increases in the need for irrigation. In Petrolina, Brazil, where

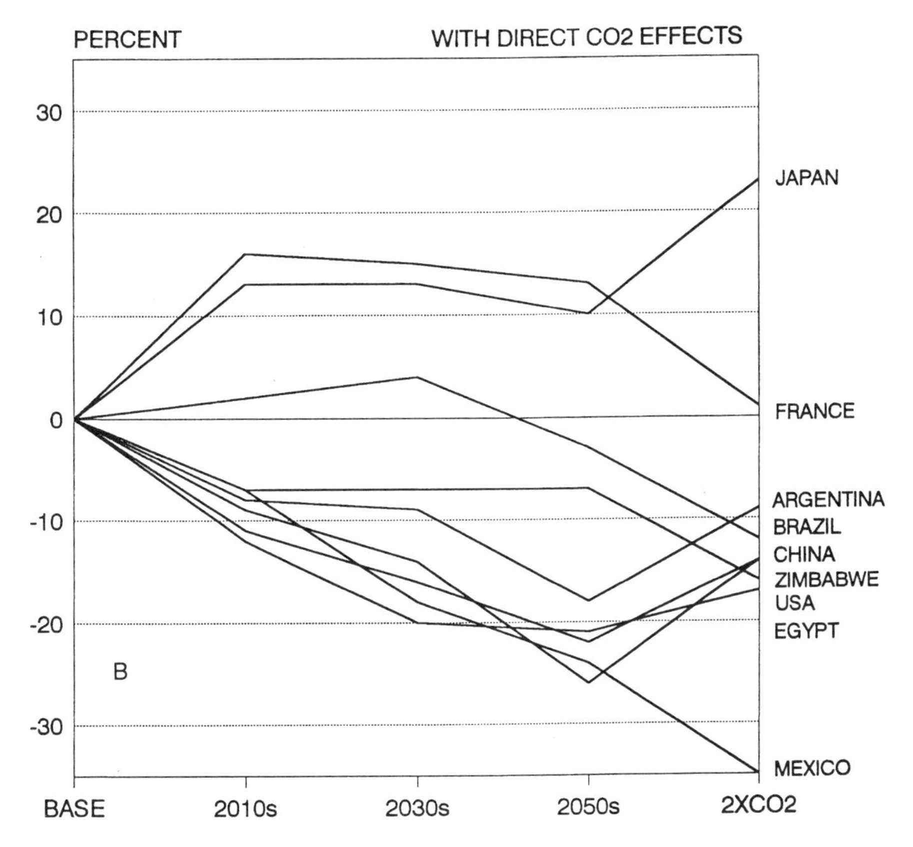

Figure 5. (*Continued.*)

soybeans are currently produced under rainfed conditions, soybean response to the UKMO climate change was negative even with direct CO_2 effects, but positive with full irrigation. Climate change may increase demand for irrigation water for crops already under irrigation and may encourage installation of new irrigation systems, if economic resources are available. Currently only about 15% of the world's agriculture is under irrigation. The potential problems associated with increased irrigation as an adaptive strategy are the questionable availability of water resources, the associated costs, and the environmental problems of soil salinization and water pollution.

Fertilizer

An increase in the amount of fertilizer applied can compensate in some cases for yield losses caused by climate change. Studies in Uruguay and Mexico used the nitrogen module in the crop models to test adaptation to climate change via fertilizer applications. In Uruguay, the response of barley to nitrogen

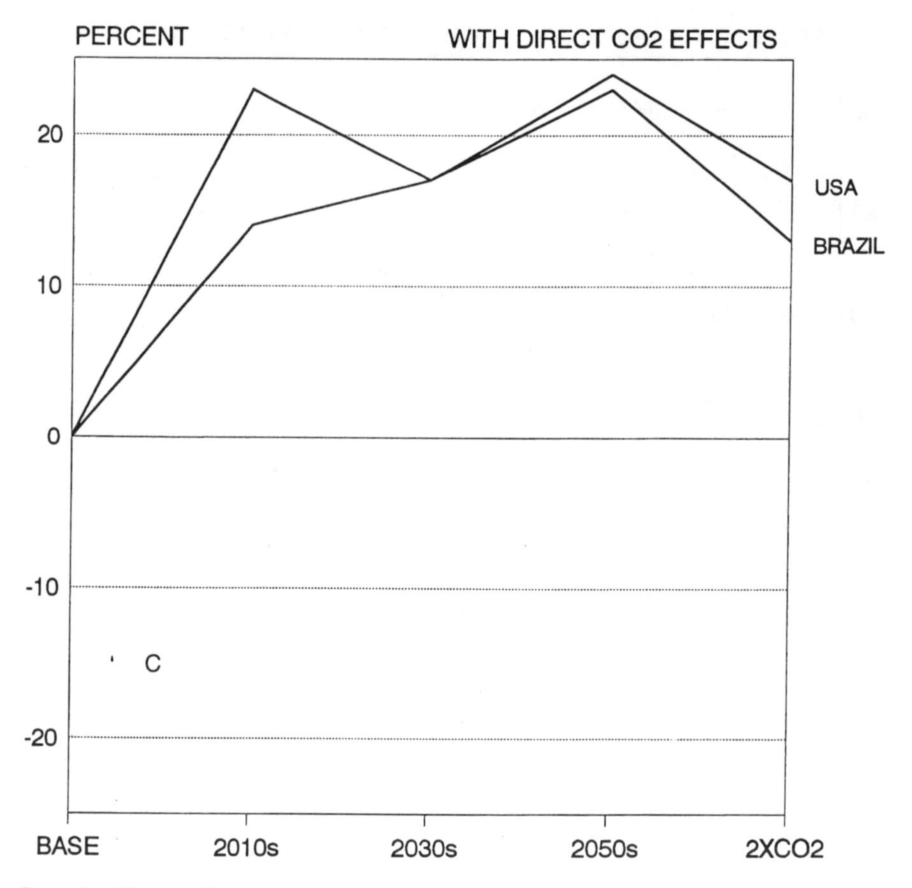

Figure 5. (*Continued.*)

fertilizer under baseline and UKMO conditions was compared. For the baseline runs the currently available cultivar was used at the normal planting date. The same cultivar was used for the UKMO runs, but sown at an earlier date with the physiological effects of increased CO_2. The planting date was changed for the UKMO runs, because the warmer temperatures of the scenario resulted in a shorter growing season. Four nitrogen fertilizer rates were used. Consequently, the baseline maximum yield was more than 1 t ha^{-1} higher than the corresponding yield under UKMO conditions. Also, the amount of N fertilizer needed to attain the maximum grain yield under UKMO was 2.6 times larger than the amount required to attain the same yield under current climatic conditions. These results may have significant implications for future fertilizer use under climate change conditions at the high end of the range of predicted warming, and indicate an important area for future research.

Crop variety

Since most regions are predicted to experience substantial increases in growing season temperature, country participants tested substitution of existing varieties with higher thermal requirements for currently grown varieties. In Mexico, use of existing cultivars produces slightly higher yields with the GISS climate change scenario (with the direct effects of CO_2), but does not overcome the negative climate change effects. Some researchers also tested hypothetical new varieties in the crop model simulations, a technique useful for establishing new breeding objectives. The crop models also allow testing of differing crop types, such as winter and spring wheats. In the former USSR, a comparison of winter and spring wheat simulations indicates that winter wheat will respond more favorably under the GCM climate change scenarios tested.

Cropping area

With further analysis of crop-climate classification, estimation of changes in cropping systems may be made. In China, climate change as projected by the three GCM scenarios would bring significant shifts in the rice cropping pattern, based on the extension of the growing period and increased thermal regime during the rice growing season (Gao et al., 1987). The regions where triple, double, and single rice crops per year could be grown would move northward and there would be increased sowing of *indica* rice now grown in southern China, replacing the current *japonica* types.

Adaptation results

The adaptation studies conducted by project participants suggest that ease of adaptation to climate change is likely to vary with latitude (Figure 6). With the existing pool of cultivars and current resources of water and fertilizer, agricultural adaptation to climate change seems likely in high and mid-latitude countries, but seems out of reach for nations in the low latitudes. In tropical and semi-tropical regions, especially semi-arid zones, the temperature changes suggested by the GCM scenarios tested are problematic. While soil moisture deficits can easily be made up with simulated automatic irrigation, the economic and environmental costs of establishing irrigation systems can be high.

Sources of uncertainty

Uncertainties in the crop yield modeling study include fundamental uncertainties in projecting the exact nature and geographic distribution of climate change; the assumptions embedded in the crop models; and the difficulty of estimating future technological improvements in agriculture.

The GCM projections used to develop the climate change scenarios were for mean changes only. Changes in climate variability were not simulated, even

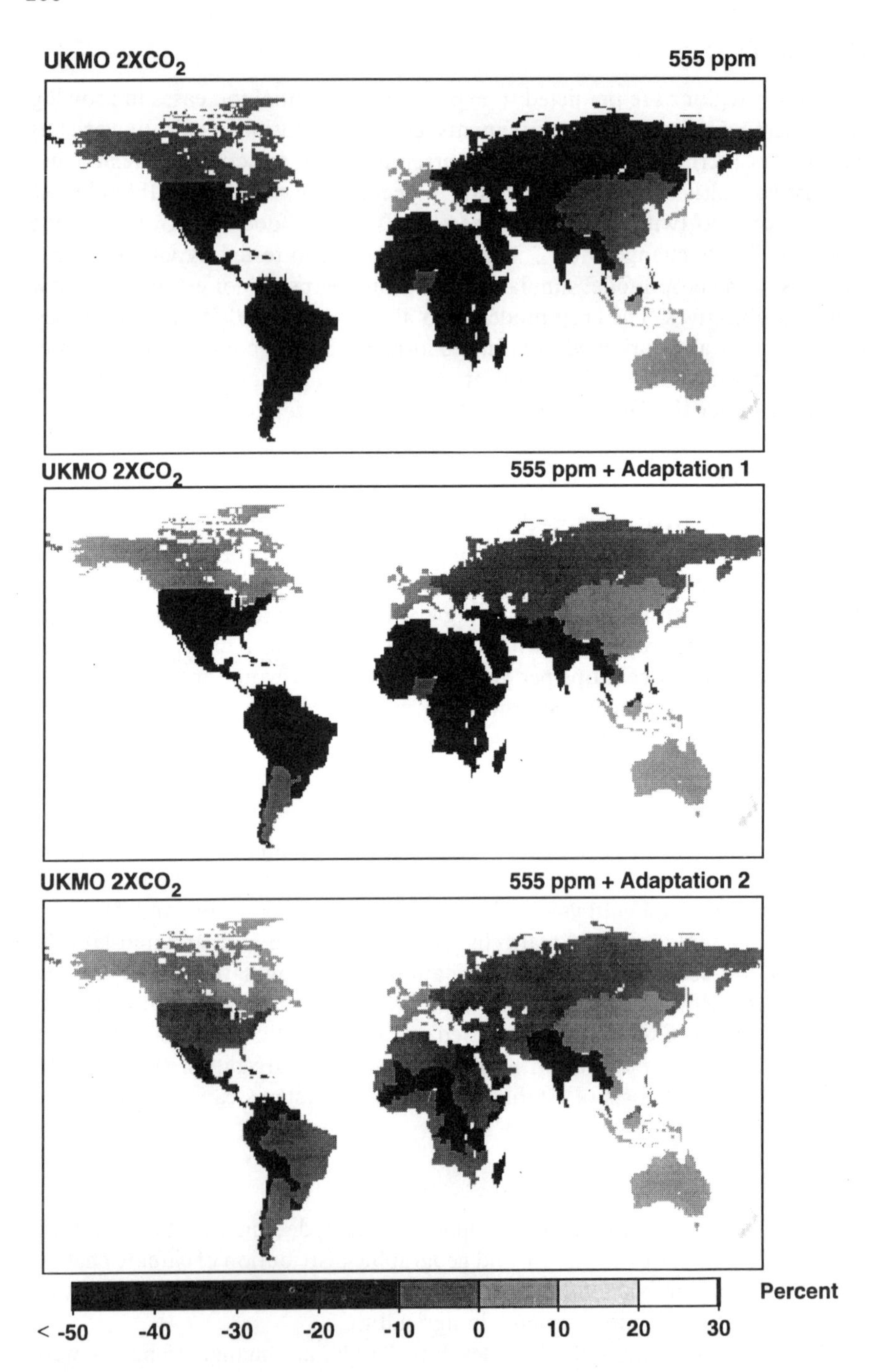

Figure 6. Estimated change in average grain yield (wheat, rice, coarse grains, and protein feed) under two levels of adaptation for the UKMO climate change with direct CO_2 effects.

though such changes could have significant effects on crop yields. Furthermore, the observed climate data used to create the climate change scenarios were taken from differing numbers of years and were of differing quality in the various country analyses. It is important, therefore, to view the results of the study as exploratory, rather than as predictive of actual impacts.

With regard to the use of the crop models in the study, there are several caveats. The generalized soil characteristics used as inputs do not encompass all the wide variety of global agricultural soils. Furthermore, the resetting of the water content of each soil profile (in most cases to full) at the beginning of each cropping season leads to underestimation of the impacts of changes in the hydrological cycle on crop production. Varying levels of nitrogen fertilization were not considered in most of the country studies, exceptions being Argentina, Uruguay and Mexico. Consistently high levels of fertilization are especially unrealistic in developing countries. Limited (often only one) cultivars were simulated at each site, although it is common practice to plant several cultivars in many cropping systems, which might respond differently to climate change.

Finally, the crop models simulate the current range of available agricultural technologies. They do not include potential improvements in such technology, although they may be used (as shown in this study) to test the effects of some potential improvements, such as improved varieties and irrigation schedules.

Conclusions and future research needs

Climate change induced by increasing greenhouse gases is likely to affect crop yields differently from region to region across the globe. Under the climate change scenarios adopted in this study, the effects on crop yields in mid- and high-latitude regions appeared to be less adverse than those in low-latitude regions. However, the more favorable effects on yield in temperate regions depended to a large extent on full realization of the potentially beneficial direct effects of CO_2 on crop growth. Decreases in potential crop yields are likely to be caused by shortening of the crop growing period, decrease in water availability because of higher rates of evapotranspiration, and poor vernalization of temperate cereal crops. When adaptations at the farm level were tested, such as change in planting date, switch of crop variety, changes in fertilizer application and irrigation, compensation for the detrimental effects of climate change was found to be more successful in developed countries.

Future research needs include determining how countries, particularly developing countries, can and will respond to reduced yields. More detailed adaptation studies in many different locations will help to address this need. In order to minimize possible adverse consequences to climate change worldwide, the agricultural sector should be encouraged to continue to develop crop breeding and management programs for heat and drought conditions (these will be immediately useful in improving productivity in marginal environments today). Another important activity is to enlarge, maintain, and screen crop genetic resources at established seedbanks. Resilience of the agricultural production sector also depends on improved use of systems for monitoring weather, soil moisture, nutrient requirements, and pest infestations. Finally, strong communication links among the agricultural research, production, and policy sectors are essential.

References

Acock B, Allen L H Jr (1985) Crop responses to elevated carbon dioxide concentrations. Pages 33–97 in Strain B R, Cure J D (Eds.) Direct effects of increasing carbon dioxide on vegetation. US Department of Energy. DOE/ER-0238. Washington, DC.

Adams R M, Rosenzweig C, Peart R M, Ritchie J T, McCarl B A, Glyer J D, Curry R B, Jones J W, Boote K J, Allen L H Jr (1990) Global climate change and US agriculture. Nature 345(6272):219–22.

Allen L H Jr, Boote K J, Jones J W, Jones P H, Valle R R, Acock B, Rogers H H, Dahlman R C (1987) Response of vegetation to rising carbon dioxide: Photosynthesis, biomass and seed yield of soybean. Global Biogeochemical Cycles 1:1–14.

Callaway J N, Cronin F J, Currie J W, Tawil J (1982) An analysis of methods and models for assessing the direct and indirect impacts of CO_2–induced environmental changes in the agricultural sector of US Economy. Pacific Northwest Laboratory, Battelle Memorial Institute. PNL-4384. Richland, Washington.

Cure J D (1985) Carbon dioxide doubling responses: A crop survey. Pages 33–97 in Strain B R, Cure J D (Eds.) Direct effects of increasing carbon dioxide on vegetation. US Department of Energy. DOE/ER-0238. Washington, DC.

Cure J D, Acock B (1986) Crop responses to carbon dioxide doubling: A literature survey. Agricultural and Forest Meteorology 38:127–145.

Fischer G, Frohberg K, Keyzer M A, Parikh K S (1988) Linked national models: a tool for international food policy analysis. Kluwer Academic Publishers, Dordrecht, The Netherlands.

FAO (1988) 1987 Production Yearbook. Food and Agriculture Organization of the United Nations, Statistics Series No. 82, Rome, Italy.

France J, Thornley J H M (1984) Mathematical models in agriculture. Butterworths, Boston and London.

Gao L, Lin L, Jin Z (1987) A classification for rice production in China. Agricultural and Forest Meteorology 39:55–65.

Godwin D, Ritchie J T, Singh U, Hunt L (1989) A User's Guide to CERES-Wheat v2.10. International Fertilizer Development Center. Muscle Shoals, Alabama, USA.

Hansen J, Russell G, Rind D, Stone P, Lacis A, Lebedeff S, Ruedy R, Travis L (1983) Efficient three-dimensional global models for climate studies: Models I and II. Monthly Weather Review 111(4):609–662.

Hansen J, Fung I, Lacis A, Rind D, Russell G, Lebedeff S, Ruedy R, Stone P (1988) Global climate changes as forecast by the GISS 3-D model. Journal of Geophysical Research 93(D8):9341–9364.

IPCC (1990a) Climate change: the IPCC scientific assessment. Houghton J T, Jenkins G J, Ephraums J J (Eds.) Intergovernmental Panel on Climate Change. Cambridge University Press, Cambridge, UK.

IPCC (1990b) Climate change: the IPCC impacts assessment. Tegart W J McG, Sheldon G W, Griffiths D C (Eds.) Intergovernmental Panel on Climate Change. Australian Government Publishing Service, Canberra, Australia.

IPCC (1992) Climate change 1992. The supplementary report to the IPCC scientific assessment. Houghton J T, Callander B A, Varney S K (Eds.) Intergovernmental Panel on Climate Change. Cambridge University Press, Cambridge, UK.

IPCC (1996a) Climate change 1995: the science of climate change. Houghton J T, Meira Filho L B, Callander B A, Harris N, Kattenberg A, Maskell K (Eds.) Intergovernmental Panel on Climate Change. Cambridge University Press, Cambridge, UK.

IPCC (1996b) Agriculture in a changing climate: impacts and adaptation. Pages 427–467 in Watson R T, Zinyowera M C, Moss R H (Eds.) Climate change 1995. Impacts, adaptations and mitigation of climate change: scientific-technical analyses. Contribution of Working Group II to the Second Assessment Report of the Intergovernmental Panel on Climate Change. Cambridge University Press, Cambridge, UK.

International Benchmark Sites Network for Agrotechnology Transfer (IBSNAT) (1989) Decision Support System for Agrotechnology Transfer Version 2.1 (DSSAT v2.1). Department of Agronomy and Soil Science, College of Tropical Agriculture and Human Resources, University of Hawaii, Honolulu, Hawaii, USA.

Jones C A, Kiniry J R (1986) CERES-Maize: a simulation model of maize growth and development. Texas A&M Press, College Station, Texas, USA.

Jones J W, Boote K J, Hoogenboom G, Jagtap S S, Wilkerson G G (1989) SOYGRO v5.42: soybean crop growth simulation model. User's guide. Department of Agricultural Engineering and Department of Agronomy, University of Florida, Gainesville, Florida, USA.

Kane S, Reilly J, Tobey J (1991) Climate change: economic implications for world agriculture. US Department of Agriculture. Economic Research Service, AER-No. 647.

Kimball B A (1983) Carbon dioxide and agricultural yield. An assemblage and analysis of 430 prior observations. Agronomy Journal 75:779–788.

Lamb P J (1987) On the development of regional climatic scenarios for policy-oriented climatic impact assessment. Bulletin of the American Meteorological Society 68:1116–1123.

Leemans R, Solomon A M (1993) Modeling the potential change in yield and distribution of the earth's crops under a warmed climate. Climate Research 3:79–96.

Manabe S, Wetherald R T (1987) Large-scale changes in soil wetness induced by an increase in CO_2. Journal of Atmospheric Science 44:1211–1235.

Martin R J, Salinger M J, Williams W M (1990) Agricultural industries in climate change: impacts on New Zealand. Ministry for the Environment, Wellington, New Zealand.

NDU (1980) Crop yields and climatic change for the year 2000. Volume 1. Lesley F, McNair J (Eds.) National Defense University, Washington, DC.

Nix H A (1985) Agriculture. Pages 105–130 in Kates R W (Ed.) Climate impact assessment. SCOPE 27. John Wiley & Sons, Chichester, UK.

Otter-Nβcke S, Godwin D C, Ritchie J T (1986) Testing and validating the CERES-wheat model in diverse environments. AgGRISTARS YM-15-00407. Johnson Space Center No. 20244, Houston, Texas, USA.

Parry M L, Carter T R, Konijn N T (Eds.) (1988) The impact of climatic variations on agriculture. Volume 1. Assessments in cool temperate and cold regions. Volume 2. Assessments in semi-arid regions. Kluwer Academic Publishers, Dordrecht, The Netherlands.

Pearman G (1988) Greenhouse: planning for climate change. CSIRO, Canberra, Australia.

Peart R M, Jones J W, Curry R B, Boote K, Allen L H Jr (1989) Impact of climate change on crop yield in the southeastern USA. In Smith J B, Tirpak D A (Eds.) The potential effects of global climate change on the United States, Appendix C. Report to Congress. US Environmental Protection Agency, EPA-230-05-89-050, Washington, DC.

Ritchie J T, Otter S (1985) Description and performance of CERES-wheat: a user-oriented wheat yield model. Pages 159–175 in Willis W O (Ed.) ARS wheat yield project. Department of Agriculture, Agricultural Research Service, ARS-38. Washington, DC.

Ritchie J T, Singh U, Godwin D, Hunt L (1989) A user's guide to CERES-maize v2.10. International Fertilizer Development Center. Muscle Shoals, Alabama, USA.

Rogers H H, Bingham G E, Cure J D, Smith J M, Surano K A (1983) Responses of selected plant species to elevated carbon dioxide in the field. Journal of Environmental Quality 12:569–574.

Rosenberg N J, Crosson P R (1991) Processes for identifying regional influences of and responses to increasing atmospheric CO_2 and climate change: the MINK project. An overview. Resources for the future. Department of Energy, DOE/RL/01830T-H5, Washington, DC.

Rosenzweig C (1990) Crop response to climate change in the Southern Great Plains: A simulation study. Professional Geography 42:20–39.

Rosenzweig C, Parry M L (1994) Potential impact of climate change on world food supply. Nature 367:133–138.

Singh U, Ritchie J T, Godwin D (1993) A User's Guide to CERES-Rice. International Fertilizer Development Center, Muscle Shoals, Alabama, USA.

Smit B (1989) Climatic warming and Canada's comparative position in agricultural production and trade. Pages 1–9 in Climate change digest. CCD 89–01. Environment Canada.

Smith J B, Tirpak D A (Eds.) (1989) The potential effects of global climate change on the United States. Report to Congress. U.S. Environmental Protection Agency. EPA-230-05-89-050. Washington, DC.

UK Department of the Environment (1991) The potential effects of climate change in the United Kingdom. United Kingdom Climate Change Impacts Review Group. HMSO, London.

UN (1996) World population prospects: the 1996 Revision. United Nations Department for Economic and Social Information and Policy Analysis Population Division, New York, USA.

Warrick R A, Gifford R M, Parry M L (1986) CO_2, climatic change and agriculture. Assessing the response of food crops to the direct effects of increased CO_2 and climatic change. Pages 393–473 in Bolin B, Dos B R, Jager J, Warrick R A (Eds.) The greenhouse effect, climatic change, and ecosystems. SCOPE 29. John Wiley & Sons, New York, USA.

Wigley T M L (1987) Climate scenarios. Prepared for the European Workshop in Interrelated Bioclimate and Land Use Changes. National Center for Atmospheric Research. NCAR 3142–86–3.

Wilson C A, Mitchell J F B (1987) A doubled CO_2 climate sensitivity experiment with a global climate model including a simple ocean. Journal of Geophysical Research 92(13):315–343.

World Food Institute (1988) World food trade and US agriculture, 1960–1987. Iowa State University, Ames, Iowa, USA.

Evaluation of land resources using crop models and a GIS

F.H. BEINROTH[1], J.W. JONES[2], E.B. KNAPP[3], P. PAPAJORGJI[2] and J. LUYTEN[2]

[1]*Department of Agronomy and Soils, University of Puerto Rico, Mayaguez, Puerto Rico 00608, USA*

[2]*Department of Agricultural and Biological Engineering, University of Florida, Gainesville, Florida 32611, USA*

[3]*CIAT, AA 6713, Cali, Colombia*

Key words: decision support system, geographic information system, land evaluation, land use planning, crop simulation models, climate change

Abstract

The evaluation of land resources has been greatly facilitated by recent advances in information science and computer technology. We describe an effort to move land evaluation into the information age that resulted in AEGIS, the Agricultural and Environmental Geographical Information System. Although the crop models in the DSSAT are basically one-dimensional, agriculture occurs in time and space. It was a logical step to link DSSAT with a geographic information system. The chapter outlines the evolution of the various versions of AEGIS that culminated in the a version for Windows, AEGIS/WIN. Sample applications of AEGIS are presented in three case studies: the feasibility of small irrigation projects in the Andes of Colombia; the evaluation of alternative cropping systems for former sugarcane land in southern Puerto Rico; and assessment of the possible impact of climate change on regional crop production in the southern USA. The accuracy of output from AEGIS is conditioned by limitations of the crop models and by the quality of soil and weather data and their spatial and temporal variability. Nevertheless, AEGIS is a useful tool for expanding the scope of analysis of the DSSAT from a point to an area. The field of systems-based land evaluation is progressing at a rapid pace, and future versions of AEGIS could be greatly enhanced by taking more advantage of the spatial modeling power of a geographic information system.

Introduction

Evaluation of land resources for agricultural purposes has been practiced for more than 4000 years. The past two decades have been a time of rapidly evolving methodology in land evaluation. The art and science of land appraisal has been transformed through the use of tools and techniques of the information age, particularly geographic information systems (GIS) and modeling. There has also been a profound shift in attitudes towards land evaluation. Agricultural overproduction in western Europe, for example, poses quite different problems from those prevailing in many developing countries, where there is a desperate need to match food production with population growth. An area of increasing involvement by land resource specialists is environmental impact assessment.

G.Y. Tsuji et al. *(eds): Understanding Options for Agricultural Production, 293–311*

There is a consensus that modern land resource evaluation must be able to assess the potential of land-use patterns and predict impacts and performance under different policy or management options. Generating such assessments and predictions obviously requires land evaluation methods that involve the interaction of GIS, crop simulation and yield prediction, and environmental impact statements associated with given policy and management scenarios.

In this chapter we describe how the principal product of the IBSNAT project, the Decision Support System for Agrotechnology Transfer (DSSAT) and its resident crop simulation models, can be employed in the process of land evaluation by linking it to a GIS. Consequently, our discussion is focused primarily on the evaluation of land resources for crop production, although related environmental aspects are also considered. The underlying rationale for this effort was that while crop models are point specific, agriculture occurs in space and time. Coupling DSSAT with a GIS was thus a logical step that expands the scope of analysis of DSSAT from a site to an area.

Principles and methods of land evaluation

Reference background

Land, according to a widely used definition by FAO (1976), is 'an area of the earth's surface, the characteristics of which embrace all reasonably stable, or predictably cyclic, attributes of the biosphere vertically above and below this area including those of the atmosphere, the soil, and underlying geology, the hydrology, the plant and animal populations, and the results of past and present human activity, to the extent that these attributes exert a significant influence on present and future uses of the land by man.'

Land is thus viewed as areas that comprise biophysical and environmental characteristics which are or may be of importance to human use (Davidson, 1992). The FAO definition excludes purely economic and social characteristics from the concept of land on the rationale that these form part of the socio-economic context (FAO, 1976). The Soil Survey Division Staff (1993), on the other hand, considers the word land to imply 'attributes of place and other factors besides soil', which is a more inclusive and flexible concept. Irrespective of whether socio-economic characteristics are included in the definition of land or not, they are nonetheless decisive criteria in land evaluation. They are time-specific, however, whereas most physical land features are of a more permanent nature. Agricultural and other managed landscapes, of course, are dynamic systems that are shaped to varying degrees by human activity. Although the rate of change in biophysical conditions is normally much slower than that of socio-economic aspects, it may be accelerated by practices such as terracing, fertilization and liming, and drainage. In any event the intimate linkage between the biophysical aspects of managed land and human activities is obvious.

Land resource evaluation is concerned with the assessment of land performance for specified purposes. The process involves the execution and interpretation of basic surveys of soils, climate, vegetation and other aspects of land in terms of the requirements of alternative forms of land use within the physical, economic and social context of the area considered. The Soil Survey Division Staff (1993) also places emphasis on the importance of parameters other than those related to soil, and states that 'planners must consider place, size of area, relation to markets, social and economic development, skill of land users, and other factors.'

Principles of land evaluation

Land evaluation is based on the following fundamental principles (FAO, 1976): (1) Land suitability is assessed and classified with respect to specified kinds of use. (2) Evaluation requires a comparison of the benefits obtained and the inputs needed on different types of land. (3) A multi disciplinary approach is required. (4) Evaluation is made in terms relevant to the physical, economic, and social context of the area concerned. (5) Suitability refers to use on a sustained basis. (6) Evaluation involves comparison of more than a single kind of use.

Nearly two decades after they were formulated, these principles remain valid and relevant today. The process of land evaluation usually involves the following steps (Soil Survey Division Staff, 1993):

(1) Assembling information about the soils and the landscapes in which they occur.
(2) Modeling other necessary soil characteristics from the soil data.
(3) Deriving inferences, rules, and guides for predicting soil behavior under specific land uses.
(4) Integrating these predictions into generalizations for the map unit.

The function of land evaluation is '... to guide decisions on land use in such a way that the resources of the environment are put to the most beneficial use for man, whilst at the same time conserving those resources for the future' (FAO, 1976).

Methods of land evaluation

Scientific approaches to land appraisal depend on inventories of soils, land, and climate information, and knowledge of crop requirements. The development of land capability schemes during the 1930s in the USA signaled the beginning of efforts that culminated in the publication of a handbook (Klingebiel and Montgomery, 1961). This system was adopted by the then Soil Conservation Service of the US Department of Agriculture and became known as the USDA Land Capability Classification. The system classifies soils for mechanized production of field crops commonly cultivated in the USA but does not apply to crops that require little cultivation, such as nuts, or flooded crops, such as rice. It also cannot be used for farming systems that depend on primitive implements and extensive hand labor (Soil Survey Division Staff, 1993). Although the system has some conceptual and operational shortcomings, it is still widely used today.

The publication of 'A Framework for Land Evaluation' by FAO (1976) marked a major methodological breakthrough. The framework provides a set of guidelines for an ecological analysis whereby land mapping units are evaluated with reference to defined land uses which also incorporate social, economic and technological criteria. A set of soil and non-soil properties are synthesized in land qualities and the area-specific land qualities are compared with the biophysical and management requirements of specific land uses. Classes of land suitability are established based on the degree to which the intrinsic land qualities meet the requirements of the specified land use.

Another approach to land appraisal for crop production is the designation and delineation of management groups that comprise soils that require similar kinds of practices to achieve acceptable performance. Such attempts include the Fertility Classification System (Sánchez et al., 1982) and methods to group soils by productivity ratings or soil potential indices (Soil Survey Division Staff, 1993).

All of the aforementioned methods have one characteristic in common, they depend on conventional 'pencil and paper' approaches to data management and interpretation. Advances in information science and computer technology, however, now present new ways to address issues of land evaluation. Prominent among these are relational database management systems, GIS, crop simulation models, soil erosion and hydrology models, and rule-based systems.

One such effort is the Automated Land Evaluation System (ALES) developed at Cornell University. ALES is based on the FAO methodology and is itself a framework within which evaluators build their own models. By itself, the system contains no knowledge. Rather, it is a shell which provides a reasoning mechanism and compels the evaluator to express inferences using this mechanism. ALES is thus a computerized realization of the FAO framework and also can be thought of as a model of expert judgement, i.e., the codification in a constrained form of the inferences already present in the minds of an expert (Rossiter, 1990).

The development of GIS has had a considerable impact; a GIS consists of a set of computer-based tools that permit the processing, storage, retrieval, transformation and display of spatially referenced data. There are several attractions of GIS for use in land evaluation (Davidson, 1992):

(1) The ability to produce output in the precise form required by clients.
(2) The capacity to integrate data sets from a wide range of sources including digitized maps and remote sensing imagery.

(3) The potential for mapping change through monitoring programs.
(4) The capability to interact with models to produce outputs in the form of landscape simulations under different policy scenarios, or predicted crop or financial yield under different management strategies.

A GIS that incorporates appropriate predictive capabilities can explore a wide variety of 'what if' questions, which is of paramount importance to agricultural and environmental planning and management and helps to reduce uncertainty in decision making. Prediction of change and the ability to map the results using a GIS comprise the current research frontier in land evaluation. Recognizing the potential of this technology, limited GIS capabilities have been built into the latest version of DSSAT v3 (Thornton et al., 1997) with a public-domain mapping module, spatial interpolation routines, and custom software to operate with IDRISI (Eastman, 1990) and ARC/VIEW (ESRI, 1994). Similarly, users of ALES now have the option to include an interface in their software that links it to IDRISI (Eastman, 1990). The resulting system is called ALIDRIS. In the following sections we describe software whose development was associated with the IBSNAT project.

The AEGIS family

The original AEGIS

AEGIS (Agricultural and Environmental Geographic Information System) was initiated to explore how a GIS can be linked to crop models and, by implication and extension, to DSSAT. To address this question, we selected three different agroenvironments in western Puerto Rico on the basis that the environmental dissimilarity results in correspondingly divergent crop yields and management requirements that should be clearly reflected in the output of AEGIS.

Detailed soil maps at a scale of 1:20,000 produced by the USDA Natural Resources Conservation Service were digitized for each of the three areas using PC ARC/INFO Version 3.4D (ESRI, 1992). The maps, each comprising about 3000 ha, delineate a total of 88 map units, mostly phases of 38 soil series that represent seven of the 11 soil orders recognized in Soil Taxonomy (Soil Survey Staff, 1975). A simple rule-based system was employed to identify the 67 map units suitable for bean production. The soil parameters required to run the DSSAT crop models and the Universal Soil Loss Equation (Wischmeier and Smith, 1978) were derived from the soil survey reports and series-specific analytical data available for 28 of the 38 series. For the remaining series, surrogate data were estimated using analogue procedures (Beinroth, 1990). Combining field and laboratory data, soil data files were created for each of the 67 polygons. To enhance user friendliness, these units were aggregated into 12 generic soil groups.

The climate in the region ranges from humid to subhumid to semiarid tropical. Historical weather data recorded at four representative stations were used. Long-term solar radiation data were not available but time sequences were estimated with a stochastic default procedure developed for Puerto Rico (Esnard et al., 1994). A land-use coverage was generated from processed LANDSAT imagery.

AEGIS was designed to estimate crop yields, management requirements and environmental impact for various land-use strategies. Users can select combinations of crops, crop varieties, planting dates, irrigation, and nitrogen fertilization. AEGIS simulates these scenarios and generates tables and thematic maps for yields, biomass, runoff and nitrogen leaching. The system also allows the creation, modification and storage of outputs for subsequent analysis and display.

The original AEGIS consisted of the following components (Calixte et al., 1992, Papajorgji et al., 1994b):

(1) A set of spatial databases of soil, weather, and land use attributes in a relational database management system.
(2) A relational database management system, dBaseIV (Ashton Tate, 1990) for the user interface, control of simulation models, and analysis of simulated results.
(3) A crop model, BEANGRO v1.1 (Hoogenboom et al., 1990) written in FORTRAN.
(4) A simple expert system for assessing land suitability for bean production.
(5) A geographical information system (PC ARC/INFO) for creating and displaying thematic maps based on user-selected simulations.
(6) The seasonal analysis program of DSSAT.
(7) A menu interface that facilitates the interaction of users with the system.

In the first version of AEGIS, the concept of a land-use plan was limited. A land-use plan could consist of a single crop and a single management practice so that comparisons of crop performance over different land units could be made. A more general interpretation of land-use plans allows for several crops, each having different management to represent the variability that may exist in a given region. Each plan that is created may be saved, along with simulated results, for later re-analysis and comparison with other land-use plans. This capability, now available in both PC and SUN workstation versions of AEGIS, allows a high level of flexibility for land-use evaluation.

The prototype AEGIS demonstrated the technical feasibility of linking crop models to a GIS and indicated the kinds of issues that the system can address. Its application is limited, however, to the three regions of Puerto Rico for which it was developed. This limitation results from the use of precomposed maps of the regions, the use of region names in the code, and the lack of a framework for extending the system to other areas (Papajorgji et al., 1994a).

Generic versions of AEGIS

Various versions of AEGIS were developed to overcome limitations of the original prototype. AEGIS–2 (Papajorgji et al., 1993, 1994b), developed with

PC ARC/INFO (ESRI, 1992), has access to all crop simulation models in DSSAT v2.1. In addition, it is generic; any region can be installed if the coverages have the required polygon attributes and soil and weather data are available for all defined areas of the map coverages. The required attributes for each polygon are weather station code, soil map unit, and soil group. Daily weather data files for the defined areas must be available for running the crop models, and soil profile data must be available for each soil in DSSAT v2.1 file formats. After selecting weather zones and soil groups to include in the analysis, users then specify the crop, variety, and management details to be analyzed. Simulations are automatically performed for a number of years, results are averaged, and averages are linked back to the coverage so that they can be displayed as thematic maps. Users can save simulated results as a spatial plan and create new plans for comparison. Detailed or aggregated results of the simulations can be printed as tables.

AEGIS–2.5 (Papajorgji et al., 1994a) is similar to AEGIS–2 in most respects. However, it was developed to utilize DSSAT v2.5, which incorporated crop models with the ability to study climate change and CO_2 effects on crops as described by Rosenzweig and Jones (1990) and Peart et al. (1989). In this version, users can specify increases or decreases in temperature, solar radiation, precipitation, and atmospheric CO_2 concentration to estimate the impact of those changes on production over a region. This version allows for the use of weather files from General Circulation Models (Smith and Tirpak, 1989).

Generic workstation and PC versions for DSSAT v3

DSSAT v3 (Tsuji et al., 1994) consists of major changes in all aspects of the crop models, databases, and functions compared with DSSAT v2.1, which was used in earlier versions of AEGIS. Two generic AEGIS versions were developed for different computer platforms, utilizing new features of DSSAT v3 and its crop models. AEGIS/WIN v3.0 (Engel and Jones, 1995) and AEGIS+ (Luyten et al., 1994), developed for use in the Microsoft Windows environment on PCs and in the Unix environment on workstations, respectively, have been documented and are available for general use. Both are designed for use with vector coverages using ARC/INFO formats.

AEGIS/WIN and AEGIS+ can be used for any region such as a piece of land on a farm, region, or other spatial scale. For a coverage, different crop management practices can be assigned to different fields or homogenous land areas or polygons. These versions of AEGIS allow one to evaluate the effects of various practices arranged over the spatial scale to estimate crop production, explore land-use options, and determine possible environmental impacts. The user interfaces on these systems are used to perform the following functions:

(1) Create thematic maps based on either polygon characteristics or features of a coverage such as soil type, total area, weather station, and climate

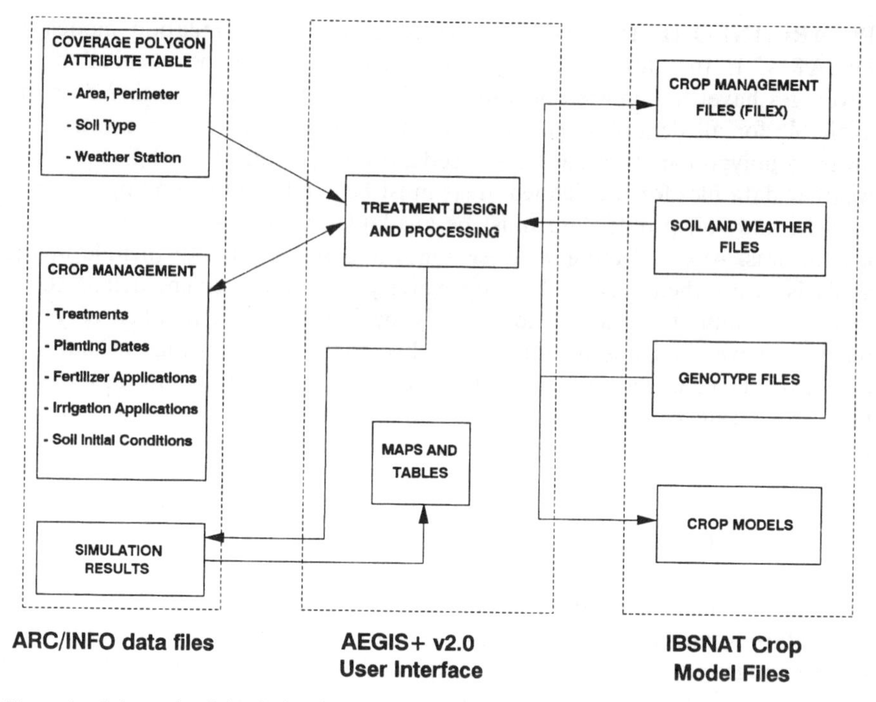

Figure 1. Schematic of the design for AEGIS+ v2.0 (Luyten et al., 1994).

zones, or land-use characteristics and features such as the crops and varieties being grown.

(2) Define and modify various crop management practices, including crop, variety, planting management, irrigation, fertilizer, and residue management. In the PC version (AEGIS/WIN), these modifications are made in the DSSAT v3 environment, whereas in the AEGIS+ workstation version, these changes are made using a form presented on the workstation screen.

(3) Select and unselect polygons for simulation. In AEGIS+, users can assign crops and management practices to polygons on a displayed map, whereas this must be done in DSSAT v3 in the AEGIS/WIN version.

(4) Perform simulations and examine results of several simulated factors. Simulation results can be graphically presented in thematic maps. Statistical results can be computed and displayed in tables and graphs.

(5) Print maps, graphs, and tables.

Figure 1 shows the overall design schematic for the Unix workstation version. PC AEGIS/WIN uses DSSAT v3 with no modifications to any of its components. However, the workstation version uses only the crop models from DSSAT v3 and associated data files and definition files. The crop models had

Table 1. System requirements of AEGIS/WIN (Engel and Jones, 1995).

HARDWARE:
* IBM compatible Personal Computer 80386 or higher
* Hard disk with over 30 megabytes (MB) of free space for DSSAT v3 and ArcView 2.0. Required disk space will increase with the addition of new coverages and weather data to AEGIS/WIN.
* 8 MB random access memory (RAM)
* 12 MB swapfile (virtual memory on the hard disk)
* Microsoft compatible mouse

SOFTWARE:
* MS-DOS or PC-DOS 3.1 or higher
* MS-Windows 3.1 or higher
* DSSAT v3 (Tsuji et al., 1994)
* ArcView 2.0 (ESRI, 1994)

to be modified slightly (Luyten et al., 1994) for running on the SUN workstation on which AEGIS+ was developed.

Data requirements of AEGIS

The coverages of AEGIS require general information such as map unit symbols, generic groupings of map units if appropriate, and weather station identification. To run the crop models in DSSAT, the user must provide soils, daily weather, and genotype data for each polygon. The quality of AEGIS output obviously depends on the accuracy of the map unit delineations, the precision of the predictions of the crop and other simulation models and their validation for the region in which they are applied, and the adequacy of the input data required by the models. Complete, reliable and site-specific primary data are often not available, resulting in a major obstacle to the successful and confident application of AEGIS. This is particularly true for soil, solar radiation and plant genetic data (Jones et al., 1994), forcing the user to make do with surrogate or estimated data. Default procedures for data generation have been discussed by Beinroth (1990) for soil data and by Geng et al. (1988), Geng and Auburn (1987), and Richardson and Wright (1984) for weather data. Running AEGIS with default parameters inevitably increases the errors associated with system output, however.

Computer system requirements for AEGIS/WIN v3 and AEGIS+ v2.0 are shown in Tables 1 and 2.

Application of AEGIS

Using AEGIS to evaluate the feasibility of small irrigation projects for watersheds in the Andes of Colombia

In developing countries, there is a general need to identify data and tools for efficient, objective and comprehensible regional land-use planning to support

Table 2. System requirements of AEGIS+ (Luyten et al., 1994).

HARDWARE:
* SUN SPARC workstation
* Disk with over 30 MB of free space. The required disk space increases considerably when new coverages and weather files are added.

SOFTWARE:
* SUN OS version 4.1 or higher
* OpenWindows version 3.0 or higher
* ARC/INFO version 6.1 or higher (ESRI, 1991)
* DSSAT v3 crop models (Hoogenboom et al., 1994), recompiled to run under the SUN workstation

multi-party negotiation and consensus building. This study tested the AEGIS+ v2.0 modeling software for its appropriateness in assessing the potential effects of small irrigation projects on the regional water balance within a small mountain catchment in southwestern Colombia. Conflicting goals of within-watershed and offsite water demands are likely to become increasingly contentious issues as societies attempt to keep pace with the food and water demands of increasing populations.

The specific region chosen for this proof-of-concept analysis was the 3200 ha Río Cabuyal sub-catchment nested within the 100,000 ha Río Ovejas watershed. Certain critical assumptions were made which affect the reliability and acceptance of the analysis. No climate station exists within the Río Cabuyal sub-catchment, although several stations lie within a close radius. Consequently, long-term daily weather sequences were generated using the weather analysis program of DSSAT v3 (Pickering et al., 1994), and interpolated using an inverse square function. Model parameters for the four soil mapping units in the sub-catchment, all classified as Andisols, were taken from previously published soil surveys, an historical database of more than 1000 samples made available from the state agricultural department, and corroborating field measurements.

Hillside environments are characterized by variable topography which may severely limit crop management decisions such as machinery use and irrigation options. A digital terrain model was created from existing 1 : 10,000 topographic maps using ARC/INFO which allowed for the interactive selection of slope class intervals. For the analyses presented, three class intervals were chosen: 0 to 7%, 7 to 30%, greater than 30% slopes. The potential for irrigation was restricted to the first slope class within the watershed. Estimates of river water flow at upper, mid, and lower locations in the Río Cabuyal were made using accepted, empirical, cost-effective field procedures.

For this study, production estimates and economic yield predictions were of secondary importance to water balance calculations. As such, drybean (Phaseolus vulgaris) was chosen as a proxy for a high-value crop with a short season and shallow roots. Assumptions of lesser importance for the conclusions of the study relate to choice of variety, seeding and fertilization rates, residue

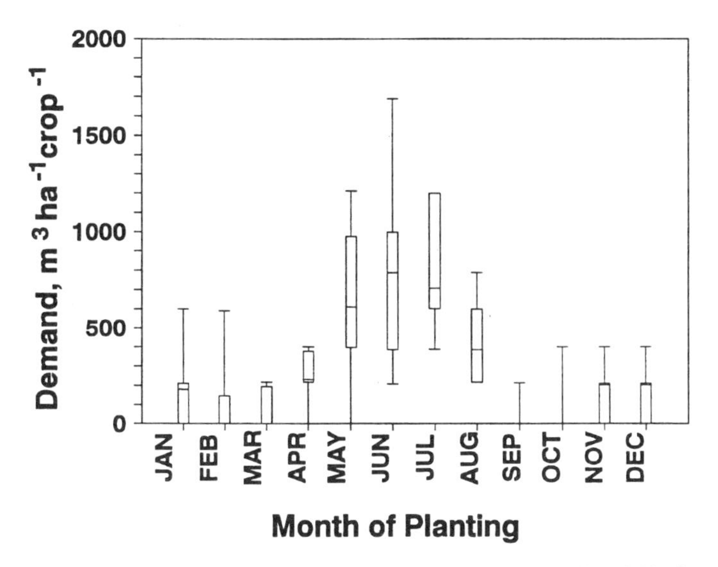

Figure 2. Frequency distributions and percentiles (10, 25, 50, mean, 75 and 90) for total crop demand per hectare for irrigation water. Seeding was on the first day of each month shown.

management, and non-limiting pest and disease control. In a typical environment of negotiation and consensus building, the analysis could be broadened by taking full advantage of crop and management practices including irrigation method and automatic irrigation scheduling strategies which are relevant to this analysis.

An important advantage of the GIS link provided by AEGIS is the capability to define accurately geographic space and access data directly from maps and images. Overlaying climate, soil and slope coverages resulted in the creation and definition of 993 polygons or 'fields' within the 3200 ha Río Cabuyal subcatchment. The area of noncontiguous 'fields' varied from less than 500 m^2 to several hectares. Some of this area, however, is roadway, homestead and watercourse, and can be 'de-selected' as never being available for agriculture.

During an interactive, compromise planning exercise, stakeholders can define conditions and select fields for inclusion in one or more planned irrigation projects. The tradeoffs in terms of water partitioning for domestic reserves, irrigation demand and downstream use and concomitant temporal variability can then be assessed using AEGIS. For example, one of many possible initial plans calls for irrigation of all fields with slopes between 0 and 7%, and irrigation could be gravity fed rather than pumped to the fields. Figure 2 presents summary output of seasonal demand for irrigation by month of planting for the predominant soil mapping unit. The box plots are based on AEGIS+ tabular output which summarizes seasonal minimum, average and maximum demand by 'field' and by cropping strategy. Ten years of crop production were simulated for each month of planting to obtain the variability

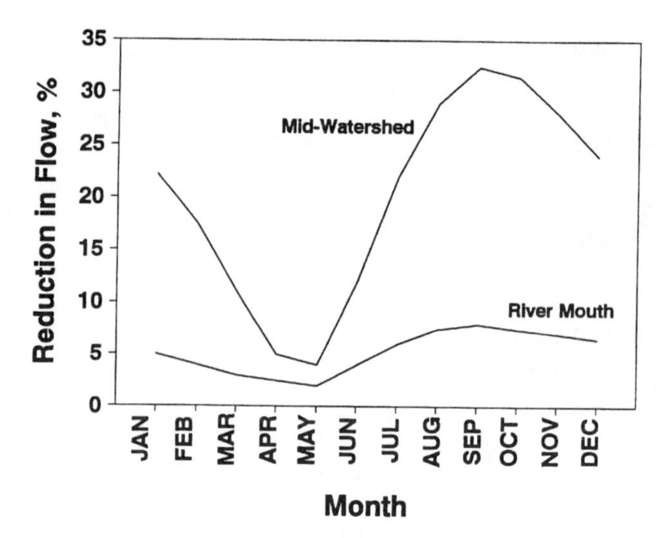

Figure 3. The expected percent reduction in river flow at two points in the watershed for the one year in ten of greatest demand for a benchmark of 100 ha total area under irrigated crops.

shown. Irrigation demand is greatest for crops planted in May with growing periods of June, July and August, the driest months of the year. Irrigation demand is low for September planting because October, November, and December have higher rainfall.

Figure 3 shows the expected percent reduction in river flow at two points (mid-watershed and river mouth) for the one year out of ten in which irrigation demand was greatest, assuming that a total of 100 ha were irrigated. Percent reduction in river flow peaked in September when river flow was at its lowest and the maximum (one out of ten years) irrigation demand for September was a large percentage of that flow.

Using the approach outlined in this example, initial positions about resource planning can be quantified and presented for discussion. During discussions, new positions can be formulated and simulated in an iterative manner until a consensus is reached. AEGIS has been demonstrated to stakeholders in the region and they agree that data collection requirements are well within the capabilities of government agencies in most Latin American countries, and that the software might usefully and economically be applied for ex ante impact analysis and compromise planning addressing multiple objectives of land and water use.

Assessing agricultural potential of previous sugar cane land in Puerto Rico

The south central coastal valley of Puerto Rico is one of the most fertile areas of the island, used in the past for sugar cane production. As a result of the depressed sugar market and the arrival of substitute sweeteners, sugar cane is no longer attractive to farmers of the region. Much of the land previously devoted to sugarcane is now idle and does not contribute to the economy of the island. Vegetable crops are now being grown in the valley. Growers are using substantial quantities of inorganic fertilizer and pest control chemicals on these crops. Many of these lands are near rivers and the coast. Development of new types of agriculture on these lands could create important economic activity for these areas, but could also pose risks to surface and ground water quality. Agricultural planners and policy makers need to have information on the likely impacts of alternative agricultural uses of these lands previously devoted to sugar cane production.

AEGIS was used by Hansen et al. (1997) as a framework for analyzing the possible tradeoffs between production and economics of proposed land use in a small watershed in southern Puerto Rico. This study area is a gently sloping valley between the Jacaguas and Canas rivers. In this pilot study, ARC/INFO coverages were developed for soil, weather and land use in the area. These coverages were overlaid and linked with AEGIS/WIN. Data for candidate crops were assembled (sugar cane, maize, bean, sorghum, tomato, and soybean). Figure 4 shows results for tomato yield and nitrogen leaching over this watershed. Because of their high market value, tomatoes are likely to gain in importance in this watershed. Hansen et al. (1997) showed that tomato double cropped with cereal crops would increase profits for land use and would probably decrease risks of erosion and nitrogen leaching, compared with a number of alternative cropping systems.

Assessing possible impacts of climate change on regional crop production

Agricultural production is vulnerable to variability in weather conditions. Projections of global climate change have prompted a number of studies concerning the possible impacts on agriculture. Most of these studies have used crop models to quantify possible changes in crop yields if temperatures and precipitation change as projected by General Circulation Models. These studies have typically selected specific sites over some spatial scale for analysis. Current weather data are used to simulate crop performance under existing climate conditions. Modified weather data are then used to simulate crop performance under changed conditions at the selected sites. Differences are then attributed to the changes in weather imposed on the models. Many of these studies have used the DSSAT v2.1 or v3 crop models (e.g., Adams et al., 1990; Curry et al., 1990; Rosenzweig et al., 1995). A natural extension of these studies was to use GIS for organizing the spatial weather and soils data for use in the simulation analyses.

AEGIS/WIN and AEGIS+ were used to study climate change in two states in the South-Eastern USA: Florida and Georgia (Peart et al., 1995). This study had as its goal the preparation of state coverages based on county polygons for use in climate change studies. Soil maps and climate zone maps were overlaid with county maps to determine the most predominate soil in each county and the weather station to use in each county. One difficulty in this

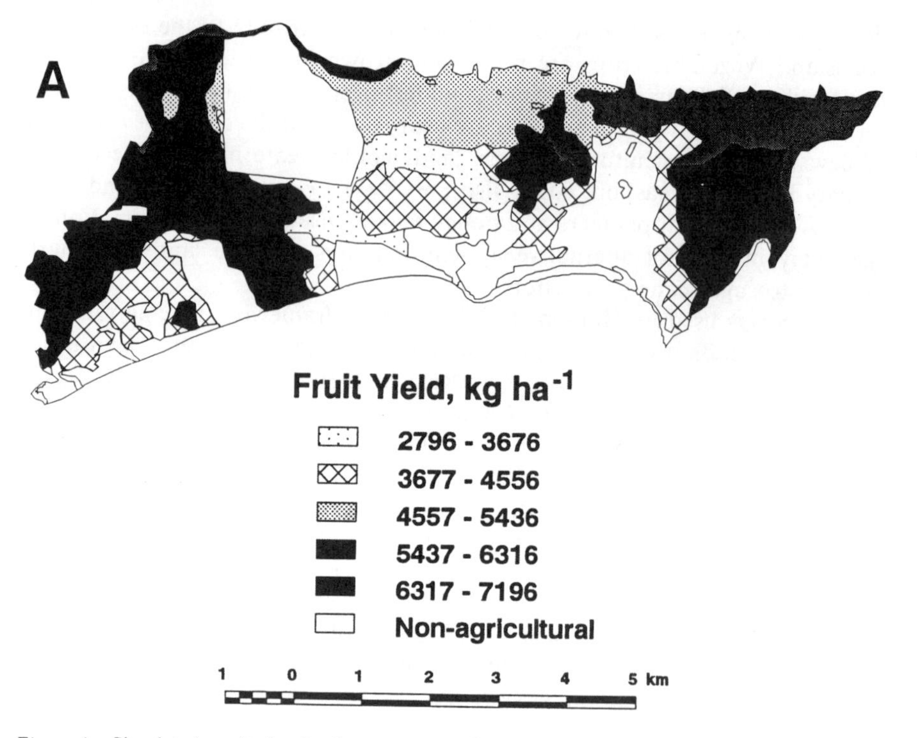

Figure 4. Simulated results for the Jacaguas watershed using AEGIS/WIN. (A) Simulated tomato yields over the area. (B) Simulated nitrogen leaching, averaged over 20 years (from Hansen et al., 1997).

study was the creation of meaningful coverages at the state level for climate change analysis and the development of soil and weather data for running the simulations. Analyses were performed by each of the AEGIS versions to demonstrate their functionality. Figure 5 shows the spatial variability of simulated soybean yield averages for Georgia if the same cropping practices were used for the entire state for current climate conditions and for a 3°C temperature increase. The differences visually displayed on the maps are entirely due to the assumed 3°C temperature increase as there were no assumed changes in rainfall or atmospheric CO_2 concentrations for this analysis. It was found that the linkage between crop simulation models and GIS provided by AEGIS makes it convenient to conduct analyses of climate change effects, to obtain averages and spatially-weighted averages, and to interpret results.

Outlook and conclusions

The utility of GIS technology for land resource evaluation will inevitably be judged by the quality of the final product. Errors may be caused by imperfec-

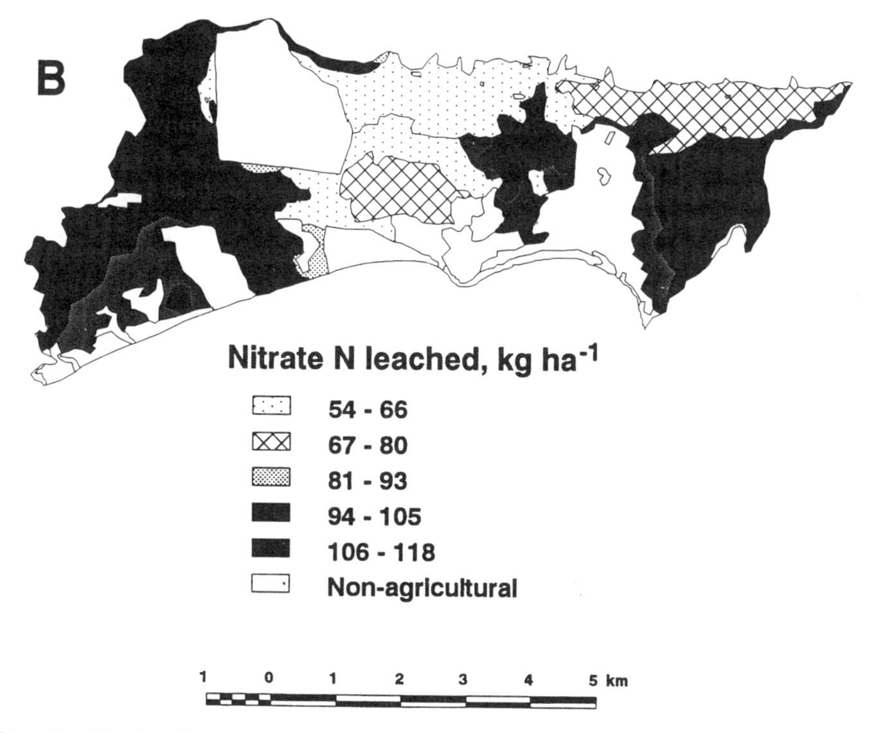

Figure 4. (*Continued.*)

tions in the tools that comprise the system. A more common source of error is probably in the database. Soil variability, in particular, is an area of much concern. It is an inherent property of soils that they vary in space and variability exists within individual pedons, in taxonomic units within mapping units and within mapping units. Conventionally, soil and landscape continua are divided into classes to reduce complexity and to create more manageable units which are typified by a central concept and defined by a set of discriminating criteria. Mapping units at any scale will always include soils that do not accord with the mapping unit. In the USA the target for map unit purity was set at 85% (Soil Survey Division Staff, 1993). That goal is rarely attained, however, and purity values of 50% or less are not uncommon. This implies that an evaluation based on the characteristics of the central concept of the map unit may only apply to half of cases. However, purity may not be a serious problem because sub-dominant soils in the unit are likely to be closely related to the named soil series and thus may have properties that lead to similar effects on land use. Nevertheless, situations exist where inclusions in the map unit are strikingly different from the defining soil; in such cases errors in evaluation may be substantial.

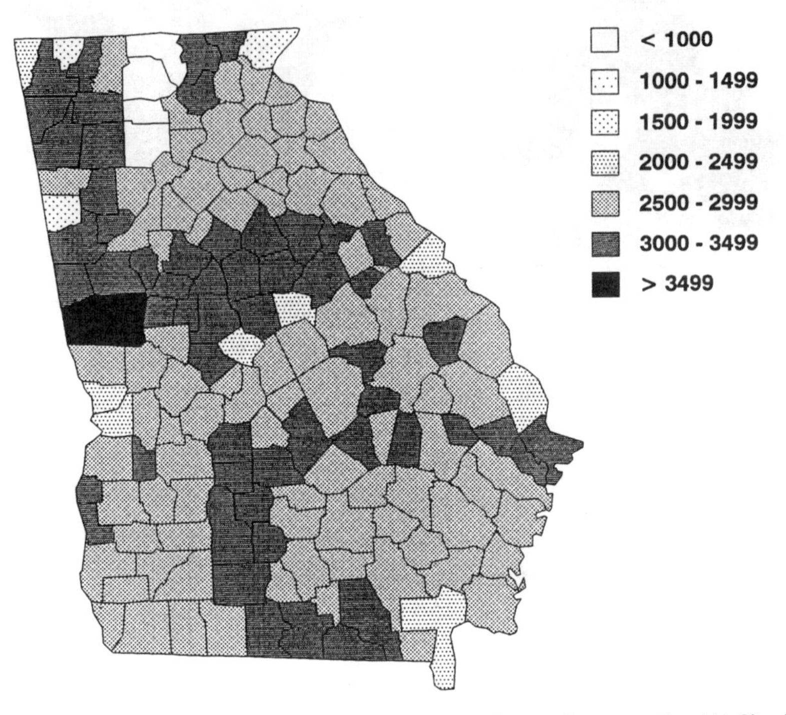

Figure 5. Example results from the use of AEGIS+ in climate change studies. (A) Simulated soybean yields for Georgia under current weather conditions. (B) Simulated soybean yields for Georgia under assumed future weather that is hotter by 3°C with current rainfall and CO_2 conditions.

An alternative approach is not to generalize information within mapping units, which always causes loss of information, but to store the survey data for each sampled point and then use spatial interpolation techniques. With the wide availability of Global Positions Systems (GPS) the precise longitude, latitude and altitude can now be routinely and almost instantaneously determined in the field. GIS integrated with digital elevation models, which portray the continuous variation of relief over space, Landsat Thematic Mapper images, and geological and vegetation data, have also been used successfully to enhance the definition of taxonomic soil components within mapping units (Amen et al., 1994). Fuzzy set theory has been applied to land resource evaluation (Burrough, 1989); this is a generalization of set theory known as Boolean algebra, and offers a useful alternative to existing methodology, but it is not yet widely used. Another avenue for research in land evaluation is the application of artificial

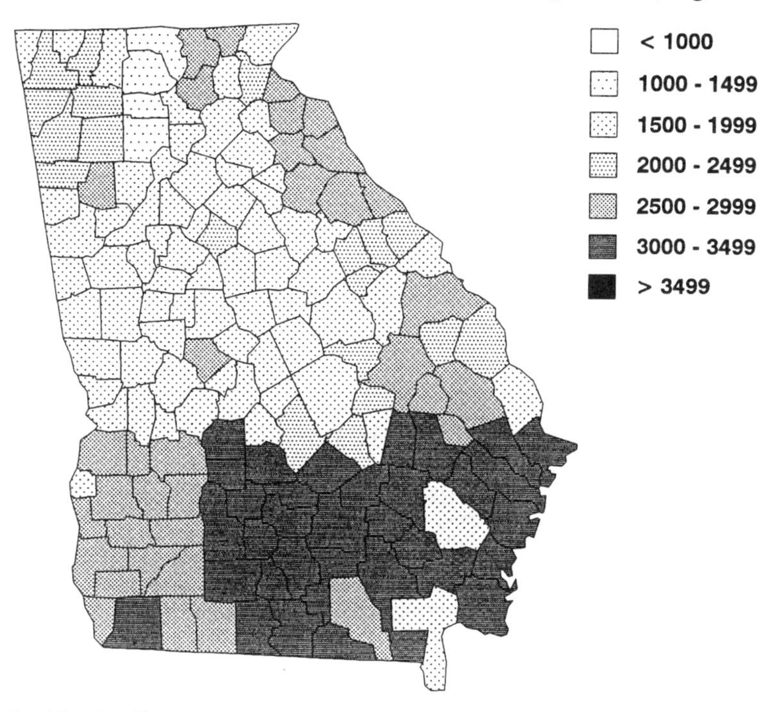

Figure 5. (*Continued.*)

neural networks that can deal with incomplete or inadequate data (Davidson, 1992).

Land resource planning and management becomes increasingly complex, especially because temporal variability in climatic variables and risk prediction must now be taken into consideration. It is essential that biophysical, economic and social factors be integrated in these approaches. Scientific and technological advances will make this approach more feasible but there will always be the need for reliable land resource data as they provide the factual underpinning for realistic analysis and synthesis that may foster a more sensitive approach to the wise use of the land.

Given the rapid advances in information science and the increasing sophistication of the tools and techniques of the information age, AEGIS will undoubtedly look crude within a short time. It was a pioneering effort, nevertheless, to link the DSSAT models to a GIS and thereby expand the scope of analysis from a site to a region.

References

Adams R M, Rosenzweig C, Peart R M, Ritchie J T, McCarl B A, Glyer J D, Curry R B, Jones J W, Boote K J, Allen L H Jr (1990) Global climate change and US agriculture. Nature 345:219–224.

Amen A E, Blaszcynski J, Page D, Jeffey J (1994) Soil survey enhancement on rangeland and wildland areas. Agronomy Abstracts P 336.

Ashton-Tate Corporation (1990) dBASE IV for developers. Ashton-Tate, Torrance, California, USA.

Beinroth F H (1990) Estimating missing soil data: how to cope with an incomplete data base. Agrotechnology Transfer 11:10–11. IBSNAT Project, Department of Agronomy and Soil Science, University of Hawaii, Honolulu, Hawaii, USA.

Burrough P A (1989) Fuzzy mathematical methods for soil survey and land evaluation. Journal of Soil Science 40:477–92.

Calixte J P, Beinroth F H, Jones J W, Lal H (1992) Linking DSSAT to a geographic information system. Agrotechnology Transfer 15:1–7. IBSNAT Project, Department of Agronomy and Soil Science, University of Hawaii, Honolulu, Hawaii, USA.

Curry R B, Peart R M, Jones J W, Boote K J, L H Allen, Jr (1990) Response of crop yield to predicted changes in climate and atmospheric CO_2 using simulation. Transactions of the American Society of Agricultural Engineers 33(4):1383–1390.

Davidson D A (1992) The evaluation of land resources. John Wiley and Sons, New York, New York, USA.

Eastman J R (1990) IDRISI: a grid-based geographic analysis system. Graduate School of Geography, Clark University, Worcester, Massachussetts, USA.

Engel T, Jones J W (1995) AEGIS/WIN version 3.0: User's Manual. Agricultural and Biological Engineering Department, University of Florida, Gainesville, Florida, USA.

Esnard A M, Pérez-Alegría L R, Beinroth F H, Goenaga R (1994) Generation of solar radiation values for use in weather files of crop simulation models. Journal of Agriculture, University of Puerto Rico, Volume 78 (1–2):33–44, Mayaguez, Puerto Rico.

ESRI (1991) ARC/INFO User's Guide. Applications programming language reference. Environmental Systems Research Institute, Inc. Redlands, California, USA.

ESRI (1992) PC ARC/INFO reference manuals, version 3.4D+ Environmental Systems Research Institute, Inc. Redlands, California, USA.

ESRI (1994) ArcView, the geographic information system for everyone. Environmental Systems Research Institute, Redlands, California, USA.

Food and Agriculture Organization (FAO) (1976) A framework for land evaluation. FAO Soils Bulletin no. 32, FAO, Rome, Italy.

Geng S, Auburn J (1987) Weather simulation models based on summaries of long-term data. Pages 237–254 in International Symposium on Impact of Weather Parameters on the Growth and Yield of Rice, 7–10 April, 1986. IRRI, Manila, Philippines.

Geng S, Auburn J, Brandstetter E, Li B (1988) A program to simulate meteorological variables: documentation for SIMMETEO. Agronomy Report No. 204, Univ. California, Davis Crop Extension, Davis, California, USA.

Hansen J W, Beinroth F H, Jones J W (1997) Systems-based land use evaluation at the South Coast of Puerto Rico. Applied Engineering in Agriculture (in press).

Hoogenboom G, White J W, Jones J W, Boote K J (1990) BEANGRO v.1.0: Dry bean crop growth simulation model user's guide. Department of Agricultural Engineering, University of Florida, Gainesville, Florida, USA.

Hoogenboom G, Jones J W, Wilkens P W, Batchelor W D, Bowen W T, Hunt L A, Pickering N B, Singh U, Godwin D C, Baer B, Boote K J, Ritchie J T, White J W (1994) Crop models. Pages 95–244 in Tsuji G Y, Uehara G, Balas S (Eds.) DSSAT version 3, Volume 2. Department of Agronomy and Soil Science, University of Hawaii, Honolulu, Hawaii, USA.

Jones J W, Hunt L A, Hoogenboom G, Godwin D C, Singh U, Tsuji G Y, Pickering N B, Thornton P K, Bowen W T, Boote K J, Ritchie J T (1994) Input and Output Files. Pages 1–95 in Tsuji G Y, Uehara G, Balas S (Eds.) DSSAT version 3, Volume 2. Department of Agronomy and Soil Science, University of Hawaii, Honolulu, Hawaii, USA.

Klingebiel A A, Montgomery P H (1961) Land-capability classification. Agricultural Handbook 210, US Department of Agriculture, Soil Conservation Service, Washington, DC, USA.

Leonard R A, Knisel W G, Still D A (1987) GLEAMS: Groundwater Loading Effects of Agricultural Management Systems. Transactions of the American Society of Agricultural Engineers 30(5):1403–1418.

Luyten J C, Jones J W, Calixte J P, Hoogenboom G, Negahban B (1994) AEGIS+ version 2.0: User's and Developer's Manual. Research Report AGE 94–1. Agricultural Engineering Department, University of Florida, Gainesville, Florida, USA.

Papajorgji P, Jones J W, Calixte J P, Bienroth F H, Hoogenboom G (1993) A generic geographic decision support system for estimating crop performance. Pages 340–348 in Proceedings, Integrated Resource Management and Landscape Modification for Environmental Protection. American Society of Agricultural Engineers, St. Joseph, Michigan, USA.

Papajorgji P, Jones J W, Peart R M, Curry R B (1994a) Using crop models and geographic information systems to study the impact of climate change in the Southeastern USA. Soil and Crop Science Society of Florida Proceedings 53:82–86.

Papajorgji P, Jones J W, Hoogenboom G, Calixte J P (1994b) Exploring concepts for linking GIS and crop models. Pages 598–602 in Proceedings, Computers in Agriculture 1994. American Society of Agricultural Engineers, St. Joseph, Michigan, USA.

Peart R M, Jones J W, Curry R B, Boote K J, Allen L H (1989) Impact of climate change on crop yield in the southeastern USA: A simulation study. Pages 1–54 in The Potential Effects of Global Climate Change on the United States. Appendix C, Agriculture, Volume 1. US Environmental Protection Agency Report 230–05–89–053.

Peart R M, Curry R B, Jones J W, Boote K J, Allen L H Jr (1995) Integrated systems for evaluating impacts of changing climate on water used by agriculture in the southeast region. Final Report to SE Regional Climate Center, Agricultural and Biological Engineering Department, University of Florida, Gainesville, Florida, USA.

Pickering N B, Hansen J W, Jones J W, Wells C M, Chan V K, Godwin D C (1994) WeatherMan: A utility for managing and generating daily weather data. Agronomy Journal 86:332–337.

Richardson C W, Wright D A (1984) WGEN: A model for generating daily weather variables. Publication ARS–8. US Department of Agriculture, Washington, DC, USA.

Rosenzweig C, Jones J W (1990) Climate change crop modeling study. US Environmental Protection Agency, Office of Policy, Planning, and Evaluation, Washington, DC, USA.

Rosenzweig C, Ritchie J T, Jones J W, Tsuji G Y, Hildebrand P (1995) Climate change and agriculture: analysis of potential international impacts. ASA Special Publication Number 59, American Society of Agronomy, Madison, Wisconsin, USA.

Rossiter D G (1990) ALES: A framework for land evaluation using a microcomputer. Soil Use and Management 6:7–20.

Sánchez P A, Couto W, Boul S W (1982) The fertility capability soil classification system: interpretation, applicability, and modification. Geoderma 27:283–309.

Smith N P, Tirpak D A (Eds.) (1989) The potential effects of global climate change in the United States, Appendix C, Agriculture. US Environmental Protection Agency, Washington, DC, USA.

Soil Survey Division Staff (1993) Soil survey manual. US Department of Agriculture Handbook 18, US Government Printing Office, Washington, DC, USA.

Soil Survey Staff (1975) Soil taxonomy – a basic system of soil classification for making and interpreting soil surveys. US Department of Agriculture Handbook 436, US Government Printing Office, Washington, DC, USA.

Thornton P K, Booltink H W G, Stoorvogel J J (1997) A computer program for geostatistical and spatial analysis of crop model outputs. Agronomy Journal (in press).

Tsuji G Y, Uehara G, Balas S (Eds.) (1994) DSSAT version 3. Department of Agronomy and Soil Science, University of Hawaii, Honolulu, Hawaii, USA.

Wischmeier W H, Smith D D (1978) Predicting rainfall erosion losses. US Department of Agriculture Handbook 537, US Government Printing Office, Washington, DC, USA.

The simulation of cropping sequences using DSSAT

W.T. BOWEN[1], P.K. THORNTON[2] and G. HOOGENBOOM[3]

[1]*Centro Internacional de la Papa/International Fertilizer Development Center, Apartado 1558, Lima 100, Peru*

[2]*International Livestock Research Institute, P.O. Box 30709, Nairobi, Kenya*

[3]*Department of Biological and Agricultural Engineering, University of Georgia, Griffin, Georgia 30223, USA*

Key words: rotation, management, model, soil, water, nitrogen

Abstract

The crop models that are distributed with DSSAT version 3 permit the simulation of long-term cropping sequences. This added capability enhances the value of crop growth models as a tool for analyzing the long-term consequences of management practices on a site-specific basis. The models can be used to predict changes and detect trends in biophysical indicators such as crop yield, nitrogen uptake, soil carbon levels, and nitrate leaching. In this way, management practices that are potentially sustainable or unsustainable can be identified. In the systems approach to research, simulation tools remain in constant need of critical evaluation and refinement. The models, which currently provide reasonable simulation of single-season processes, require a better understanding of long-term soil processes and how these processes are affected by management.

Introduction

Agricultural science today is facing a critical dilemma. At the same time that it is expected to help achieve a sustainable growth in food production, it is also being asked to conduct research with a dwindling allocation of resources. To confront this challenge, agricultural scientists need to place a greater emphasis on the efficient organization of research and the knowledge that it generates. One approach to gaining improved efficiency is through the integration of research activities with the construction and application of dynamic simulation models.

Following advances in computer technology and accessibility, models of soil and plant systems have become increasingly valuable instruments for assimilating knowledge gained from experimentation. Their use within a research program has the potential to increase efficiency by emphasizing process-based research, rather than the study of site-specific net effects. Consequently, a modeling approach lends structure to a research program, helping to focus on the quantitative description of soil and plant processes. This information can then be used to predict how the system might respond to different environmental and management factors. A modeling approach also provides a dynamic, quantitative framework for multidisciplinary input.

If it is to increase the efficiency of research, the modeling process must become a truly integrated part of the research process. Experimentation and

G.Y. Tsuji et al. *(eds): Understanding Options for Agricultural Production, 313–327*

model development need to proceed jointly; whereas new knowledge is used to refine and improve models, models are used to identify gaps in knowledge, thereby helping to set research priorities. To be most effective, the modeling approach requires a regular evaluation of progress and continual refinement of objectives and priorities. It also requires a commitment to the development of software and the establishment of data standards that facilitate a functional understanding of how soil and plant systems work.

By integrating experimental and modeling activities, several multidisciplinary teams of scientists have been able to assemble comprehensive models that provide quantitative estimates of crop production under a wide range of soil, weather, and management conditions (Whisler et al., 1986; Penning de Fries et al., 1989; Jones and Ritchie, 1991; Ritchie, 1991; Jones, 1993). Constructed primarily for predicting yield during a single growing season, these models usually describe plant growth on a daily basis at the process level, such as carbon assimilation, partitioning, phenology, and water and nitrogen uptake. Attention to this level of detail has resulted in crop growth models that realistically simulate the sensitivity of growth and development to changes in solar radiation, temperature, photoperiod, and water and nitrogen availability. While these models have helped to improve our understanding of crop, soil, weather, and management interactions in the field, their application has been mostly limited to simulating processes during the course of a single growing season. In general, they have not been applied to the simulation of long-term cropping systems.

Efforts to address the sustainability of cropping systems demand simulation tools that go beyond a single season to include the realistic simulation of crops grown in sequence. To be useful, such tools must incorporate a quantitative understanding of how the soil resource changes with time because of weather and the effects of management, and how such changes have an impact on crop production. Interest in predicting the long-term productivity of cropping systems has resulted in the construction of a few models designed to simulate certain soil processes over long periods of time (Shaffer, 1985; Parton et al., 1987; Williams, 1990). To make them applicable to a wide range of cropping systems, most of these models contain generic plant growth models that describe growth and development processes in less detail than many of the single-season crop models referred to earlier. The limited sensitivity of the generic growth models has sometimes meant grain or biomass yields were not realistically simulated in all years when compared with long-term experimental data (Steiner et al., 1987; Paustian et al., 1992; Parton and Rasmussen, 1994; Probert et al., 1995). In addition, the generic growth models are usually non-crop specific and often need local calibration.

This brief mention of existing crop production models points to an obvious contrast in present capabilities. Some models are able to mimic the sensitivity of plant growth and development during a single season, without the capability to simulate crops grown in sequence; other models, on the other hand, are able to simulate long-term cropping sequences, but have less sensitive and robust plant growth components. Of course, there are historical reasons for the apparent contrast in model capabilities, among which could be included earlier limitations in computing power as well as differences in the objectives for developing specific models. Today, however, ever more powerful computers and broader objectives are driving the development of improved simulation tools for analyzing both the long-term productivity of different agricultural systems and their potential impact on the environment.

One approach to the development of improved tools would be to expand the capabilities of the more detailed single-season growth models to include the long-term simulation of soil-related processes. In a significant step toward this goal, the crop growth models that are distributed with the DSSAT version 3.0 software (DSSAT v3) have been linked together to enable them to simulate crops grown in a rotation or a continuous sequence. This new feature of DSSAT v3 allows a user to simulate a specific cropping sequence for any number of years using either measured or statistically generated weather. Also included in DSSAT v3 is an analysis program for studying long-term trends and variability in simulated output using a combination of graphical and statistical procedures (Thornton et al., 1995).

The crop sequencing option in DSSAT v3 now provides an operational tool for simulating selected sequences of crops, although further development and testing are needed to improve the simulation of long-term soil processes. Recognizing this need, we believe that implementation of the sequencing capability contributes a basic structure of detailed plant growth simulators to which more comprehensive and accurate descriptions of soil processes can be added as our understanding of these processes improves.

In this chapter, we describe how the crop models have been linked to simulate cropping sequences and provide some simple examples of how this capability might be used to evaluate the long-term productivity of a cropping system. We also discuss opportunities for model improvement, particularly in the long-term simulation of soil processes.

The DSSAT crop models

The DSSAT v3 originally contained five separate models for simulating the growth of 11 different crops (Table 1). Although these models have been developed by different groups of researchers and institutions, there has been a coordinated effort to standardize input and output data formats (Jones et al., 1994) and to implement the same soil water and nitrogen balance in each model (Hoogenboom et al., 1994). Therefore, all of the models contain similar subroutines for describing soil-related processes and for reading and writing data using the same variable names. The models themselves remain separate entities because crop growth and development processes continue to be described differently in each model.

Table 1. Principal crop models released as part of the DSSAT version 3.0 (Hoogenboom et al., 1994).

Crop model	Crops simulated
CERES-Generic	Maize, wheat, barley, millet, sorghum
CERES-Rice	Rice (upland and flooded rice)
CROPGRO	Soybean, peanut, dry bean
SUBSTOR-Potato	Potato
CROPSIM-Cassava	Cassava

Model linkage for simulating cropping sequences

The DSSAT v3 crop models were developed for simulating the growth of annual crops during a single season. Their similar structure, however, has allowed the use of these same models for the long-term simulation of cropping sequences with minimal modification to the code. To facilitate the running of the models in sequence, new subroutines were added to each model that permit the passing of relevant variables from one model to the next in a temporary file named 'TMP.DAT' (Figure 1). When the cropping sequence option is specified, this temporary file is written out at the end of one model run and then read at the beginning of the next model run. A separate 'driver' program was also developed to control the order in which the crop models are run. The driver program reads the order of the cropping sequence from an experimental details file at the beginning of the simulation, then continues running the respective models for the number of years specified in the same experimental details file.

The simulation of a cropping sequence requires the continuous calculation of soil-related processes on a daily basis, including the days when no crop is growing. Therefore, most of the variables passed in the temporary file are those needed for the continuous simulation of soil water, carbon, and nitrogen processes (Table 2). Passing variables this way permits a seamless long-term simulation of soil-related processes, even though different crop models are being run. For example, the daily variation in extractable (plant available) water for a simulated soybean–winter wheat–soybean rotation is shown in Figure 2. This simulation was obtained by running the CROPGRO and the CERES crop models in sequence, with initial soil conditions – volumetric soil water content, organic carbon, and inorganic nitrogen in each layer of the soil profile – specified for only the first model in the sequence. Succeeding model runs then started with the simulated values calculated for the last day of the previous model run, which were passed in the temporary 'TMP.DAT' file.

Analysis of cropping sequence simulations

When simulating a cropping sequence, the user will usually not be interested in examining daily differences in soil water, inorganic nitrogen, biomass accu-

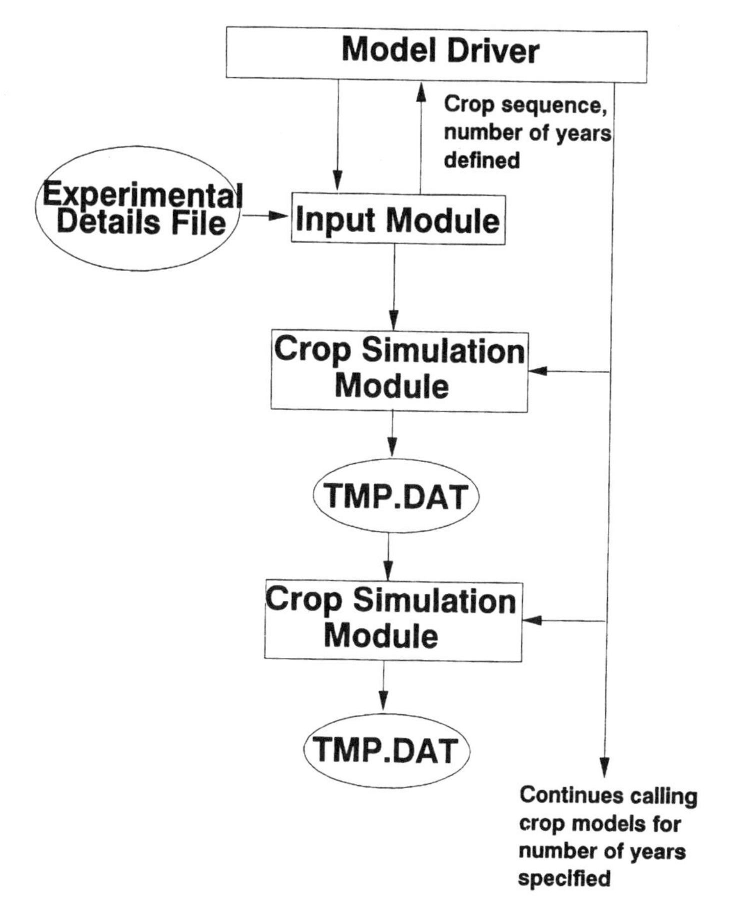

Figure 1. Links between the model driver, input module, and crop simulation models to simulate continuous cropping sequences in DSSAT v3.

mulation, or any other output generated on a daily basis, although such an analysis is possible (Figure 2). Frequently of greater interest will be the determination of any trends in end-of-season output such as yield, nitrogen uptake, or soil organic carbon levels. The tendency for such variables to change with time in a consistent direction will define the trend, and it is this trend that can be used to estimate the potential sustainability of a defined cropping sequence and management system. For example, a 30-year simulation might show yields decreasing with time, thus indicating the cropping system as simulated is not likely to be sustainable. Likewise, a positive trend or no trend in yield might indicate the system is sustainable, at least from a biophysical perspective (Lynam and Herdt, 1989).

To facilitate the analysis of long-term trends, a software program was developed to read simulated output, provide summary statistics, and present the

Table 2. Principal variables passed in the temporary file used to link DSSAT version 3.0 models.

Variable	Description
YRSIM	Date at the end of the previous model run
NREP	Model run number
STOVRL	Weight of aboveground plant residue (kg dry matter ha^{-1})
APTNPL	Amount of N in the aboveground plant residue (kg N ha^{-1})
RTWTL	Weight of roots in the soil profile (kg dry matter ha^{-1})
RTWTNL	Amount of N in the roots (kg N ha^{-1})
DEPMAX	Maximum soil depth (cm)
NLAYR	Number of soil layers
SW(L)	Soil water content of layer L (cm^3 cm^{-3})
ESW(L)	Extractable soil water content for soil layer L (mm)
PESW	Extractable soil water in the total profile (cm)
FOM(L)	Plant residues in soil layer L (kg dry matter ha^{-1})
FON(L)	Amount of N in plant residues in soil layer L (kg N ha^{-1})
HUM(L)	Stable humic fraction material in soil layer L (kg C ha^{-1})
NHUM(L)	N in the stable humic fraction in soil layer L (kg N ha^{-1})
BD(L)	Bulk density of soil layer L (g cm^{-3})
PH(L)	pH of soil layer L
OC(L)	Soil organic C content in soil layer L (%)
NO3(L)	Nitrate-N content in soil layer L (mg kg^{-1})
NH4(L)	Ammonium-N content in soil layer L (mg kg^{-1})
RSEED	Random number seeds for weather generation

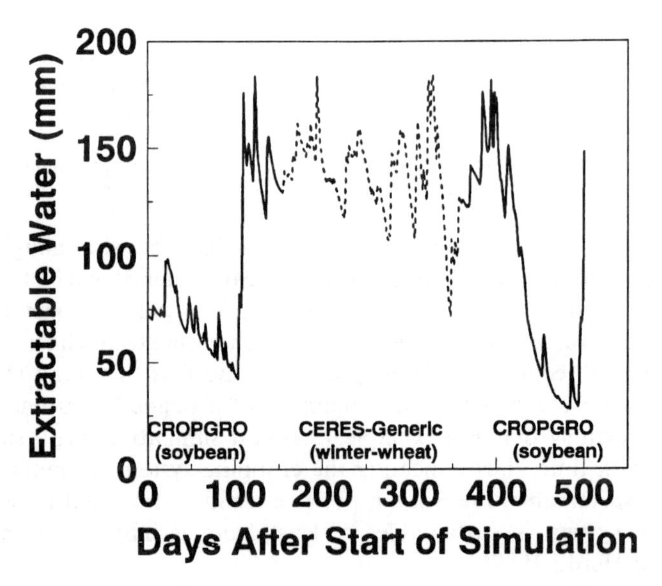

Figure 2. Variation in extractable water obtained from simulating a soybean–winter wheat–soybean rotation for a site in Georgia, USA.

data in both tabular and graphical forms (Thornton et al., 1994; Thornton et al., 1995). In order that sustainability may be analyzed from an economic as well as a biophysical perspective, this program also performs an analysis of net monetary returns or gross margins on the simulated output using product prices and production costs set by the user. Since future costs and prices are not known with certainty, the user can choose to specify their variability according to a normal distribution; alternatively, a single, fixed cost or price may be assigned.

The analysis program is particularly useful when a cropping sequence is simulated using more than one series of synthetic weather, which can be generated using a statistical weather generator coded into each crop model (Hoogenboom et al., 1994). Unless a cropping sequence is being simulated to compare results with observed data, in which case there exists only one series of relevant measured weather data with which to run the models, there is no unique series of weather data that can be assigned to the simulation of a hypothetical sequence. For a given site, any number of possible weather patterns might prevail, and there is no way of predicting a specific weather pattern. Therefore, to account for the uncertainty in expected weather, the DSSAT v3 sequencing option allows the user to run the same cropping sequence several times with different combinations of synthetic weather. In this way, probability distributions for different model outputs can be obtained.

The capability to simulate and analyze a cropping sequence under equally plausible weather scenarios provides a powerful tool for estimating the full range of probable outcomes, as well as for separating out any trends or variability due to weather. Depending on the site, variability in weather can have a substantial impact on expected outcomes, particularly dry matter production. To illustrate, Figure 3 shows a time-series graph of grain yield obtained from the simulation of a continuous maize-fallow rotation for 50 years. The graph shows two different lines representing the expected yield using different 50-year runs of synthetic weather (only two runs are shown for ease of presentation, although many more could be simulated). Yields are shown to decrease because the simulation assumed no external source of nitrogen was added, so the system eventually degrades as the soil nitrogen supply is depleted. When the two runs in Figure 3 are compared, we see that yields were different in most years of the sequence because weather inputs varied.

By making more runs of the same sequence with a greater number of weather scenarios, it is possible to obtain a distribution of expected yields which can be used to estimate probability levels. For example, ten runs (i.e., ten different weather scenarios) of the same 50-year maize-fallow rotation could provide sufficient replication to obtain cumulative probability function (CPF) plots for grain yield at different times during the rotation. This is illustrated in Figure 4, which shows the CPF plots for grain yield at five years and then 49 years after the start of the rotation. Clearly, for all weather scenarios, yields are expected to decrease as the rotation continues with no external sources of extra nitrogen.

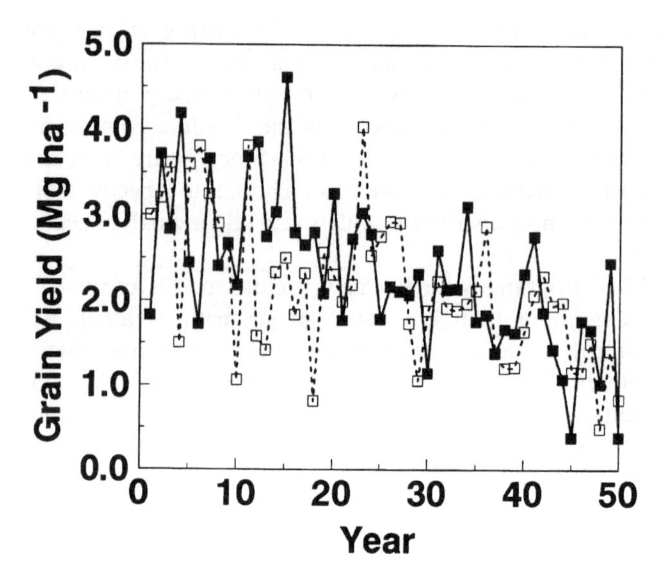

Figure 3. Expected grain yields obtained from simulating a maize–fallow rotation (assuming no external sources of N) using two different weather scenarios generated for a site in central Brazil.

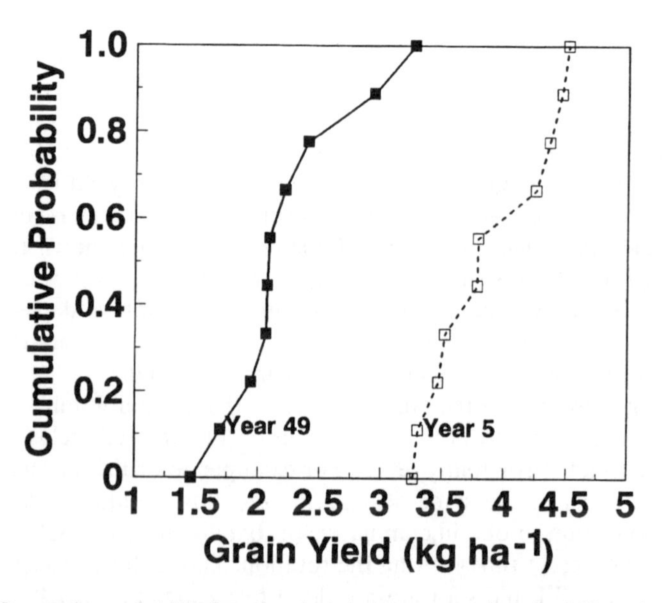

Figure 4. Cumulative probability function plots showing expected grain yields at 5 and 49 years after the start of a maize–fallow rotation (assuming no external sources of N) simulated using ten different weather scenarios generated for a site in central Brazil.

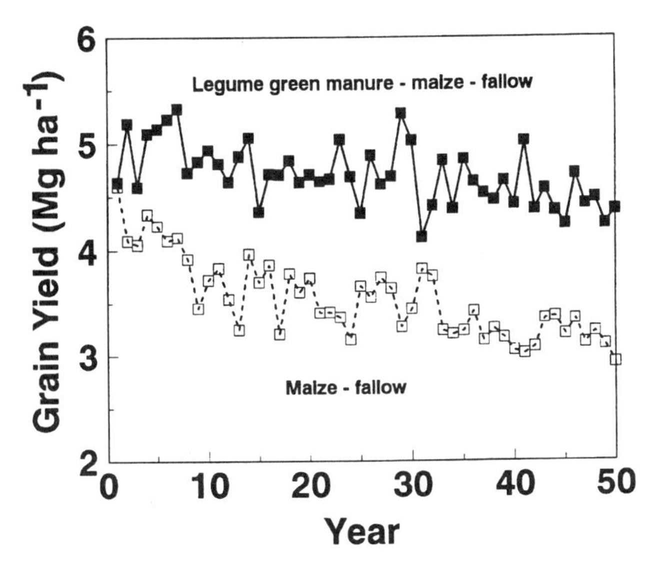

Figure 5. Expected grain yields obtained from simulating a 50-year short-duration legume green manure–maize–fallow rotation compared with a maize–fallow rotation with no N inputs for a site in central Brazil.

In the DSSAT v3 sequence analysis program, separate runs of the same sequence using different weather inputs are analyzed as replications (Thornton et al., 1995). For the example cited above involving ten different runs of a 50-year cropping sequence, the analysis program would provide statistics on the simulated output assuming there were ten replications. Simulated results can then be presented as tables with means, standard deviations, and maxima and minima, or they can be plotted as box plots, CPF plots (Figure 5), mean-variance diagrams, coefficient of variation plots, or variance plots. In general, to obtain more stable estimates of output means and variances, it is suggested that hypothetical sequences be simulated using at least ten replications of synthetic weather (Thornton et al., 1994).

Potential application examples

Examples of how the DSSAT v3 crop sequencing option and the analysis program might be used to investigate the long-term productivity of different cropping systems have been offered by Singh and Thornton (1992), Bowen et al. (1993a), Jones et al. (1993), and Thornton et al. (1995). Produced when the sequencing capability and other enhancements (such as the subroutines for N_2-fixation and decision modules for automatic planting and harvesting) were being added to the DSSAT v3 models, these examples were meant to provide only a few simple illustrations of the types of simulations and analyses that might be performed using the improved model capabilities.

Singh and Thornton (1992) provided an early demonstration of how simulated yield distributions might be obtained using replicated weather scenarios. The basis for their simulation study was the comparison of a continuous maize system with nitrogen fertilizer applied versus a soybean-maize rotation with no nitrogen fertilizer applied for a site in Florida. In a similar comparison of continuous maize and soybean-maize rotations with different nitrogen management practices for a site in Brazil, Jones et al. (1993) examined changes in grain yield and soil organic carbon obtained with replicated 60-year simulations. This last simulation study used a modified version of the soil nitrogen model, which was based on the way soil carbon and nitrogen pools are represented in the CENTURY model (Parton et al., 1987). Following development of the sequence analysis program, Thornton et al. (1995) showed how the program might be used to investigate the long-term stability and profitability of a simulated cropping system.

The simulation study described by Bowen et al. (1993a) provides a practical example of how the crop sequencing option and other model capabilities might be used in formulating and testing hypotheses. It also serves to demonstrate how the models may be used to guide further field research. The purpose of the simulation study was to determine the feasibility of growing a short-duration legume green manure each year as a nitrogen source for continuous maize in the wet-dry tropics of central Brazil. The models were thought to be an appropriate tool for such an analysis since previous studies have shown they can provide realistic estimates of nitrogen made available from decomposing leguminous residues (Bowen et al., 1993b; Quemada and Cabrera, 1995). Using the automatic planting and harvesting decision modules, the simulation was set up through the experimental details file so that the models automatically plant a legume at the beginning of the rainy season when soil moisture conditions are appropriate, incorporate the legume when it reaches its flowering stage, and then plant maize. Employing the automatic planting and harvesting decision modules means that planting and incorporation dates will vary from year to year depending on weather conditions. This same rotation of a short-duration legume followed by maize was simulated continuously for 50 years using replicated weather scenarios. With the replicated sequences it was possible to obtain output distributions for analyzing the following: (1) the earliest dates the legume might be planted so that it received adequate rainfall; (2) how long the legume might grow before it reached the flowering stage; (3) how much dry matter and nitrogen from both N_2-fixation and soil nitrogen uptake might be in the legume when it was to be plowed in at flowering; (4) how much nitrogen might be made available to succeeding maize; and (5) how much grain might the maize crop yield. The mean grain yields obtained from a 50-year simulation of this system are shown in Figure 5. For comparison purposes, Figure 5 also shows the mean yields obtained from simulating a continuous maize system with no short-duration legume or other nitrogen inputs.

The simulation results from this example reveal that a short-duration legume could provide a significant, and perhaps sustainable, source of nitrogen for maize when plowed in as a green manure, although the amount of nitrogen would not be sufficient for maximum yields. After demonstrating the long-term feasibility of such a cropping system, the next step might be to grow one or more legumes in the field to determine if the simulated dry matter and nitrogen production levels are realistic for short-duration growth. Applied this way, the simulation models provide a valuable tool for testing hypotheses and screening management options, thus indicating which components of a proposed cropping system might need further study in the field.

Opportunities for improvement

The development of reliable models for simulating cropping sequences is a continuing activity. Initially, the challenge was to create a mechanism for long-term simulation without constructing a new model or rewriting present versions of the DSSAT v3 models. As noted above, this was accomplished through the creation of a driver program that controls which models are called according to a cropping sequence defined by the user. To provide for a continuous simulation, each model was modified so that relevant variables can be passed between models using a temporary file. Concurrent with this activity was the development of graphical and statistical programs for displaying and interpreting simulation data. Together, these activities have resulted in the implementation of an easily accessible option for the simulation and analysis of many different cropping sequence and management alternatives.

Now that the mechanics of sequencing the DSSAT v3 crop models have been addressed, and software developed for conveniently examining output using graphical and statistical methods, much more emphasis needs to be placed on improving the knowledge base presently incorporated into the models, specifically that related to the description of long-term soil processes. Although there has been a continuous effort to test and improve the crop growth components of the DSSAT v3 models (e.g., Grimm et al., 1993; 1994), less has been done to improve on the simulation of soil-related components. If the capability to run the crop models in sequence is to provide realistic simulations of the long-term consequences of alternative management systems, then better submodels for describing soil-related processes must be incorporated.

Simulating soil productivity

The soil resource is a critical component of productive cropping systems. If not managed properly, its capacity to supply water and nutrients becomes limiting, thus resulting in a decrease in potential yield. Although many factors impact on the long-term productivity of a cropping system, any decrease in yield with time can often be traced to a loss in soil productivity. This loss may come about for any number of reasons, including the loss of soil by erosion, a decrease in soil organic matter, a decrease in the supply of an essential nutrient, or the build up of a toxic element.

Since a decrease in soil productivity is often gradual, with changes in soil properties occurring over many seasons, the true nature of these changes can best be quantified through long-term field experiments and monitoring studies. Such studies, however, cannot be conducted for all possible combinations of soil, crop, weather, and management factors that merit evaluation. Because simulation models can include innumerable combinations of these factors, they have the potential to be practical tools for the site-specific evaluation of alternative cropping systems and management practices. Before they can be used reliably, however, such models must be shown through rigorous testing to provide realistic estimates of how crops, weather, and management interact to change the soil resource, and how these changes in turn affect crop production. Such testing requires good long-term field data, and this underlines the importance of well-conducted field experiments for both model construction and model evaluation (Acock and Acock, 1991).

Although the DSSAT v3 models have been linked to simulate cropping sequences, their reliability in simulating long-term soil processes has yet to be demonstrated. This is partly the result of the sequencing capability being added only relatively recently, and also because there is a scarcity of long-term data sets sufficiently complete for conducting appropriate tests of model assumptions. At some point, the DSSAT v3 models must undergo more rigorous testing with long-term field data. Such testing will undoubtedly reveal weaknesses in the modeling of long-term soil processes, both in terms of processes inadequately described by the models as well as critical processes simply not included in the present version of the models.

The DSSAT v3 models were originally developed to simulate a limited number of soil processes (water, carbon and nitrogen dynamics), during a single season. Because of the emphasis on simulating a single growing season, many of the processes known to affect soil productivity during longer periods of time were not included. Of the processes that were included, fairly simple descriptions were often found to be adequate when simulating soil-related processes during only one growing season. Such simple descriptions, however, are not likely to be sufficiently accurate for long-term simulation. For example, even though the use of a one-pool model to describe the carbon and nitrogen dynamics of soil organic matter (SOM) was shown to provide realistic estimates of nitrogen released as nitrate from SOM during one growing season (Bowen et al., 1993b), a one-pool model may not be appropriate for predicting SOM dynamics for longer periods of time. The heterogenous nature of SOM and differences in component contributions to the soil nitrogen supply will undoubtedly require implementation of a multi-pool model for describing both carbon and nitrogen dynamics (Duxbury et al., 1989).

The long-term productivity of a cropping system can be influenced by many soil-related processes. Too numerous and too complex to be incorporated entirely in any one model, these processes can at least be reduced to a subset of those most likely to have an impact on long-term yields. Such a subset should contain, at a minimum, those processes that define water and nitrogen availability; long-term simulation models, to be useful, need to include accurate descriptions of any process or property of the soil that influences the availability of water and nitrogen with time.

Closer examination of the DSSAT v3 crop models reveals that many of the processes needed to accurately simulate the long-term availability of water and nitrogen across a wide range of conditions are presently lacking. As noted above, there is a need to include a multi-pool SOM submodel. Also needed is the capability to distinguish between plant residues left on the surface and those incorporated in the soil; the models presently simulate the decomposition of residue assuming it is all incorporated, with no accounting made of the different dynamics driving the decomposition of residue left on the surface and the impact such surface residue might have on the water balance. Other important processes presently left out include those related to soil erosion and tillage. Further analysis might also point to the need to include a description of how changes in soil reaction, particularly a build-up of soil acidity, might occur because of the management of nitrogen fertilizer inputs, and how such acidity might affect crop production. The retarded leaching of nitrate in the subsoil of variable charge soils is another process that field data indicates is important and should be included in the models (Bowen et al., 1993b).

Ways to incorporate these major processes within the framework of the DSSAT v3 crop models as well as other closely related models are being investigated. Descriptions of some of these processes within a similar crop model framework (APSIM) have already been made (Probert et al., 1995; McCown et al., 1996).

Error propagation problems

The simulation of long-term cropping systems has the potential to provide inaccurate predictions of system behavior as time progresses from the start of the simulation. Such inaccurate predictions may arise because of errors in the mathematical assumptions underlying process descriptions, inaccurate model parameters, or inaccurate inputs. While these errors or inaccuracies might prove inconsequential when simulating one growing season, they can propagate and result in substantial error when simulating several consecutive growing seasons. It is this potential for error propagation that demands reliable experimental data be used to critically test and evaluate the output provided by long-term simulation.

Conclusions

Continued experimentation, monitoring studies, and access to analysis tools such as crop simulation models are needed to secure the long-term productivity of cropping systems. When pursued jointly, modeling, experimentation, and monitoring activities provide a powerful approach to improving our understanding of how soil, crop, weather, and management factors interact, at a given site, to influence the sustainability of a cropping system.

To assist in the modeling of different cropping systems, a basic framework for simulating and analyzing long-term cropping sequences has been assembled using the crop models released with DSSAT v3. The framework is flexible enough to easily accommodate the inclusion of new or improved crop models as well as new and improved routines for describing soil-related processes. Although it presently lacks sufficiently robust algorithms for dealing with some important soil-related processes such as soil erosion and tillage, the framework represents significant progress in providing a flexible tool for investigation and hypothesis testing using well-validated annual crop models. Continuing research activities coupled with further model development and testing are bound to result in an ever more useful framework for analyzing the complexity of agricultural systems.

References

Acock B, Acock M C (1991) Potential for using long-term field research data to develop and validate crop simulators. Agronomy Journal 83:56–61.

Bowen W T, Jones J W, Thornton P K (1993a) Crop simulation as a potential tool for evaluating sustainable land management. Pages 15–21 in Kimble J M (ed.) Utilization of soil survey information for sustainable land use. USDA-SCS, Washington, USA.

Bowen W T, Jones J W, Carsky R J, Quintana J O (1993b) Evaluation of the nitrogen submodel of CERES-Maize following legume green manure incorporation. Agronomy Journal 85:153–159.

Duxbury J M, Smith M S, Doran J W, Jordan C, Szott L, Vance E (1989) Soil organic matter as a source and a sink of plant nutrients. Pages 33–67 in Coleman D C, Oades J M, Uehara G (Eds.) Dynamics of soil organic matter in tropical ecosystems. NifTAL Project, University of Hawaii, Honolulu, Hawaii.

Grimm S S, Jones J W, Boote K J, Hesketh J D (1993) Parameter estimation for predicting flowering dates of soybean cultivars. Crop Science 33:137–144.

Grimm S S, Jones J W, Boote K J, Herzog D C (1994) Modeling the occurrence of reproductive stages after flowering for four soybean cultivars. Agronomy Journal 86:31–38.

Hoogenboom G, Jones J W, Wilkens P W, Batchelor W D, Bowen W T, Hunt L A (1994) Crop model user's guide. DSSAT Version 3.0. Vol. 2. University of Hawaii, Honolulu, Hawaii.

Jones J W (1993) Decision support systems for agricultural development. Pages 459–471 in Penning de Vries F W T, Teng P, Metslaar K (Eds.) Systems approaches for agricultural development. Kluwer Academic Publishers, the Netherlands.

Jones J W, Ritchie J T (1991) Crop growth models. Pages 63–89 in Hoffman G J, Howell T A, Solomon K H (Eds.) Management of farm irrigation systems. The American Society of Agricultural Engineers, St. Joseph, Michigan.

Jones J W, Bowen W T, Boggess W G, Ritchie J T (1993) Decision support systems for sustainable agriculture. Pages 123–138 in Ragland J, Lal R (Eds.) Technologies for sustainable agriculture in the tropics. American Society of Agronomy, Special Publication 56, Madison, Wisconsin.

Jones J W, Hunt L A, Hoogenboom G, Godwin D C, Singh U, Tsuji G Y, Pickering N B, Thornton P K, Bowen W T, Boote K J, Ritchie J T (1994) Input and output files. Pages 1–94 in Tsuji G Y, Uehara G, Balas S (Eds.) DSSAT version 3, Vol. 2-1. University of Hawaii, Honolulu, Hawaii.

Lynam J K, Herdt R W (1989) Sense and sustainability: sustainability as an objective in international agricultural research. Agricultural Economics 3:381–398.

McCown R L, Hammer G L, Hargreaves J N G, Holzworth D P, Freebairn D M (1996) APSIM: a novel software system for model development, model testing and simulation in agricultural systems research. Agricultural Systems 50:255–271.

Parton W J, Schimel D S, Cole C V, Ojima D S (1987) Analysis of factors controlling soil organic matter levels in Great Plains grasslands. Soil Science Society of America Journal 51:1173–1179.

Parton W J, Rasmussen P E (1994) Long-term effects of crop management in wheat-fallow: II. CENTURY model simulations. Soil Science Society of America Journal 58:530–536.

Paustian K, Parton W J, Persson J (1992) Modeling soil organic matter in organic-amended and nitrogen-fertilized long-term plots. Soil Science Society of America Journal 56:476–488.

Penning de Vries F W T, Jansen D M, ten Berge H F M, Bakema A (1989) Simulation of ecophysical processes of growth in several annual crops. Simulation Monograph Series, PUDOC, Wageningen, The Netherlands and IRRI, Los Banos, Philippines.

Probert M E, Keating B A, Thompson J P, Parton W J (1995) Modelling water, nitrogen, and crop yield for a long-term fallow management experiment. Australian Journal of Experimental Agriculture 35:941–950.

Quemada M, Cabrera M L (1995) CERES-N model predictions of nitrogen mineralized from cover crop residues. Soil Science Society of America Journal 59:1059–1065.

Ritchie J (1991) Specifications of the ideal model for predicting crop yields. Pages 97–122 in Muchow R C, Bellamy J A (Eds.) Climate risk in crop production: models and management for the semiarid tropics and subtropics. CAB International, Wallingford, UK.

Singh U, Thornton P K (1992) Using crop models for sustainability and environmental quality assessment. Outlook in Agriculture 21:209–218.

Shaffer M J (1985) Simulation model for soil erosion-productivity relationships. Journal of Environmental Quality 14:144–150.

Steiner J L, Williams J R, Jones O R (1987) Evaluation of the EPIC simulation model using a dryland wheat-sorghum-fallow crop rotation. Agronomy Journal 79:732–738.

Thornton P K, Wilkens P W, Hoogenboom G, Jones J W (1994) Sequence analysis. Pages 67–136 in Tsuji G Y, Uehara G, Balas S (Eds.) DSSAT version 3.0. Vol. 3-2. University of Hawaii, Honolulu, Hawaii.

Thornton P K, Hoogenboom G, Wilkens P W, Bowen W T (1995) A computer program to analyze multiple-season crop model outputs. Agronomy Journal 87:131–136.

Whisler F D, Acock B, Baker D N, Fye R E, Hodges H F, Lambert J R, Lemmon H E, McKinion J M, Reddy V R (1986) Crop simulation models in agronomic systems. Advances in Agronomy 40:141–208.

Williams J R (1990) The erosion-productivity impact calculator (EPIC) model: a case history. Philosophical Transactions of the Royal Society, London, Series B 329:421–428.

Risk assessment and food security

P.K. THORNTON[1] and P.W. WILKENS[2]
[1] *International Livestock Research Institute (ILRI), P.O. Box 30709, Nairobi, Kenya*
[2] *International Fertilizer Development Center (IFDC), P.O. Box 2040, Muscle Shoals, Alabama 35662, USA*

Key words: risk, variability, enterprise, utility, Bernoullian, mean-variance analysis, stochastic dominance, mean-Gini dominance, forecasting

Abstract
Risk and uncertainty affect all decisions and result in inefficiencies in the agricultural sector as well as insecurity in food supply in many countries. Carefully validated crop models have a role to play in addressing these issues, because they can be used to isolate and quantify risk associated with weather variability and economic risk arising from uncertainty in the costs and prices of production. The methodologies for risk assessment in the DSSAT software are reviewed and illustrated. These are based on Bernoullian utility theory and stochastic efficiency criteria. At the household level, the models have been used to derive site- and season-specific crop management practices that reduce risk in southern Africa. At the regional level, a pilot information system has been constructed that links crop models, satellite-derived estimates of rainfall in real time, and a geographic information system, to obtain millet yield forecasts for districts in Burkina Faso that can be updated regularly through the growing season, to give early warning of impending production shortfalls. While work is still required to improve the methodologies used, the potential of models for providing information that can help reduce down-side risk, and thereby improve food security at the household and regional levels, is considerable.

Introduction

Risk and uncertainty pervade the environment in which agriculture is practiced. During the coming growing season, the impact of weather, pests, diseases, and prices and costs on productivity are all unknown to varying degrees. For most decision makers, the future can only be gauged in terms of their experience of the (usually recent) past. Risk is important because it affects decision making; decisions once made may turn out to be hopelessly inadequate in a highly uncertain environment. How are appropriate decisions to be made, given particular levels of information and available technology, and given that the future is highly uncertain and risky?

Risk has an impact on decisions because it may modify, sometimes very strongly, choices made from amongst a set of alternatives. Risk has been cited as a cause of slowed technology diffusion, fragmentation of land holdings, increased population growth, and price instability, for example (Walker and Ryan, 1990). Indeed, farm management has much to do with the management of risk, with impacts on food security at both household and regional levels.

G.Y. Tsuji et al. *(eds): Understanding Options for Agricultural Production, 329–345*

Informed decision making under risk involves a marriage of two strands of information – the decision maker's expectations about what is likely to happen in the future, and the decision maker's preferences. How these two strands are combined to form a logical framework for rational choice is the task of decision theory. Many different theories have been proposed and tested, but all are inadequate to some extent. People often have difficulty in formalizing their expectations concerning the future, even when based on extensive past experience. Expectations have to be derived not in terms of point prediction into the future ('yield in this field will be 4.3 t ha^{-1} next season') but in terms of a probability distribution ('our expectation of yield in this field next season is distributed normally with a mean of 4.3 t ha^{-1} and a standard deviation of 1.2 t ha^{-1}'). Simulation models offer one way of deriving such outcome distributions. If expectations are difficult to derive, then personal preferences are often even harder to quantify. At present, models of human behavior, where they exist at all, are generally inadequate.

In terms of agricultural research, the policy objective of food security is synonymous with yield stability (Walker and Ryan, 1990). At the household level, food security for the family may relate to an adequate income level, dependent not only on production stability but also on off-farm employment opportunities, for instance, but even here a large portion of the food security problem can be viewed as a problem in risk management. The year-to-year variation in food production, whether brought about by weather or pests and diseases, for example, has a direct impact on food security, as can seasonality, be it climatic, economic, or sociocultural (Gill, 1991). In this chapter, the methods used in the DSSAT software package to carry out risk assessment are briefly reviewed. Some applications of the crop models to risk and food security issues are described, from the enterprise to the regional and national levels, and their limitations discussed.

Risk management

Expectations

Part of managing risk in agriculture consists of coping with production variability between years. At the household level, it may be critical for the farmer to minimize the fluctuations in household income over time or to maintain or increase a particular wealth level and nutritional status; at the national level, governments have to ensure adequate supply of food to the populace in all sectors of society. The crop models can be excellent tools for assessing the production variability associated with weather for various strategies. In comparing two varieties at a site, for example, the same simulation experiment can be replicated using a number of different historical weather years. Some of these years will be more conducive to crop production than others. This process of replication produces a distribution of yields for that environment. The probability distribution of yield (or any other output that the model is capable of producing) then becomes a proxy for the investigator's expectation of what may happen next season – it is assumed that next year constitutes a sample drawn at random from the unknown outcome distribution, which can be approximated using the model and recent weather conditions. Usually there will be no basis for assuming that conditions next year are going to be fundamentally different from the last few years.

This is a general method of deriving an output distribution. Historical weather data may be used in the simulation experiment, or a statistical generator may be used that produces weather records with similar statistical characteristics as the historical weather data for the site. In either case, the initial conditions of the model are reset at the start of each season. Crop performance is thus replicated over a single growing season, and there is no carry-over from one year to the next; each season is completely independent of any other. Weather years could be chosen at random to produce the same output distribution – the results of this seasonal analysis would be the same.

There are two other methods of running the crop models that are of use. First, crop simulations can be updated during a growing season, using up-to-date weather data, the output of interest being the movement of the outcome distribution through time as more and more 'unknown' weather is replaced with historical weather. Second, the models can be used to investigate crop rotations or sequences. The user specifies a sequence treatment or rotation (such as maize followed by bean) and this is then simulated for as many years as the user specifies. The simulation experiment starts from a user-specified set of initial conditions, but from thereon, the outputs from one model become the inputs for the next, in terms of soil moisture and nutrient status. In this way, the performance of a crop rotation over long periods of simulated time can be assessed. In addition, the entire crop sequence can be replicated using another set of generated weather data, to indicate the variability associated with the crop sequence.

These three modes of running crop models are summarized in Table 1, in terms of the sources of variability that may be of interest to the investigator, the time horizon involved, whether carry-over effects from one season to another are simulated (if not, then the simulation returns to the specified initial conditions at the start of each season), and a published example of a study involving each. The differences between these three modes are worth clarifying, because each is appropriate to a different type of problem, but all can provide objective output distributions that can be used to assess risk and crop yield stability in different environments.

Preferences

How to deal with decision-makers' preferences? Of all the frameworks that have been proposed to deal with decision making under risk, that based in

Table 1. Three modes of running crop models.

Mode	Sources of variation	Time horizon	Carry-over simulated?[a]	Example
SEASONAL				
1 Single	Between treatments Between replicates	Next season	No	Singh et al. (1993)
2 Updated[b]	Between treatments Between replicates Between updates	This season in progress	No	Thornton and Dent(1984)
SEQUENCE				
3 (General)	Between treatments[c] Between replicates Between sequence years	The next n seasons ($n > 1$)	Yes	Bowen et al. (1993)

[a] Refers to the carry-over of soil-water, nutrient and carbon status from one crop to the next in the sequence or rotation.
[b] Updated seasonal simulations cannot currently be carried out directly in the DSSAT v3.0 (although with a simple driver program they can be done outside the DSSAT shell).
[c] Sequence treatments cannot currently be directly compared using the analysis program in the DSSAT v3.

Bernoullian utility theory has perhaps received the most attention. This posits the existence of a utility function that relates outcomes to the particular decision-maker's utility or level of satisfaction or happiness that he or she feels about particular outcomes. For the utility function $U(x)$, x is associated with monetary gains or losses, although conceptually utility functions can be constructed for any other factors that affect choice in a particular situation. The utility function can be thought of as a mapping or weighting function; its purpose is to transform money outcomes into utility outcomes. Different decision-makers' attitudes to risk can be encoded in a utility function, and so a decision problem for a particular farmer (and, hence, for a particular attitude to risk) can be solved in the knowledge that the solution takes account of that farmer's preference for certain outcomes over other outcomes. It appears that most decision-makers are risk averse; in the Bernoullian framework, this is equivalent to saying that incremental additions to wealth produce diminishing utility. Thus $1,000 is 'worth' more to a risk-averse individual who has no money than it would be if the same person had $100,000.

Utility functions are difficult to measure without bias, however, and the evidence as to whether utility explains farmers' decision making behavior a great deal better than other frameworks is not unequivocal. Instead of having to derive a utility function for each and every decision-maker, it is possible to make general statements about risk by using stochastic efficiency criteria. These criteria are particularly suited to the analysis of simulation model output, and, while based in utility theory, differ in the assumptions that are made about the decision-maker's attitude to risk. The application of stochastic efficiency criteria involves a pair-wise comparison of random variables that relate to financial gains and losses. The outcome of the analysis is an efficient set, which contains a subset of treatments that are superior (there may be one only, but sometimes there will be more than one efficient treatment) in the sense that they would be preferred by decision-makers whose risk attitudes conform to the assumptions of the analysis (Anderson et al., 1977).

Risk assessment procedures in the DSSAT

In Version 2.1 of the DSSAT (IBSNAT, 1989), the user can set up seasonal strategies (different varieties, planting dates, or fertilizer application schedules, for instance) and simulate these using historical weather data or weather records generated stochastically using WGEN (Richardson, 1985) or WMAKER (Keller, 1987). Simulation results can then be analyzed using a set of fixed prices and costs specified by the user. Efficient strategies are then identified on the basis of two stochastic efficiency criteria, mean-variance analysis and stochastic dominance (Table 2). There are a number of differences in Version 3 of the DSSAT (Tsuji et al., 1994). For running seasonal experiments, the weather generator WMAKER has been replaced by SIMMETEO (Geng et al., 1988). In seasonal analyses, the two-stage stochastic dominance routines have been replaced with another stochastic efficiency criterion, mean-Gini dominance (Table 2). The program also allows price and cost variability to be taken into account, so that analyses can isolate and quantify both weather-related and price-related risk. Users specify price and cost distributions, and these are combined algebraically with model-simulated output distributions to produce a distribution of gross margin or net return per hectare. In both released versions of the DSSAT, the seasonal analysis programs assume that returns per hectare can be used as a proxy for changes in total wealth. While this assumption is not strictly valid except for a particular class of utility functions (and it may be a reasonable assumption only rarely), it greatly facilitates treatment comparison and interpretation. In addition, the yield, cost and price distributions are assumed to be independent; there is no direct way in the DSSAT of taking account of negatively correlated yield and price distributions, for example.

For the analysis of crop sequences, no direct comparison of different sequence treatments is possible in DSSAT Version 3, but economic analysis of any one sequence may be carried out, with price and cost variability included in the analysis if required. The program also can perform a probability analysis of the sequence (where replications have been carried out): a fixed threshold or level of returns per hectare is specified by the user, and the probability of the sequence failing to generate this threshold is calculated and plotted. Changes in this probability over time may be indicative of instability in the production system.

Risk assessment at the enterprise level

The type of risk assessment possible with the DSSAT can be illustrated with an example (from Tsuji et al., 1994). A sample simulation file that is distributed with the DSSAT v3 package is a soybean experiment run at Gainesville, Florida, where the experimental variable is the amount of irrigation water applied to the crop. There are six treatments, a control (rain-fed) and five treatments involving different irrigation triggers; when the water content of the soil profile reaches a particular percentage of its total capacity, then irrigation is applied automatically. This simulation experiment is replicated using 20 years of weather data for Gainesville.

Table 3 summarizes the results of this simulation experiment (an example of run mode 1, Table 1). Using a set of illustrative costs and prices, the impact of large amounts of irrigation on monetary returns per hectare is clear (Figure 1A). If the amount of irrigation water applied in the six treatments is compared with the yields obtained, it is apparent that much of the water applied in treatments 2, 3, and 4 is wasted; only a small amount of supplemental irrigation is required to alleviate most of the water stress that may occur in most years. The same can be seen in terms of the number of irrigation applications made

Table 2. Stochastic efficiency criteria used in the DSSAT v2.1 and v3.0.

Criterion	Decision rule[a]	Form of $f_A(x)$ and $f_B(x)$	Form of $U(x)$	Power
Mean-variance	A dominates B if $E(A) = E(B)$ and $V(A) < V(B)$ or if $V(A) = V(B)$ and $E(A) > E(B)$	Normally distributed	Quadratic	Weak-medium
Stochastic dominance	*A* dominates *B* by first-order SD if $F_A(x)$ $F_B(x)$ for all x with at least one strong inequality	Any	$U_1(x) > 0$	Medium
	A dominates *B* by second-order SD if $G_A(x) \leq G_B(x)$ for all x with at least one strong inequality	Any	$U_1(x) > 0$ $U_2(x) < 0$	Strong
Mean-gini	*A* dominates *B* if $E(A) \geq E(B)$ and $E(A) - \Gamma(A) \geq E(B) - \Gamma(B)$	Any	Any	Strong

[a] Pair-wise comparison of risky prospects *A* and *B* with probability density function $f(.)$ described by mean $E(.)$, variance $V(.)$, gini coefficient $\Gamma(.)$ (half the value of Gini's mean difference: the absolute expected difference of a pair of randomly selected values), and cumulative distribution function $F(.)$ and cumulative cdf $G(.)$. The utility function $U(x)$ has first derivative $U_1(x)$ and second derivative $U_2(x)$.

Table 3. Results of a soybean simulation experiment involving five automatic irrigation triggers (expressed as a percentage of total soil-water holding capacity) at Gainesville, Florida, replicated over 20 different weather years.

Treatment	Yield kg ha^{-1}		Irrigation applications	Irrigation applied mm	Gross margin[a] \$ ha^{-1}		$E(x) - \Gamma(x)$[b]
	mean	st.dev			mean	st.dev	
1 RAINFED	3135	444	0.0	0.0	623	237	487
2 AUTOMATIC 90%	3529	144	22.2	239.7	356	191	244
3 AUTOMATIC 70%	3528	146	10.8	170.5	534	187	424
4 AUTOMATIC 50%	3540	123	6.6	135.1	607	187	497
5 AUTOMATIC 30%	3506	110	3.1	77.5	669	189	558
6 AUTOMATIC 10%	3228	287	0.6	19.1	638	192	526

[a] Product price = \$320 t^{-1} (cv = 10%), base production costs = \$345 ha^{-1}, irrigation cost = \$0.50 mm^{-1}, irrigation cost per application = \$12.50.
[b] $E(x)$ is the mean, (x) is the gini coefficient, of the gross margin distribution.

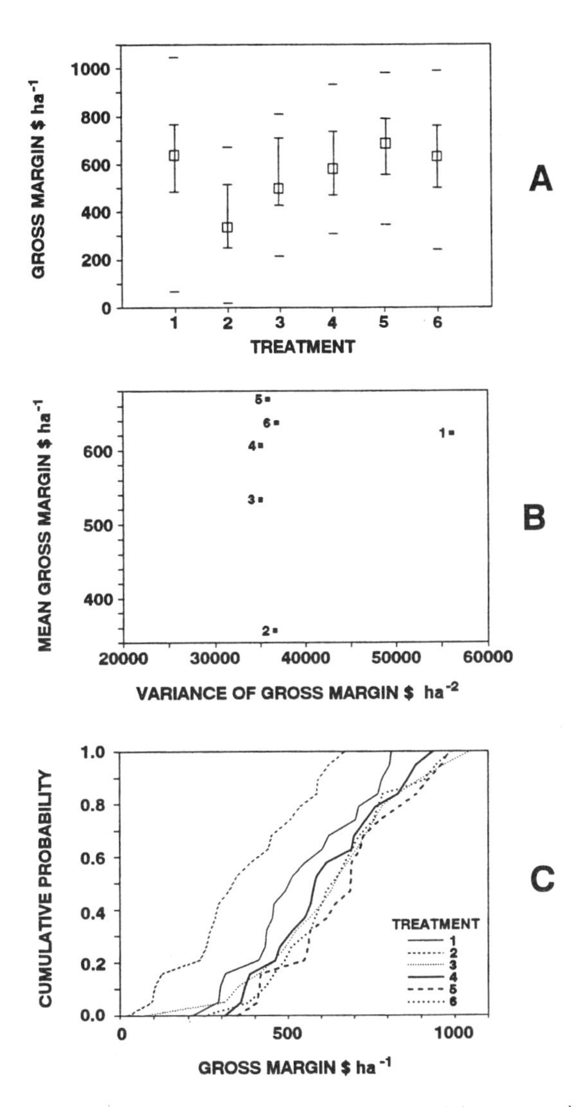

Figure 1. Analysis of a six-treatment simulation experiment involving automatic irrigation thresholds on soybean in Gainesville, Florida, replicated using 20 years' of weather data. (A) Box plot of gross margin, showing the 0th, 25th, 50th, 75th and 100th percentile of each treatment's gross margin distribution. (B) Mean gross margin plotted against its variance by treatment. (C) Cumulative distribution functions of gross margin by treatment.

in each treatment. It can be concluded that the cost of irrigation water per mm, the cost of each irrigation application, and the limited yield response to heavy irrigation applications, combine in such a way that the most profitable option of the six considered is to apply irrigation at a threshold of 30% of field capacity.

For strategy analysis, the mean-Gini dominance calculations are shown in Table 3. The dominant strategy (treatment 5) also has the highest mean monetary return. In mean-Gini dominance, this will always be the case, but it will sometimes happen that the variability associated with this 'high-mean' treatment will be so great that it will be joined in the efficient set by another treatment with lower mean but lower variability. In such a case, no decision can be made with respect to which of the efficient treatments is preferable without knowing more about the preferences and attitudes of the particular decision-maker. Mean-variance analysis can be carried out simply by inspecting Figure 1B. Treatment 5 has the highest mean, but its variance is greater than the variances for treatments 3 and 4, so according to the E-V rules, treatments 3, 4 and 5 are all E-V efficient. For purposes of comparison, inspection of the CPF plot of monetary returns for the six treatments (Figure 1C) shows that none of the treatments is dominant by first-order stochastic dominance, since the CPF for treatment 5 crosses the CPFs of treatments 1 and 6. Further analysis shows that treatment 5 is efficient according to second-order stochastic dominance.

Such analyses can provide useful information concerning the possible suitability of particular strategies or treatments of a simulation experiment. Treatments may involve different crops, varieties, planting dates and densities, nitrogen and water management, or any other management factors to which the crop models are sensitive. Costs and prices can be changed readily, allowing sensitivity analysis to be performed. The running and analysis of seasonal simulation experiments provides a powerful tool for assessing site-specific management recommendations that can help to minimize production risk. In highly variable rainfall environments, however, there is also a need to tailor crop management recommendations to the current season. It has been shown in a number of environments in the tropics and subtropics that indicators such as the date of start of the rains can contain a great deal of information about the coming season. This information can be exploited with crop models, and forms the basis of response farming (Stewart, 1991).

The data of Sivakumar (1988), for example, show a strong relationship between date of onset of the rains and length of the growing season at many sites in the southern Sahelian and Sudanian climatic zones of West Africa. The implications of this relationship for crop and disaster planning are underlined in Sivakumar (1990), where he shows that in longer growing seasons (characterized by earlier-starting rains), there is generally enough moisture to establish a second crop of cowpea for hay after millet, the traditional first crop, has been harvested. A similar relationship was exploited in the studies of McCown et al. (1991) and Keating et al. (1993) concerning maize production in a semi-arid region of Kenya. A modified version of the crop model CERES-Maize V2.1 (Ritchie et al., 1989) was used to examine the feasibility and value of modifying agronomic practices in response to a seasonal forecast of rainfall. The forecast was based on date of onset of the rainy season and cumulative rainfall in the early part of the season.

The strong relationship between date of onset of the rains and total seasonal rainfall (or length of the growing season) was also demonstrated for a number of sites in the mid-altitude maize growing region of central Malawi (Thornton et al., 1995), using a simple water balance model and proxies for the start and end of the growing season based on those of Jones (1987). Again, this simple information can be exploited using the crop model to provide management recommendations based on a small number of season types characterized by the starting date of the rains. Simulated maize yield distributions varied substantially for three distinct, equally-likely season types (early-, normal- and late-starting rains) for a site near the town of Kasungu in central Malawi. The economically optimal amount of nitrogen fertilizer varied also, from 90 kg ha^{-1} for early-starting seasons to 30 kg ha^{-1} for the late-starting seasons, based on 1993 costs and prices. Apart from helping to assess season-specific management inputs, this type of analysis could also be used to help select maize cultivars for a region based on the length of the growing season. In a late-starting season, for example, it may be better for farmers to use smaller amounts of fertilizer or lower planting densities than they would in earlier-starting seasons, or it may be better to grow shorter-duration maize cultivars or shorter-duration crops altogether.

The monetary value of this type of information can be estimated by comparing different scenarios under various management decision rules (Thornton and MacRobert, 1994). Such analyses are important for determining whether it is worthwhile fine-tuning management recommendations and what the demand for such information is likely to be at the farm level. The true value of this kind of information is not easy to estimate, however, because the implications of very poor years on a household's income levels will be much greater than their apparent probability of occurrence, especially in subsistence systems. Expected utility is not an appropriate framework for assessing the benefits of risk reduction measures in such situations.

Yield stability at the regional level

Information on yield stability at the regional level is of critical importance in many environments. Early warning of a poor season, for example, can give policy makers some time to take appropriate action to combat likely food shortages. The crop models can be used to provide such information using spatial databases of soils and weather data within the framework of a Geographic Information System (GIS). Simulations can be updated through the season, resulting in successive modifications of the simulated distribution of yield or other outputs of interest (run mode 2, Table 1). This means that the crop models can be run at regular intervals throughout the growing season to

Table 4. Dekadal rainfall totals at Dori, Burkina Faso.

Date	Day of year	Dekad	Rainfall (mm)[a]	
			1986	1990
01–10 June	152–161	16	9.5	21.7
11–20 June	162–171	17	36.2	16.1
21–30 June	172–181	18	3.7	27.6
01–10 July	182–191	19	10.1	7.4
11–20 July	192–201	20	4.5	72.1
01–31 July	202–212	21	49.2	43.2
01–10 August	213–222	22	99.1	56.2
11–20 August	223–232	23	14.8	106.4
21–31 August	233–243	24	38.0	20.3
01–10 September	244–253	25	28.8	21.7
11–20 September	254–263	26	5.7	5.4
21–30 September	264–273	27	10.3	30.0
Total			309.9	428.1

[a] Long-term average annual rainfall from 1963 to 1990: 470.8 mm.

produce production forecasts in the following way. For each date of forecast, the crop model uses weather data acquired from climatological stations and meteorological satellites up to the current date. Weather from the forecast date to the end of the growing season is unknown, but the crop models can be run for several sequences of daily weather data obtained from statistical weather generators that are based on historical data for each site. In this way, a probability distribution of simulated crop yield is obtained, with an expected value (the mean of the replicates) and a confidence interval. This confidence interval is associated with the interactions of crop growth and the weather experienced up until the day of forecast and the uncertainty of weather thereafter. For forecasts made in this way early in the growing season, the confidence interval may be wide, because most of the weather to be experienced during the growing season is as yet unknown. As the season progresses, however, subsequent production forecasts use a higher proportion of observed weather data and less unknown weather. The resultant confidence intervals become smaller and the expected production estimates better.

To illustrate the way in which a yield distribution can be modified through a growing season, some simulations were carried out using CERES-Millet in Dori in northern Burkina Faso (14°2′N, 0°2′W, elevation 276 m, average annual rainfall 470 mm). The soil at this site is a Typic Natraqualf. For all simulations, millet was assumed to be planted on day 160 (9 June); in this environment the crop matures in about 100 days. Dekadal or ten-day totals were calculated for 1986 and 1990 (Table 4) from the historical rainfall records for Dori. (The reason for using dekadal rainfall data is that the only realistic way of obtaining weather data in real time for Burkina Faso is from satellite remote sensing; at present these data are provided only on a dekadal basis.) For each year, the

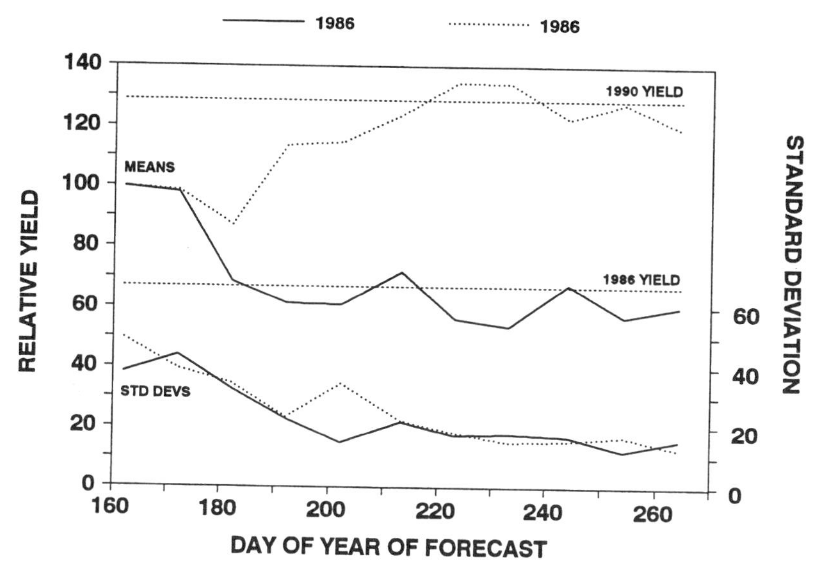

Figure 2. Simulated millet yield distributions for conditions in Dori, Burkina Faso, derived at various forecast dates using historical dekadal rainfall totals up to the day of forecast and probabilistic weather data thereafter, for the 1986 and 1990 seasons.

millet model was then run for 11 simulation dates; these would correspond to 'forecast' dates in the situation where the model was being run through the current growing season. Up to the day of forecast, dekadal totals were used, split up probabilistically according to which days within the dekad were simulated to be wet. From the day of forecast onwards, rainfall was simulated probabilistically until the end of the growing season. Thus for these replicated millet simulations, dekadal total rainfall for the start of the season was identical (its distribution within the dekad changed between replicates, but the dekadal amounts were identical), whereas rainfall from the date of forecast to the end of the season varied between replicates in terms of both its amount and distribution.

The results of simulations replicated 20 times for each of 11 forecast dates, corresponding to the start of dekads 17 to 27, are shown in Figure 2, for the two contrasting years, 1986 and 1990. Yields are expressed relative to the mean of the yield distribution obtained when all the weather is generated probabilistically (corresponding to the pre-season expectation of yield, when nothing is yet known about the current season). As the forecast date progressed, the mean simulated millet yield decreased in 1986 but increased in 1990, while the standard deviations of the yield distributions decreased in both years as the amount of probabilistic weather in the simulations decreased. The variation in yield obtained for later forecast dates was due purely to the different distributions of rainfall within each dekad – the dekadal and seasonal rainfall totals were identical between replicates for each year.

Rainfall at Dori in 1986 was 67% of the long-term average (Table 4). In Seno, the district in which Dori is situated, the observed average district-wide yield of millet in 1986 was 237 kg ha^{-1}, compared with a long-term average from 1984 to 1992 of 352 kg ha^{-1} (FEWS, 1994). The 1986 yield was thus only 67% of normal. The simulated yield deviation from normal agrees well with this value – final average yield at Dori was simulated to be about 70% of normal in 1986. From Figure 2, by the fourth forecast date (dekad 20, day-of-year 192), the forecast average deviation was already at the 60% level, giving early indications that the season was likely to be poor. This was indeed the case in 1986. In 1990, by contrast, rainfall was more plentiful, although still less than average; the district average yield was 29% above the long-term average, however (454 kg ha^{-1}). By day-of-year 213, the mean simulated millet yield was 20% above the average, giving a reasonably early indication that the season was going to be better than average.

If reasonably accurate indications of final yields could be given 4 or 6 weeks before crop maturity, this would give decision-makers some time in which to take appropriate action. A prototype system has been constructed to generate this type of information on a district basis in Burkina Faso, using estimates of dekadal rainfall for all of the country from regular satellite imagery of cold cloud duration converted to rainfall estimates (Thornton et al., 1997). The prototype system uses IDRISI (Eastman, 1992), an inexpensive GIS that can be run on a PC, together with a simple interface to the millet model, databases, and satellite images. District-level simulations can be carried out, in which the model is run for all unique combinations of soil type and weather conditions that occur in the district. The area planted to millet in the district can be estimated from a crop use intensity coverage, allowing production to be estimated and compared with historical area and production figures. The final result from all the simulations is an estimated percentage seasonal deviation from normal for the district under study (Figure 3). The prototype needs much field testing, but the potential for providing timely production information at a district level would appear to be considerable.

Prospects and conclusions

The increasing pressures on natural resources brought about by population growth and land degradation will bring the food security issue into ever sharper relief in the future. Providing information to decision-makers that can help, even in a limited way, to combat risk and ensure food security is thus a research area that warrants intense activity. Food security issues are being addressed at a number of levels in the agricultural system hierarchy in different ways. At the regional and national levels, a wide variety of crop models and techniques

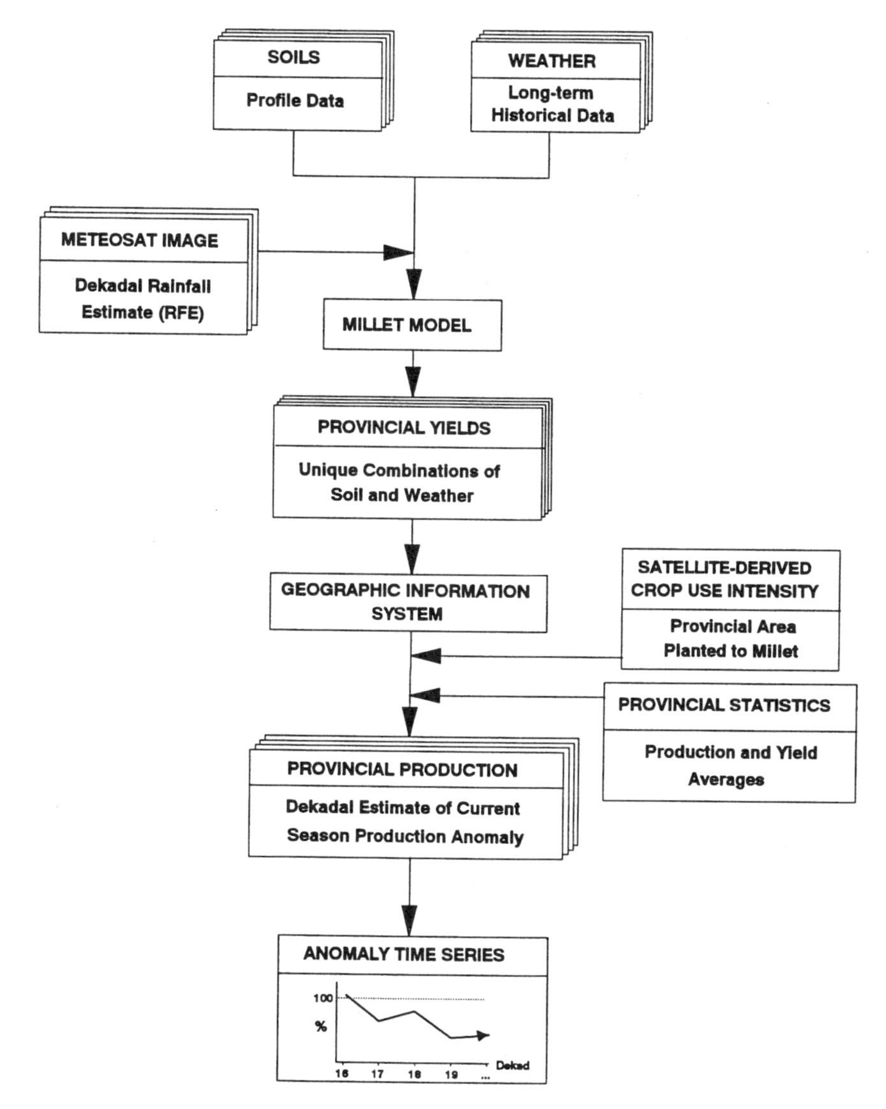

Figure 3. Input data and information flow in a prototype district-level millet yield forecasting system for Burkina Faso (from Thornton et al., 1997).

can be used for yield forecasting (Jones et al., 1995). GIS and simple energy models are being used to study human carrying capacity in Senegal, allowing possible ways of increasing food security to be explored (Moore et al., 1994). Similarly, Veeneklass et al. (1994) used models and interactive multiple goal linear programming to help develop land use options in a region of Mali that fit in with government policy objectives. Such work could help to achieve food security at the regional level by identifying feasible land use options that minimize household production risk, for instance.

A somewhat different approach is being taken at the farm level, involving the coupling of biophysical models with models of the household. For example, an integrated model of the food system in a region of Zimbabwe (Gundry, 1994) is being constructed with the aim of carrying out comparative analyses of the factors that affect the nutritional status of households in the area, to help identify those areas where households are most at risk. Such a framework could be used to assess proposed policy interventions aimed at promoting food security, and to assist in targeting food aid distribution. The approach of Hansen (1995) to modeling the farm household involves a coupling of crop models with decision models, in which enterprise options can be assessed for their 'survivability' over the medium term (6 to 10 years) in the face of price risk and natural resource degradation in the form of soil erosion. The food security problem has to be set within a framework that considers the household as an entity operating within socioeconomic and cultural as well as biophysical subsystems, so the importance of household models is paramount. There is, however, substantial imbalance between the predictive capabilities of many biophysical models and current socioceonomic models, and much work will be needed before this is rectified even partially.

As for all models and applications, the integrity of the information produced by crop models to study risk and food security depends on sound models, sound validation procedures, and sound input data. The input data required to run such models can pose substantial problems, as can objective testing. The real potential of models in the study of food security issues has yet to be tapped, however.

References

Anderson J R, Dillon J L, Hardaker J B (1977) Agricultural decision analysis. Iowa State University Press, Ames, Iowa, USA.

Bowen W T, Jones J W, Thornton P K (1993) Crop simulation as a tool for evaluating sustainable land management. Pages 15–21 in Kimble J M (Ed.) Utilization of soil survey information for sustainable land use. USDA, Soil Conservation Service, National Soil Survey Center, Washington, USA.

Eastman J R (1992) IDRISI Version 4.0 technical reference manual. Clark University, Worcester, Massachusetts, USA.

FEWS (1994) AGMAN, an agricultural database manager, version 2.00. Famine Early Warning System Project, Arlington, Virginia, USA.

Geng S, Auburn J S, Brandstetter E, Li B (1988) A program to simulate meteorological variables: documentation for SIMMETEO. Agronomy Progress Report 204, Department of Agronomy and Range Science, University of California, Davis, California, USA.

Gill G J (1991) Seasonality and agriculture in the developing world. Cambridge University Press, Cambridge, UK.

Gundry S (1994) An integrated model of the food system in a region of Zimbabwe. Institute of Ecology and Resource Management, University of Edinburgh, Edinburgh, UK.

Hansen J W (1995) A systems approach to characterizing farm sustainability. PhD thesis, Department of Agricultural Engineering, University of Florida, Gainesville, Florida, USA.

IBSNAT (1989) Decision Support System for Agrotechnology Transfer (DSSAT) V2.1. Department of Agronomy and Soil Science, University of Hawaii, Honolulu, Hawaii, USA.
Jones P G (1987) Current availability and deficiencies in data relevant to agro-ecological studies in the geographic area covered by the IARCs. Pages 69–83 in Bunting A H (Ed.) Agricultural environments. CAB International, Wallingford, UK.
Jones P G, Thornton P K, Hill P (1995) Agrometeorological models: crop yield and stress indices. Proceedings of the EU/FAO Expert Consultation on Crop Yield Forecasting Methods, Villefranche-sur-Mer, 24–27 October, 1994.
Keating B A, McCown R L, Wafula B M (1993) Adjustment of nitrogen inputs in response to a seasonal forecast in a region of high climatic risk. Pages 233–252 in Penning de Vries F W T, Teng P S, Metselaar K (Eds.) Systems approaches for agricultural development. Kluwer, Dordrecht, The Netherlands.
Keller A A (1987) Modeling command area demand and response to water delivered by the main system. Unpublished PhD dissertation, Agricultural and Irrigation Engineering Department, Utah State University, Utah, USA.
McCown R L, Wafula B, Mohammed L, Ryan J G, Hargreaves J N G (1991) Assessing the value of a seasonal rainfall predictor to agronomic decisions: the case of response farming in Kenya. Pages 383–410 in Muchow R C, Bellamy J A (Eds.) Climatic risk in crop production: models and management for the Semi-arid Tropics and Subtropics. CAB International, Wallingford, UK.
Moore D G, Tappan G G, Howard S M, Lietzow R W, Nadeau A, Renison W, Olsson J T, Kite R (1994) Geographic modeling of food production from rainfed agriculture: Senegal case study. Pages 133–158 in Uhlir P E, Carter G C (Eds.) Crop modeling and related environmental data. Monograph Series Volume 1, CODATA, Paris, France.
Richardson C W (1985) Weather simulation for crop management models. Transactions of the American Society of Agricultural Engineers 18(5):1602–1606.
Ritchie J T, Singh U, Godwin D C, Hunt L A (1989) A user's guide to CERES maize V2.10. IFDC, Muscle Shoals, Alabama, USA.
Singh U, Thornton P K, Saka A R, Dent J B (1993) Maize modelling in Malawi: an tool for soil fertility research and development. Pages 253–273 in Penning de Vries F W T, Teng P, Metselaar K (Eds.) Systems approaches for agricultural development. Kluwer, Dordrecht, The Netherlands.
Sivakumar M V K (1988) Predicting rainy season potential from the onset of rains in southern Sahelian and Sudanian climatic zones of West Africa. Agricultural and Forest Meteorology 42:295–305.
Sivakumar M V K (1990) Exploiting rainy season potential from the onset of rains in the Sahelian zone of West Africa. Agricultural and Forest Meteorology 51:321–332.
Stewart J I (1991) Principles and performance of response farming. Pages 361–382 in Muchow R C, Bellamy J A (Eds.) Climatic risk in crop production: models and management for the Semi-arid Tropics and Subtropics. CAB International, Wallingford, UK.
Thornton P K, Dent J B (1984) An information system for the control of *Puccinia hordei*. Agricultural Systems 15:209–224.
Thornton P K, MacRobert J F (1994) The value of information concerning near-optimal nitrogen fertilizer scheduling. Agricultural Systems 45:315–330.
Thornton P K, Hansen J W, Knapp E B, Jones J W (1995) Designing optimal crop management strategies. Pages 333–351 in Bouma J, Kuyvenhoven A, Bouman B A M, Luyten J C, Zandstra H G (Eds.) Ecoregional approaches for sustainable land use and food production. Kluwer, Dordrecht, The Netherlands.
Thornton P K, Bowen W T, Ravelo A C, Wilkens P W, Farmer G, Brock J, Brink J E (1997) Estimating millet production for famine early warning: an application of crop simulation modelling in Burkina Faso. Agricultural and Forest Meteorology 83:95–112.
Tsuji G Y, Uehara G, S Balas (Eds.) (1994) DSSAT v3. University of Hawaii, Honolulu, Hawaii, USA.
Veeneklass F R, van Keulen H, Cisse S, Gosseye P, van Duivenbooden N (1994) Competing for limited resources: options for land use in the fifth region of Mali. Pages 227–247 in Fresco L O, Stroosnijder L, Bouma J, van Keulen H (Eds.) The future of the land: mobilising and integrating knowledge for land use options. Wiley, Chichester, UK.
Walker T S, Ryan J G (1990) Village and household economies in India's Semi-Arid Tropics. Johns Hopkins University Press, Baltimore, Maryland, USA.

Incorporating farm household decision-making within whole farm models

G. EDWARDS-JONES[1], J.B. DENT[2], O. MORGAN[1]
and M.J. McGREGOR[3]

[1] *Rural Resource Management Department, Scottish Agricultural College, West Mains Road, Edinburgh EH9 3JG, UK*

[2] *Institute of Ecology and Resource Management, University of Edinburgh, West Mains Road, Edinburgh EH9 3JG, UK*

[3] *Muresk Institute of Agriculture, Curtin University of Technology, Northam, Western Australia 6401, Australia*

Key words: words: agricultural systems, socio-economics, farm household, modeling, whole farm model

Abstract

If models of agricultural systems are to have a major impact on policy formulation and assessment then it is important to integrate the human elements of these systems with biological models in a whole farm context. This chapter outlines the structure of one such model, which integrates CERES-Maize and BEANGRO with family decision-making and demographic models in a whole farm model of a subsistence farming system. Although the development of this prototype whole farm model has demonstrated the feasibility of integrating socio-economic and biological models, further work is required before such integrations can become more widespread. Particular problems relate to model formulation, the choice of the appropriate scale for modeling farm families, data collection, and validation. Should these problems be overcome then the requirement to integrate socio-economic and biological models may have implications for the design of future crop models.

Introduction

Most ecosystems contain elements which behave in a manner analogous to that observed in agricultural situations. Thus soil-plant interactions, nutrient cycling, seed germination, insect herbivore, plant and animal growth, lactation and disease transmission are all observable in natural ecosystems, and could legitimately be studied by ecologists. The one feature of agricultural systems which clearly differentiates them from any natural system is the human element. Almost by definition all agricultural systems have at their center one or more humans who manage the natural systems in order to meet anthropocentric objectives. Given this obvious relationship between man and agriculture, it is perhaps surprising that so little effort and resources had until recently been directed at understanding and predicting the dynamics of the social elements of agricultural systems. This is unfortunate, as it is self-evident that it is the human element of farming systems which determines the success of policy and technology transfer initiatives.

G.Y. Tsuji et al. *(eds): Understanding Options for Agricultural Production, 347–365*

Awareness of the role of social systems in agriculture has, however, been increasing during the last 25 years as farmers and their families, who have always been studied by rural sociologists (Gasson and Errington, 1993), have become important in Farming Systems Research (FSR) (Gladwin, 1983; Low, 1988; Dent and McGregor, 1994) and agricultural economics (Anderson et al., 1985; Byerlee and de Polanco, 1986; Doorman, 1991). Despite this recent progress, social systems have not been analyzed as intensively or as quantitatively as other elements of the farm systems, such as livestock and crop production. Thus relatively few models exist which attempt to simulate the behavior of farm families, or to integrate their behavior with other elements of the agricultural system.

Although the traditional reductionist approach which pervades much scientific research may have been partly responsible for the omission of farm families from agricultural research agendas, there is no doubt that the apparent difficulty of studying and modeling human behavior and interactions has also been a major hurdle. As Anderson (1985) confessed, 'The complexity of family arrangements in diverse cultural settings intimidates this FSR observer into a reluctance to open this particular Pandora's box'.

Farm household economics have been the subject of econometric analyses (Caillavet et al., 1994; Singh et al., 1986), but these models tend to make some major assumptions about farm families behaving so as to maximize a single objective, usually profit. This assumption runs counter to increasing amounts of evidence which suggests that such assumptions may mask the real complexity of Anderson's 'Pandora's Box' (Carr and Tait, 1991; Casebow, 1981; Coughenour and Tweeten, 1986; Coughenour and Swanson, 1988; Gilmor, 1986; McGregor et al., 1995, 1996). For example, the inadequacy of utilizing simple economic models for modeling one facet of farm family behavior, the adoption of new technologies, was questioned by Herath et al. (1982), who demonstrated that profit maximization was probably not the major driving force behind the selection of rice varieties in Sri Lanka. Later studies of technology adoption utilized more sophisticated, multi-variate models to understand and simulate farmer behavior (Akinola, 1987; Polson and Spencer, 1992; Shakya and Flinn, 1985; Voh, 1982), and although variables such as age, farm size, education, tenure status, and the availability of credit and markets did explain significant amounts of variation in these models, no generic model for predicting technology adoption has been forthcoming. Perhaps this is not surprising, given the apparent difficulty in representing the behavior of a diverse range of humans within a model and in identifying the variables which are important in understanding their behavior.

When considering the problems of incorporating individuals from a diverse range of cultures within a model, it is apparent that a continuum of perspectives exists. At one end lies the belief that all individuals react in the same way to any given set of inputs, while at the other is the belief that all individuals are truly unique. Although both economics (Bryant, 1992) and socio-biology

(Wilson, 1975) suggest that there is some degree of commonality between individual behavior patterns, agricultural modelers are remarkable in that they assume absolute commonality between individuals, and have generally treated all individuals as similar entities (Amir et al., 1991; Ghadim et al., 1991; Wossink et al., 1992; Moxey et al., 1995). In this situation it is apparent that if variation does exist between individuals, then models which assume homogeneity in farmer behavior would not be generally applicable or transferable. In fact, there is a relatively poor correlation between the results of such models and real life, and this is increasingly being recognized by some social scientists, who maintain that significant variation in behavior exists between individuals and that the relationship between this behavior and structural variables, such as region, education, and farm type may not be as simple as previously thought (Shucksmith, 1993).

Although models which assume a high degree of commonality between individuals would seem to be invalid on theoretical grounds, models at the opposite end of the continuum, those that assume individual uniqueness, appear to be untenable on practical grounds. For example, it is essential for purely practical, computational reasons that some element of commonality between the behavior of individuals be incorporated into any generic model. It is impossible, given current technology, to develop a generic model that simulates the unique behavior of individuals. Data collection, model structuring and validation would all be impossible.

Given these facts it would seem logical to develop a model which assumed some degree of commonality in the behavior of individuals, but which recognized that the characteristics of the individuals will influence the specifics of any generalized response. Even in this situation the range of specific behavior must ultimately be limited, as an infinite array of responses is computationally impractical.

Adoption of such a modeling framework poses a series of questions regarding the representation and amount of commonality to be included in the model. For example, should the common elements constitute 50% or 95% of the decision analysis? Should common elements be based on a typology constructed around regional, farm type or socio-economic variables? The answers to these and other related questions depend largely upon the effort available for data collection. If the model is to be of practical use then the data requirements must not be too arduous. Thus pressure exists to develop a minimum socio-economic data set which reflects the important elements of similarity between farm households, while simultaneously representing important socio-cultural differences. The practical requirements of collecting such a data set place further constraints on model developers.

General aims of integrating crop and socio-economic modeling

The importance of developing models of farm households, and linking these with biological crop models in a whole farm model (WFM) was recognized by

Dent and Thornton (1988), but to date no such study has been completed. Further, such models may remain rare as many of the more difficult questions outlined above remain unanswered. Given this dearth of experience in the field a preliminary study was defined which would first consider the theory and constraints of modeling farm households; second, identify appropriate techniques for representing the decision-making of farm households in a modeling framework; and third, develop a prototype whole farm model. Thus the objectives of this preliminary work were as follows:

(a) To analyze the interactions between the social and economic systems of the farm household with the biological cropping systems of the farm.
(b) To consider, in the context of crop production, the interactions between the farm household and the wider socio-economic systems beyond the farm boundaries.
(c) To examine the feasibility of modeling the interactions identified in (a) and (b) above at the farm scale.
(d) To identify appropriate frameworks for modeling farm families and develop a prototype whole farm model which integrated a farm household model with crop models developed by the IBSNAT project.

The remainder of this chapter describes the structure of the prototype whole farm model developed in order to meet objective d) above. Other issues are discussed in Dent et al. (1995), Edwards-Jones and McGregor (1994), and Willock et al. (1995).

Identification of modeling methodology

Standard numerical algorithms were unlikely to be adequate for representing the behavior of farm families within a whole farm model. Inadequacy arose in two areas: the need to manipulate qualitative data, and the impracticality of undertaking controlled experiments to provide data for the model. In order to address these problems an expert system approach was adopted for the development of the household model. Expert systems, which seek to emulate the decision-making process of human experts (Luger and Stubblefield, 1989), have come to the fore within agriculture chiefly as advisory aids, where they emulate agricultural researchers and extension officers in recognizing and solving a variety of agricultural problems (Edwards-Jones, 1993; Edwards-Jones and McGregor, 1992). As the programming techniques utilized when building expert systems were developed out of a requirement to model human decisions, these techniques are almost pre-adapted for use in simulating farm household decision making. Their specific advantage in this situation centers around their ability to handle qualitative data and uncertainty. For these reasons it was decided to base the farm household decision model around rule-

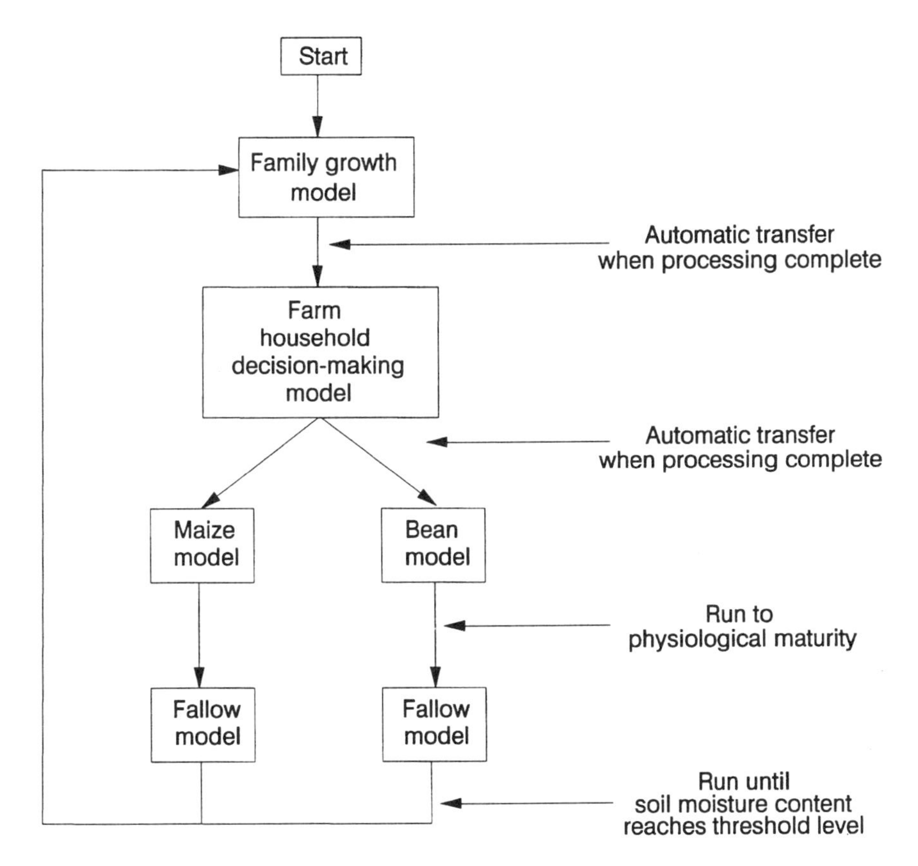

Figure 1. Sequence of module functioning within the whole farm model.

bases, and to develop the model with Prolog, a computer language specifically designed for developing expert systems (Clocksin and Mellish, 1984).

Overview of the structure and function of the whole farm model

The prototype whole farm model was developed to simulate the processes within a hypothetical subsistence farming system. The system was assumed to be based on maize and beans, dependent largely on family labor, and using limited external inputs. The whole farm model itself consisted of four main elements: the crop models, the fallow model, a family demographic model, and the household decision-making model. During simulation the focus moves between these elements in the order family demographic model, to the household decision-making model, to the crop models, and finally the fallow models, before the family demographic model is again initiated (Figure 1). The crop models integrated within the WFM were CERES-Maize (Jones and Kiniry, 1986) and BEANGRO (Hoogenboom et al., 1990).

The family demographic model

The family demographic model is based on a Leslie matrix (Leslie, 1945). This matrix assumes that only four processes govern populations: immigration, emigration, births and deaths, and that the probability of any of these occurring over a given time period can be calculated and used to model fluctuations in the population over time. Thus the 'survival' of individuals from the current time period to a future time period is specified probabilistically. This probability may vary with circumstance, and in the farm household this model varies according to the age, sex and nutritive status of the individuals. Within the WFM the family demographic model provides information on available labor, and the nutritional requirements of the family. This is achieved through the assignment of daily labor output and calorific requirements to each sex/age group in the family.

The credit adoption module

Adoption of credit can have an important impact on the behavior of small-scale farmers. Seasonal credit, which is usually paid back in full at the end of the cropping year, permits the use of improved seeds, fertilizers, pesticides and perhaps the purchase of additional labor. Long term credit supports the purchase of larger items, repayments are made over a longer period of time, and this type of loan is unlikely to be made without some collateral. Within the WFM, credit adoption is modeled as a series of rules in a two-stage process. The first stage calculates the equity status of the farmer on the basis of land tenure, farm size and cash reserves. The second stage characterizes the farmer as being either a potential credit adopter or a non-adopter. It is assumed that the uptake of credit is positively correlated with level of education and equity, and negatively with age. Thus a combination of data on age and education determine the probability of adoption, and the equity status of the farmer determines if the lending body (such as the bank) would actually lend to that individual.

The final outputs of this module are a classification of the farmer as a potential credit adopter or a non-adopter. These classifications can change with time, as both age and cash reserves change. Having categorized a farmer as a potential credit adopter, no limit is set on his borrowing; rather, it is assumed that he will only borrow money for the purchase of seed and fertilizer. This is clearly artificial, but given the simplicity of the credit adoption model it seemed a reasonable assumption.

Table 1. The classification of variables utilized to control the reassessment decision in the farm household model.

Importance class	Variable
1	Nutritional requirements of family Amount of food in store Price of inputs to cropping systems Availability of family labor
2	Financial reserves Prices received for marketed output Availability and nature of credit Availability of markets for output

The reassessment module

The presumption underlying this module is that subsistence farmers have a natural disposition to resist change. Thus if last year's harvest was satisfactory, the farmer will be unlikely to change any aspect of the farming system. The assumption here is that change may be perceived by the farmer as exposing himself to an increased risk of failure. Despite the predisposition to resist change, if there is a significant change in the endogenous or exogenous environment (i.e., either within the farm system or in the environment beyond the farm boundaries), then the model assumes that the farmer will reassess his current practice, and in view of the new situation, may implement some change to the farming system.

A major difficulty inherent in modeling this decision concerns the identification of the threshold level of change in the endogenous/exogenous environment at which a farmer seriously considers the implications of that change on his farming system. This threshold will almost certainly vary between the constituent elements of the endogenous and exogenous environments. For example, a 10% increase in output price may not stimulate a reassessment of farming practices, whereas a 5% increase in input costs may do so. In addition, variation in sociological and psychological variables such as tenureship, financial situation, stage in family cycle, personality and community characteristics, will mean that the threshold may not only vary between individuals but also over time for any given individual. While recognizing this variation, the model currently assumes the threshold for change to be constant across all situations.

Although the threshold for change is constant, the model does recognize that some variables will be more likely to initiate a reassessment than others. This is done by grouping the variables into two Importance Classes, each of which have slightly different roles in initiating a reassessment (Table 1). Thus a reassessment is automatically triggered by either a change between years of more than 5% in at least one variable from importance class 1, and/or a change of more than 5% in two or more variables from importance class 2. Some of the variables considered in this calculation are universal, i.e., they apply across all situations, whereas others are crop specific. For example, it may not be relevant to reassess the entire farming system if the only change in the endogenous/exogenous environment is an increase in the cost of inputs or outputs for one crop. Thus four of the variables are crop specific, and apply separately to maize and beans. These include amount of food in store, cost of inputs, price of outputs and availability of markets. These variables are utilized to specify whether or not any reassessment should be of the entire farming system (total reassessment) or of one element of the system (maize only or bean only).

In terms of the model cropping year, the reassessment decision occurs at pre-sowing (point d in Figure 2). The actions of the farm household decision-making model at this time are as follows. First, the family growth model is updated, then the potential for credit adoption is calculated, and finally the financial and store accounting is undertaken. The model then examines the status of all the variables given in Importance Classes 1 and 2 above and compares their current value with their value in the previous cropping year. If the difference in value between the current and previous time period is 5% or more, then a reassessment is activated. If none of the variables has changed by 5% or more, and if the management strategy of the previous year had been sufficient to feed the farm family, then the same strategy is implemented again in the current year. Depending on the variables which have exhibited change three types of reassessment are possible; total, maize only, and bean only. Each of these reassessment processes utilizes the same rule-bases, or a subset thereof, to select the perceived best variety and cropping strategy for the family in the forthcoming season.

Variety selection

One of the most important decisions that farmers make concerns crop variety selection. Breeders are continuing to develop new varieties of most crops and many of these would seem to offer significant advantages to subsistence farmers. Despite these apparent advantages new varieties may not be widely accepted in a region for cultural reasons. For example, in Latin America the color of the bean is a very important criterion in variety selection, while in Africa, some farmers suggest that many new high yielding maize varieties have an inferior taste and texture to traditional lower yielding varieties (Ashby et al., 1989a, 1989b; Negassa et al., 1991).

The selection of varieties requires farmers to consider a range of criteria including variety specific characters such as potential yield, management requirements, taste, cooking and storage qualities. Information on these criteria is held within the WFM for each variety for which there are genetic characteristics in the crop models, and these form an important input to the variety selection decision (Table 2). In addition, the individual farmers' personal experience with any variety will have a major impact on the final selection, and this personal experience is reflected within the decision-making model. The variety

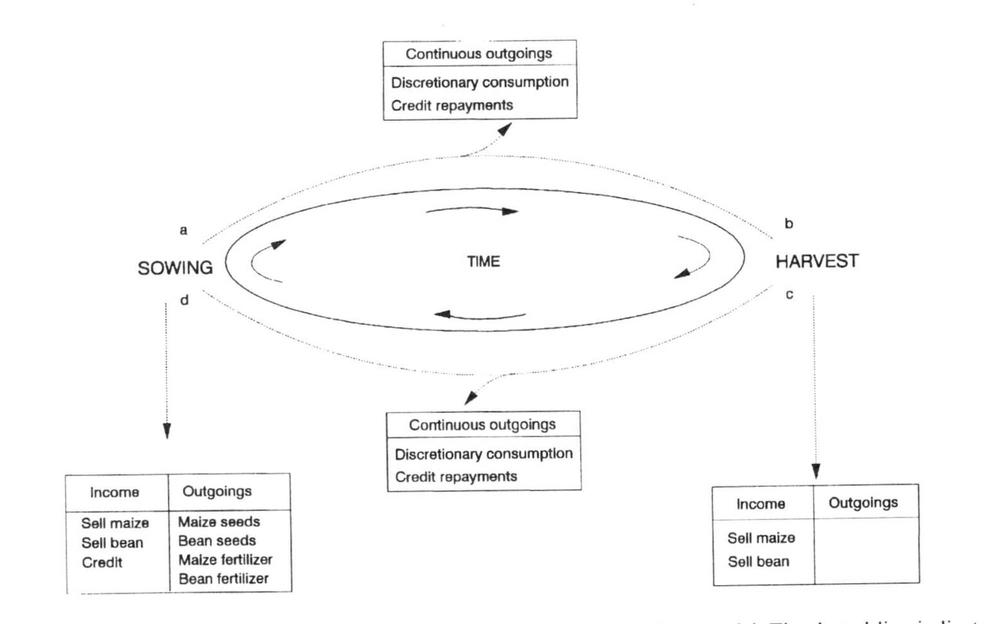

Figure 2. Timing of the different forms of financial exchange which occur within the whole farm model. The dotted line indicates financial flows, while the continuous line signifies the circular nature of the cropping season. The letters a to d identify the four available times for making and implementing decisions within the current whole farm model.

Table 2. Variety-specific information utilized by the farm household decision-making model.

Information	Units
Yield if recommended fertilizer strategy is applied	kg ha^{-1}
Yield if no fertilizer is applied	one of good, medium, poor
Storability	one of good, medium, poor
Labor requirement to achieve maximum yield with fertilizer	days
Labor requirement to achieve maximum yield without fertilizer	days
Cost of fertilizer	\$ ha^{-1}
Cost of seed	\$ per 1000 seeds

selection decision is most simple when only one crop is being reassessed and only two varieties of that crop are available for planting. The complexity of the decision increases if both crops are being reassessed, and many varieties of each are available to the farmer. Subsequent discussion of the variety selection algorithm considers the latter, more complex option.

The variety selection module functions in two main stages. First, all varieties of the required crop are passed through several initial screening rules. These rules assess whether or not the variety has the yield potential to meet the nutritional requirements of the family and whether or not financial reserves permit purchase of the required seed and fertilizer. Appropriate varieties are then passed to a second stage of selection where other agronomic, financial and cultural aspects of the varieties are considered in order to enable a final selection. This routine makes a selection of varieties based upon their overall acceptability to the farm household. Acceptability is a relative measure, which may be determined from a comparison of the perceived characteristics of each available variety with a description of the ideal characteristics of a variety in one of a range of defined situations. Ideally, this information would be obtained directly from farm households through standard social science techniques as used by Ashby et al. (1989b). The variety characteristics considered in the model were as follows:

- Financial return with different fertilizer strategies;
- Yield with different fertilizer strategies;
- Cooking and taste characteristics;
- Storage characteristics;
- Labor input required to maximize yield.

One important variable which affects the acceptability to the farm household from any given variety selection is the fertilizer strategy to be adopted. In the WFM four fertilization strategies are defined: do not apply fertilizer to either crop, apply fertilizer to maize only, apply fertilizer to bean only, and apply fertilizer to both crops. (Before running the model, the user must define these fertilizer strategies.) The final output from the variety selection module is a statement of the varieties to be grown and the appropriate fertilizer strategy to be implemented. This information is then passed to the maize and bean models.

Running the crop and fallow models

Once initialized, the crop models run to the their physiological maturity, and then automatically trigger the fallow models. These fallow models are essentially the crop models without the crop, and serve to simulate soil water and nutrient dynamics during the fallow period. However, as they do not simulate plant growth, there is no natural end point to them such as physiological maturity, and if left alone they would run indefinitely. Within the WFM it is necessary to run the fallow models up until the date of sowing the next crop. However, as the date of sowing depends crucially on the weather conditions (and perhaps on labor availability) in any particular year, the exact sowing date is not predictable from year to year. In many places sowing is timed to coincide with the annual rains, and this characteristic was used in the WFM to trigger the ending of the fallow models, and to shift control again to the family demographic model. The fallow models run until the soil moisture content reaches a stated threshold; once this is exceeded, the fallow models finish, triggering the family demographic model. This structure has proved robust and satisfactory for the current project, but the question of how control of the models may be manipulated in situations which do not have readily defined rainy seasons is unresolved.

Yield amendments

Ideally, the whole farm model would allow the farmer to implement tactical crop management decisions throughout the season (even though there may be few tactical options available to subsistence farmers). Implementing such a scheme would require the crop models to be stopped mid-run, according to some predefined rule(s), and the farmer would then be able to implement one or more predefined sets of tactical management decisions according to information available on the crop, the weather, labor availability and the market. Although such a mechanism was not implemented, it was felt that some element of agronomic management should be encompassed within the model. As many of the management actions in subsistence agricultural systems are directly related to labor availability (the extent and timeliness of weeding, pest control and harvesting), for the sake of simplicity, labor availability is taken as a proxy for good management. Within the model, therefore, if the family has sufficient labor units available to meet the entire labor demands of the crop, then the harvested yield will be the same as that given from the crop model. If, however, the family is unable to provide sufficient labor to meet the variety's requirements, then a yield reduction factor is calculated and used to amend the harvested yield.

It is also known that some varieties of crop store better than others. This is reflected in the qualitative classification of varieties into one of good, medium or poor for storability within the WFM. To reflect this variation in storability, an amendment is made to the amount of crop in store according to its storability

Table 3. Baseline conditions for testing the whole farm model.

Size of last year's maize harvest = 1000 kg
Size of last year's bean harvest = 1000 kg
Last year's maize variety = varied with test run
Last year's bean variety = varied with test run
Last year's maize fertilizer strategy = add fertilizer
Last year's bean fertilizer strategy = no fertilizer
Last year's credit availability = yes
Credit repayments this year = 0
Last year's financial reserves pre-sowing = 1500 or 450
Last years financial reserves post-sowing = 1500 or 450
Current amount of maize in store = 1000 kg
Current amount of beans in store = 1000 kg
Market price for maize = 0.5 units kg^{-1}
Market price for beans = 0.8 units kg^{-1}
Market availability for maize = varied with test run
Market availability for beans = varied with test run

class, and the amount of food stored is reduced by a given factor in accordance with the variety's storability classification.

Testing the whole farm model

Two family sizes were used during testing, one a small nuclear family and the second a large extended family. In order to start the WFM, the baseline conditions were set as detailed in Table 3. Four varieties of beans (ICTA-Ostua, Rabia de Gato, Turbo-III, C20), and three of maize (Tocoron-3, Pioneer 3541, Pioneer 3382) were made available for sowing. All of the biological characteristics of these varieties were taken directly from the DSSAT software which was also used to provide yield data with and without fertilizer. In order to obtain these data, the relevant crop models were run with the following inputs: the weather data were from Jutiapa, Guatemala, for the years 1980 to 1988; crops were planted on day-of-year 130; the planting density was 6 m^{-2} (maize) and 25 m^{-2} (bean); the row spacing was 0.5 m for maize and 0.45 m for bean; the fertilizer strategy was to apply 60 kg urea ha^{-1} at planting; the crops were rainfed, and default initial conditions were used in the soil profile.

The WFM provides 93 output variables which together represent the financial, social and agronomic condition of the farm household and crops. No single variable can represent the full complexity of change occurring in the farm system, and analysis of the output is difficult. Some results of two runs of the model are presented in Table 4, and a full set of results is presented in Edwards-Jones et al. (1994). The only differences between the two simulations whose results are shown in Table 4 relate to the size of the farm family and the availability of markets for maize and beans. In simulation (a) the family size was small and a good market for buying and selling maize and beans existed, while in simulation (b) the family was large and the market conditions for buying and selling maize and beans were poor. The results of these differ-

Table 4. Some results from the whole farm model for two different socio-economic situations.

(a) An owner occupier, working in a situation where credit is available, there is a good market for the purchase and sale of maize and beans. Initial family size is small (1 adult male, 1 adult female, 1 teenage male, 1 teenage female, 1 male child), 1 ha planted to beans, 1.2 ha planted to maize each year.

Variable	Yr1	Yr2	Yr3	Yr4	Yr5
Family labor units	34	34	34	34	34
Credit repayment ($)	0	649	1283	1283	1523
Financial reserves post harvest ($)	2	367	−688	126	−152
Seed source	buy	buy	store	buy	store
Bean variety sown	Turbo-III	Ostua	Ostua	Ostua	Ostua
Maize variety sown	Tocoron3	Tocoron3	Tocoron3	Tocoron3	Tocoron3
Bean yield (kg ha^{-1})	748	582	200	935	1603
Maize yield (kg ha^{-1})	2691	2597	2740	2620	0
Beans bought (kg)	0	0	0	0	0
Maize bought (kg)	208	0	0	0	0
Beans sold ($)	217	0	0	0	0
Maize sold ($)	0	202	60	141	0

(b) An owner occupier, working in a situation where credit is available, there is a poor market for the purchase and sale of maize and beans. Initial Family size is large (2 adult males, 2 adult female, 2 teenage males, 2 teenage females, 1 male child), 1 ha planted to beans, 1.2 ha planted to maize each year.

Variable	Yr1	Yr2	Yr3	Yr4	Yr5
Family labor units	34	34	34	34	34
Credit repayment ($)	1680	1680	1680	1680	1680
Financial reserves post harvest ($)	0	−1750	−3467	−5542	−6723
Seed source	buy	store	store	store	store
Bean variety sown	Turbo-III	Ostua	Ostua	Ostua	Ostua
Maize variety sown	Tocoron3	Tocoron3	Tocoron3	Tocoron3	Tocoron3
Bean yield (kg ha^{-1})	774	54	19	14	2
Maize yield (kg ha^{-1})	1664	2597	407	381	210
Beans bought (kg)	0	0	0	0	0
Maize bought (kg)	25	0	0	0	0
Beans sold ($)	0	0	0	0	0
Maize sold ($)	0	0	0	0	0

ences are that while in (a) the yield was sufficient to feed the family and allow some sales of excess maize, in (b) the yield in the first year of the model was insufficient to feed the family. This resulted in the family being forced to eat most of the food in store, which severely restricted the amount of farm-grown seed available to sow in year 2. As no market was functioning for the purchase of seed, the yields in year 2, and in all subsequent years, were extremely poor. It was only the continued provision of credit which kept this family alive.

Discussion

The simplified prototype whole farm model integrates farm family models with the IBSNAT crop models. The work raised many issues regarding the functioning and modeling of farm households, and some of these are currently being addressed in related research programmes (e.g., McGregor et al., 1995, 1996; Willock et al., 1995). Several of the more important issues are discussed below.

The importance of tactical decision making

All farmers are in intimate contact with their crop, and although they may be working to an overall management strategy, they clearly have to respond tactically to day-to-day problems. In the case of subsistence agriculture, response to poor germination, pest infestation or drought may be affected by factors such as a shortage of labor and/or cash. Initially, it was hoped that these tactical decision could be incorporated within the modeling framework, but several problems arose which prevented this development. First, as the crop models run for a season, even if a tactical model had been developed, then there was no way in which changes in management derived from it could be implemented. Second, as the impacts of weeds, pests and diseases were not included in the crop models at the start of the project, there was little point in attempting to model the socio-economic aspects of pest management. For these reasons an approach was adopted in which farmers only make decisions immediately prior to sowing and immediately after harvest; even then, the only decisions taken are those which are directly related to the current input variables of the crop models (such as crop, variety and fertilizer strategy). This highlights the need for further development of the crop models.

Is the farm family a single decision-making unit?

A second more conceptual problem concerns the appropriate scale for modeling farm households. It is unclear whether the model should seek to emulate the behavior and decision-making of the household as if it were a single entity, or alternatively whether each individual within the household should be modeled separately, and the final output provided by the interaction between individuals. For example, it may be argued that all family decisions are a function of the attitudes and behavior of its constituent individuals and their individual interactions with the exogenous environment. Thus in times of financial wealth, all individuals may agree to invest in a certain technology. In times of hardship, however, there may be severe disagreements within the family, as each individual may have their own relative preference for spending the limited cash. Further, the relationships between individuals may change with time, and this may affect the final household decision. For example, in the situation described above, if the children were under ten years old and still at school, then the parents' attitudes may dominate. However, the childrens' opinions may become increasingly important as they age and begin to work full time on the farm alongside their parents.

In the prototype WFM the farm household was treated as a single decision making entity, but an alternative option currently being explored is the use of a frame-based approach to modeling the farm household. In this approach, individual members of the family are represented by a frame that contains slots for specific physical, social and psychological characteristics. The frames would be controlled by rules, and would interact in order to reach any decision. This modeling approach is being used by ecologists to develop individual based models of animal populations (Judson, 1994), and would appear to be transferable to the social situation (Edwards-Jones and McGregor, 1994). Although this option is intellectually exciting and may be a closer representation of reality than the alternative of treating the family as a single decision-making entity, it does impose severe demands on data collection, which is already problematical; in addition, it may lead to a lack of balance in the overall whole farm model.

Socio-economic data and a minimum data set

As noted above, there has been a tendency, amongst those researchers who have attempted to model and analyze the behavior of farmers in a quantitative manner, to utilize data that are relatively easy to collect, such as age, farm size, education, tenure status, and the availability of credit and markets. Although these variables may explain significant amounts of variation in some models, there is a growing feeling among some researchers that analyses are not identifying the correct variables to enter into models, and that less tangible sociological and psychological variables may be more important in determining behavior than the easily collected structural variables (Dewees and Hawkes, 1988; Shucksmith, 1993). However, the identification of these psycho-social variables remains difficult, and this is hampering potentially more fruitful modeling effort. During the development of the prototype WFM, ignorance has, by and large, demanded the use of standard structural, socio-economic variables to describe the farm family. Taken together, these variables represent a minimum data set which describes the socio-economic variables required to run a farm household model. There is no doubt though that further definition of such a minimum data set (MDS) remains one of the major challenges for future work in this area. Not only is defining the MDS difficult, but also actually collecting the data specified in the MDS may be difficult. For example, if as seems likely, psycho-social variables were included within the MDS, then there would be a need to talk to a number of households in the target region, and to go through reasonably detailed questionnaires with at least one member of the family (McGregor, 1995, 1996; Willock et al., 1995). Undertaking these interviews with household members may be relatively time consuming as it may be necessary to collect a large amount of sensitive and complex data. The resource requirements of such data collection would probably result in low sample sizes with high variability, and the collation of such data into a form suitable for entry into a WFM may present severe problems to analysts.

Validation of farm household model

Validation of expert systems has always posed problems, and to date common practice includes the verification of the operation of individual modules and a series of field tests (Harrison, 1991). Validation of rule-based simulation models should be more akin to the standard validation approaches used for standard mathematical models, although as with data collection, small sample sizes and high variability may pose problems. However, several options are possible. One option is to verify the model's decision-making process against that of real farmers, which may require farmers either to discuss their own decision process or to undertake some kind of toy decision which may be observed by the analyst. Such methods have previously been used in several areas of expert system development (Edwards-Jones, 1993). A second option would be to compare model output with observed farmer behavior. For some decisions such as variety selection this should be trivial, but for other more subtle variables such as changes in diet, labor intensity and discretionary spending, it may be more difficult. These activities have not yet been carried out, but they are an obvious and essential next step.

Conclusions

The linking of social and biological data within a simplified whole farm model has been achieved. At the outset of this work the technical and modeling aspects of the work were perceived to be extremely challenging, while understanding and representing the processes within farm households were not perceived as being such a problem. This viewpoint reflected the disciplinary nature of the researchers' previous experience, and with hindsight has proved to be totally incorrect. Rapid advances in computer modeling techniques have now made the development of such models relatively simple, but meanwhile our understanding of farm household decision-making processes has not advanced greatly, despite increased work in this area.

Models that can adequately simulate farmer behavior remain rare but are still needed, although the relevance of linking these with whole farm models may vary with context. Within Europe much agricultural policy is concerned with the environment and rural community, and here there is only a limited need to link models of farm household behavior with commodity production models. More important in this context is a general understanding of the processes involved in farm household decision-making, coupled with an understanding of how variations in farm households' psychological and socio-cultural characteristics affect such processes. In developing countries, however, there is a very real need to understand how the uptake of a certain technology will affect food production. In such cases, the development of whole farm models that include commodity production models should help regional and national decision-makers understand how best to deliver certain technologies to farmers, in order that food production may be increased.

Given the rapid evolution of computing technology, the future success of whole farm modeling depends upon the success of the rural sociologists, micro-economists, psychologists and agricultural economists in providing theory and data on farm households which are suitable for inclusion in such models. It is unlikely that these groups will, unaided, collect and collate suitable data for more advanced model development; for this reason, a multidisciplinary approach will be required. Soil scientists, crop physiologists, biologists and modelers will all have a role within these teams, but it will probably be the social scientists who will highlight areas requiring further work. In developing the current model, the final output of the household decision-making module had to be directly interpretable by the crop models. This meant that output from the farm household model was restricted entirely to the input variables required by the crop models. This certainly does not reflect the situation in the field, where the farm household has to make decisions concerning many other variables that are not included in the crop models. If crop models are to be integrated with models of farm households, or indeed if they are to be useful to farmers, then it is important that they reflect the management options available in the field. Models developed by crop scientists to improve scientific understanding will not necessarily be effective tools for aiding technology transfer. To be effective in such a case, the design and development of such models may need to be influenced by the results of socio-economic field work. The management-related inputs and outputs of such models have to reflect the actual socio-economic situation in the field. In essence, the socio-economic reality may need to drive the fundamental scientific and modeling work.

References

Akinola A A (1987) An application of probit analysis to the adoption of tractor hiring service schemes in Nigeria. Oxford Agrarian Studies 16:70–82.

Amir I, Puech J, Granier J (1991) ISFARM: an integrated system for farm management: Part I – methodology. Agricultural Systems 35:455–469.

Anderson J R (1985) Assessing the impact of farming systems research: framework and problems. Agricultural Administration 20:225–235.

Anderson J R, Dillon J L, Hardaker J B (1985) Socioeconomic modelling of farming systems. In: Remenyi J V (Ed.) Agricultural systems research for developing countries. Proceedings of an international workshop held at Hawkesbury Agricultural College, Richmond, NSW, Australia, 12–15 May 1985.

Ashby J A, Quiros C A, Rivers Y M (1989a) Farmer participation in technology development: work with crop varieties. Pages 115–126 in Chambers R, Pacey A, Thrupp L A (Eds.) Farmer first. Farmer innovation and agricultural research. Intermediate Technology Publications, London, UK.

Ashby J A, Quiros C A, Rivers Y M (1989b) Experience with group techniques in work in Colombia. Pages 127–132 in Chambers R, Pacey A, Thrupp L A (Eds.) Farmer first. Farmer innovation and agricultural research. Intermediate Technology Publications, London, UK.

Bryant W K (1992) The economic organization of the household. Cambridge University Press, Cambridge, UK.

Byerlee D, de Polanco E H (1986) Farmers' stepwise adoption of technological packages: evidence from the Mexican altiplano. American Journal of Agricultural Economics 68:519–527.
Caillavet F, Guyomard H, Lifran R (Eds.) (1994) Agricultural household modelling and family economics. Developments in Agricultural Economics 10, Elsevier, Amsterdam, The Netherlands.
Carr S, Tait J (1991) Differences in the attitudes of farmers and conservationists and their implications. Journal of Environmental Management 32:281–294.
Casebow A (1981) Human motives in farming. Journal of Agricultural Economics 24:119–123.
Clocksin W F, Mellish C S (1984) Programming in PROLOG, 2nd Edition. Springer-Verlag, New York, USA.
Coughenour M C, Tweeten L (1986) Quality of life perceptions and farm structure in agricultural change: consequences for southern agriculture and rural communities. Pages 61–87 in Molnar J J (Ed.). Westview Press, Boulder, Colorado, USA.
Coughenour M C, Swanson E (1988) Rewards, values, and satisfaction with farm work. Rural Sociology 53:442–459.
Dent J B, Edwards-Jones G, McGregor M J (1995) Simulation of ecological, social and economic factors in agricultural systems. Agricultural Systems 49:337–351.
Dent J B, McGregor M J (Eds.) (1994) Rural and farming systems analysis: European perspectives. CAB International, Wallingford, UK.
Dent J B, Thornton P K (1988) The role of biological simulation models in Farming Systems Research. Agricultural Administration and Extension 29:111–122.
Dewees C M, Hawkes G R (1988) Technical innovation in the Pacific coast trawl fishery: the effects of fisherman's characteristics and perceptions on adoption behaviour. Human Organisation 47:224–234.
Doorman F (1991) A framework for the rapid appraisal of factors that influence the adoption and impact of new agricultural technology. Human Organisation 50(3):235–244.
Edwards-Jones G (1993) Knowledge-based systems for crop protection: theory and practice. Crop Protection 12:565–578.
Edwards-Jones G, Dent J B, Morgan O, McGregor M J, Thornton P K (1994) The integration of socio-economic data with biological crop models in a whole farm context. Unpublished report to the IBSNAT Project, Hawaii, USA.
Edwards-Jones G, McGregor M J (1992) Expert systems in agriculture: a history of unfulfilled potential? Pages 58–62 in Proceedings of 4th International Congress for Computer Technology in Agriculture, Versailles, France, 1–3 June 1992.
Edwards-Jones G, McGregor M J (1994) The necessity, theory and reality of developing models of farm households. In Dent J B, McGregor M J (Eds.) Rural and farming systems analysis – European perspectives. CAB, Wallingford, UK.
Gasson R, Errington A (1993) The farm family business. CAB International, Wallingford, UK.
Ghadim A A, Kingwell R S, Pannell D J (1991) An economic evaluation of deep tillage to reduce soil compaction on crop-livestock farms in Western Australia. Agricultural Systems 37:291–307.
Gilmor D A (1986) Behavioural studies in agriculture: goals, values, and enterprise choice. Irish Journal of Agricultural Economics and Rural Sociology 11:19–33.
Gladwin C H (1983) Contributions of decision-tree methodology to a farming systems program. Human Organisation 42:146–157.
Harrison S R (1991) Validation of agricultural expert systems. Agricultural Systems 35:265–286.
Herath H M G, Hardaker J B, Anderson J R (1982) Choices of varieties by Sri Lanka rice Farmers: Comparing alternative decision models. American Journal of Agricultural Economics 64: 87–93.
Hoogenboom G, Jones J W, White J W, Boote K J (1990) BEANGRO V1.0: Dry bean crop growth simulation model, user's guide. Agricultural Engineering Department and Agronomy Department, University of Florida, Gainesville, Florida, USA.
Low A (1988) Farm household-economics and the design and impact of biological research in Southern Africa. Agricultural Administration and Extension 29:23–34.
Luger G F, Stubblefield W A (1989) Artificial intelligence and the design of expert systems. The Benjamin/Cummings Publishing Company Inc. Redwood City, California, USA.
Jones C A, Kiniry J R (1986) CERES-Maize: A simulation model of maize growth and development. Texas A&M University Press, College Station, Texas, USA.
Judson O P (1994) The rise of the individual-based model in ecology. Trends in Ecology and Evolution 9:9–14.
Leslie P H (1945) On the use of matrices in certain population mathematics. Biometrika 33:183–212.

McGregor M J, Willock J, Deary I, Dent J B, Sutherland A, Grieve R, Gibson G, Morgan O (1995) Edinburgh farmer decision making study: links between psychological factors and farmer decision making. Pages 153–166 in Proceedings of the 10th International Farm Management Convention, University of Reading, 10–16 July, 1995.
McGregor M J, Willock J, Dent J B, Deary I, Sutherland A, Gibson G, Morgan O, Greive B (1996) Links between psychological factors and farmer decision making. Farm Management 9:228–239.
Moxey A, White B, Sanderson R A, Rushton S P (1995) An approach to linking an ecological vegetation model to an agricultural economic model. Journal of Agricultural Economics 46:381–397.
Negassa A, Tolessa B, Franzel S, Gedeno G, Dadi L (1991) The introduction of an early maturing maize (*Zea mays*) variety to a mid-altitude farming system in Ethiopia. Experimental Agriculture 27:375–383.
Polson R A, Spencer D S C (1992) The technology adoption process in subsistence agriculture: the case of cassava in southwestern Nigeria. IITA Research 5:12–16.
Singh I, Squire L, Strauss J (Eds.) (1986) Agricultural household models. The Johns Hopkins University Press, Baltimore, USA.
Shakya P B, Flinn J C (1985) Adoption of modern varieties and fertilizer use on rice in the eastern tarai of Nepal. Journal of Agricultural Economics 36:409–419.
Shucksmith M (1993) Farm household behaviour and the transition to post productivism. Journal of Agricultural Economics 44:466–478.
Voh J P (1982) A study of factors associated with the adoption of recommended farm practices in a Nigerian village. Agricultural Administration 9:17–27.
Willock J, Deary I, McGregor M J, Sutherland A, Dent J B, Grieve R, Gibson G, Morgan O (1995) Edinburgh study of decision making on farms: modelling the relationship between personal characteristics, attitudes and behaviour: a model for stress. Pages 208–217 in 3rd European Association of Agricultural Economics Seminar, Farmers in small-scale and large scale farming in a new perspective; objectives, decision making, and information requirements. Woudschoten, The Netherlands, 16–18 October, 1995.
Wilson E O (1975) Sociobiology: the new synthesis. Harvard University Press, Cambridge, Massachusetts, USA.
Wossink G A A, Koeijer, T J, Renkema J A, de Koeijer T J (1992) Environmental-economic policy assessment: a farm economic approach. Agricultural Systems 39:421–438.

Network management and information dissemination for agrotechnology transfer

G.Y. TSUJI

Department of Agronomy and Soil Science, 2500 Dole Street, Kr22 University of Hawaii, Honolulu, Hawaii 96822, USA

Key words: IBSNAT, network, multidisciplinary, systems analysis and simulation, DSSAT, collaborators.

Abstract

The IBSNAT project was organized as a cooperative and collaborative global network of scientists with a wide-array of disciplinary skills and interests. Outputs of the network participants were integrated and synthesized into a single, principal product, DSSAT, a decision support system software that provided a linked, computerized programming framework for crop simulation models, databases, and application programs. Utilization of project products was accomplished through annual training courses established among network participants in the United States and training workshops organized by host-country institutions and organizations at international venues. Information dissemination was accomplished principally through the printed media with technical reports and newsletters.

Introduction

The IBSNAT project was conceived to test the hypothesis that agrotechnology transfer can be more efficiently effected through systems analyses and simulation with a global network of scientists. The concept of a global collaborative network, successfully used by the Benchmark Soils Project (Swindale, 1978; Silva, 1985) to achieve project objectives, was adopted by IBSNAT as part of its strategic plan during its initial meeting at ICRISAT in 1983 (Kumble, 1984).

The rationale for a network was simple. No single institution or individual had the resources and capacity to develop new tools for the information age in a timely manner and only through a collaborative network of like-thinking and cooperative scientists could such an effort even be attempted. Participants in the network first met in 1983 and reached agreement on the 12 food crops for which models would be developed and on the minimum data set (Nix, 1984) necessary to develop and test these models.

This collective group of participants represented a global human resource, unlike those found in individual institutional settings, with knowledge and skills to synthesize decision aids and tools capable of accessing and using available information and data for a community of decision makers. These decision aids would allow decision makers to make educated choices to better

G.Y. Tsuji et al. *(eds): Understanding Options for Agricultural Production, 367–381*

match characteristics of agricultural technology and innovations to the biophysical characteristics of the land.

In combination with a systems approach, contributions from this global network of multidisciplinary collaborators were galvanized into the development of decision aids or tools with common input/output formats. These tools, by design, had compatible linking points to allow them to be integrated and merged into a single functional decision support system, DSSAT. The DSSAT software was first released for distribution in late 1989 with crop models only for wheat, maize, soybean, and peanut. That initial version was a product of a collaborative network and a systems approach and was achieved in a relatively short time.

Much has been achieved in product development and in product utilization and application since IBSNAT was inaugurated in 1983. This chapter describes the management strategy and organization of IBSNAT, and its network of collaborators. It then documents IBSNAT's efforts in information dissemination through the printed media and training programs.

Network management

A global network of benchmark sites and scientists was envisioned as necessary and essential to implement a systems research approach to agrotechnology transfer. A broader network base with the involvement of a multiplicity of disciplines was deemed critical for the project's success. Owing to the scarcity of system-oriented scientists willing and able to participate in interdisciplinary efforts, the entire international scientific community must now become the source for scientific talent. The use of interdisciplinary teams is essential because virtually all of the critical environmental problems caused by agricultural practices and policies are systems problems, not disciplinary ones (Uehara and Tsuji, 1993).

Fiscal and visionary support provided by the U.S. Agency for International Development (USAID) allowed participants in the IBSNAT network to leverage resources to achieve objectives no single individual or institution could accomplish independently. Annual workplans outlining tasks, timelines, and responsible individuals were determined collectively and were used to guide and manage the network to achieve its objectives. Measurable outputs, such as operational models installed in DSSAT, served to identify milestones achieved toward accomplishing these objectives.

System developers

The IBSNAT network of collaborators (Table 1) participated in developing, assembling and distributing a portable, computerized decision support system. The international network included a multi-disciplinary team of researchers

Table 1. Major collaborators contributing to software components of DSSAT.

Organization	DSSAT components
(1) International centers	
CIAT	CROPGRO-dry bean, CROPSIM-cassava
ICRISAT	PNUTGRO, CERES-sorghum, millet
IFDC	CERES models, CROPSIM-cassava, SUBSTOR-aroid, DSSAT, N&P modules, seasonal and sequence analysis
IRRI	Pest models, CERES-rice
(2) Universities	
Australian National University	Minimum data set
Chiang Mai University, Thailand	CERES-rice
Michigan State University	CERES-wheat, CERES-barley, CERES-sorghum, CERES-millet, SUBSTOR
University of Edinburgh	Whole-farm systems models
University of Florida	CROPGRO-soybean, peanut, pest models, WeatherMan, DSSAT
University of Georgia, Griffin	CROPGRO-dry bean, BEANGRO, DSSAT
University of Guelph	CROPSIM-cassava, genetic coefficients, DSSAT
University of Hawaii	SUBSTOR-aroid, CERES-rice, pest models, DSSAT
University of Hawaii/NifTAL	N-fixation
(3) Agencies	
CSIRO, Canberra & Griffith	Minimum data set, DSSAT
USDA/ARS/TARS, Puerto Rico	SUBSTOR-aroid
USDA/ARS, Temple, Texas	CERES-maize, minimum data set
USDA/SCS/SMSS	Soils database

and institutions involved in the development of crop simulation models and modular components of DSSAT.

During the first five years of IBSNAT, its network consisted primarily of model developers and those interested in validating models, previously referred to as data generators. Approximately 250 minimum data sets collected by collaborators with their own resources were included in the IBSNAT database. Successful calibration and validation of crop models in a range of global sites provided the necessary level of confidence to continue development of DSSAT.

System users

With the publication and release of DSSAT v2.1 in 1989, the IBSNAT network of DSSAT users grew from 50 to well over 500 representing nearly 80 countries in three years; this figure represents only the recorded number of distributed copies of version 2.1. The number of DSSAT users increased by more than 300 with the release of version 3 in 1994.

The IBSNAT network was a unique participatory network where membership was open to any individual or organization willing to share data, information, models, and experiences.

Management

Leveraging of fiscal resources from USAID with fiscal and human resources from the network of collaborators was a key element in the management of the IBSNAT network. Participants in the network, supported by their own respective institutions and organizations, collected, assembled, and contributed data sets to develop, calibrate, and validate the crop models in DSSAT and to test DSSAT itself. Their collective contributions was estimated to be worth more than 5 to 6 times the dollar amount provided by USAID (Uehara and Tsuji, 1993). Networking, with shared commitments and common goals, had a synergistic effect on participants. The key ingredient in managing and coordinating the network was trust among the participants.

Management with workplans to achieve project objectives were prioritized and adjusted annually by evaluating progress made towards projected outputs associated with each objective. Specific activities and tasks with starting and ending dates were identified along with designated responsible individuals (Kitchell, 1987). Achievements and slippages were reported as baselines of accomplishments in developing annual workplans for each project year. Milestone events representing key accomplishments under each output served as a measurable indicator of progress towards reaching project objectives.

Management team

Management of the IBSNAT project involved the coordination and facilitation of human and fiscal resources to achieve project objectives. Merging scientific visions with fiscal realities required the active interaction of IBSNAT's principal investigator, its core groups with the Technical Advisory Committee (TAC) and the Management Review Group (MRG) (IBSNAT, 1993).

Core Group. The core group included the principal investigator, a support staff, located at the University of Hawaii, and a team of subcontractors. The staff was responsible for the day-to-day coordination and monitoring of activities and tasks within the network and served to facilitate technical and logistical support to the subcontractors and to respond to requests from the user or client group of DSSAT. Subcontractors served as the technical staff and were enlisted to carry out specific tasks principally in the area of model and systems development. They included the scientists from Agricultural Research Service, USDA in Temple, Texas; the International Fertilizer Development Center, Muscle Shoals, Alabama; the International Rice Research Institute, Los Baños, Philippines; Michigan State University; the University of Edinburgh, Scotland; the University of Florida, Gainesville; the University of Guelph; and the University of Puerto Rico, Mayaguez.

Technical Advisory Committee. This committee was internationally represented and its composition was one of well-recognized scientists in their respective disciplines (IBSNAT, 1993). Amazingly, there were no absentees among them for each of the TAC meetings.

Annual technical workplans by outputs were developed by the core staff in consultation with the TAC. Workplans and associated budgets were integrated into the Project Year Workplan for review by the MRG before implementation. The TAC was not an oversight group but a participatory group in the network. Members of TAC provided the technical and visionary leadership to establish an international collaborative research network capable of making a difference in sustainable land management and agricultural development. Their vision was based on the premise that it would be more efficient to improve agroecosystem performance using a systems approach than by experimenting with the system itself. They reasoned that the development and validation of decision tools such as crop models and DSSAT would require an international team effort.

Management Review Group. The cooperative agreement between USAID and the University of Hawaii required establishing a Management Review Group to oversee and approve annual workplans. There were three members: a representative from USAID, from the University of Hawaii, and the project principal investigator.

Information dissemination

Information dissemination required the 'packaging' of information for consumption by an audience with varied backgrounds and interests. The principal ingredients in each of these packages were in the form of printed materials and software generated by collaborators. Training courses and workshops served to enhance application of products derived within IBSNAT for users and potential users. In large part, peer scientists were the major audience as reflected in the number of technical reports and publications produced. Information dissemination to a more general lay audience was accomplished through reports generated by the public media.

Publications for a general audience

Acceptance of concepts espoused by IBSNAT by a general audience was marked by reports and stories carried in both the public print and video media. These reports were prepared by individuals or organizations other than those participating in IBSNAT programs. Examples include stories written by Stephen Strauss for the *Toronto Globe and Mail* and *Technology Review* in April 1990 and by Emily Looney of the Associated Press and printed initially in the *San Diego Union* on 15 July 1991. Looney's story was subsequently reprinted in *Frontlines*, the newsletter of USAID, in October 1991. In October 1990, an Australian science newsmagazine television program, *Beyond 2000*, produced by the Australian Television Network, featured IBSNAT in a program entitled 'Global Farming' that aired initially in Australia in 1991. The *Beyond 2000* program introduces innovative ideas and concepts to general audiences and is shown globally.

Publications for a technical audience

Acceptance by scientific peers was determined by their assessment of project concepts and evaluation through tests of products and reported outcomes in technical journals and in reports of application of these products. Seven categories of published documents were produced by IBSNAT collaborators. They included papers in scientific journals, technical reports, research reports, users guides, newsletters, contributed papers in proceedings, and progress and annual reports. Each served as 'measurable indicators' of progress and accomplishments to participants' institutions and to USAID.

(1) *Journal Series.* Documents produced through this medium permitted a critical review of concepts and of results generated in filling knowledge gaps and in application of simulation models and DSSAT by peers. For IBSNAT scientists, acceptance by peers was attested to by reports and manuscripts published in scientific journals and by users as documented by the number of participants in conferences, symposia, training courses and workshops.

Much of the published materials prepared by IBSNAT collaborators documented the evolutionary stages of software or product development, its testing, and its application. Many of these appear in the reference lists for each chapter in this book. Journals selected for publishing reports and manuscripts included the *Agronomy Journal, Agricultural Systems, Transactions of the American Society of Agricultural Engineers, Phytopathology, Agricultural Administration and Extension, Agricultural Ecosystems and Environment, Experimental Agriculture, Nature,* and *Soil Science Society of America Journal.* The range of journals speaks of the multidisciplinary contributions made to the DSSAT.

(2) *Technical Reports.* This was the first series of documents published by IBSNAT. These reports were designed to establish standard procedures, methods, and formats for data collection, data storage, and data retrieval for the initial version of DSSAT (version 2.1). Five technical reports were published in all and are listed in Table 2. Principal users of these documents were those involved in collecting the minimum data sets for model validation and those interested in developing other crop models.

(3) *Research Report Series.* This series was created to document applications and case reports of crop models and DSSAT. Two publications were produced in this category. The first involved the CERES-rice model (Alocilja and Ritchie, 1988) and the second, the SUBSTOR-potato model (Griffin et al., 1993).

Table 2. Editions of IBSNAT Technical Reports.

Technical Report 1:	Experimental design and data collection procedures for IBSNAT. First edition, 1984; second edition, 1986; third edition, 1988; Spanish version, 1991.
Technical Report 2:	Field and laboratory methods for the collection of the IBSNAT minimum data set. First edition, 1990.
Technical Report 3:	Decision support system for agrotechnology transfer (DSSAT) level 1: User's guide for the minimum data set entry, version 1.1, 1986.
Technical Report 4:	Documentation for the IBSNAT data base management system, 1991.
Technical Report 5:	Documentation for IBSNAT crop model input and output files. Version 1.0, 1986; Version 1.1, 1990.

(4) *User Guides.* The User's Guide for DSSAT v2.1 (IBSNAT, 1989) was published by IBSNAT in 1989 as a 345–page document. Separate or individual user's guides for CERES-Maize v2.1 (Ritchie et al., 1989), CERES-Wheat v2.1 (Godwin et al., 1989), CERES-Rice v2.10 (Singh et al., 1993), CERES-Barley v2.10 (Otter-Näcke et al., 1991), and CERES-Sorghum/Millet v2.10 (Singh et al., 1994) were published by IFDC. A user's guide for aroids was published in 1995 (Singh et al., 1995). User's guides for SOYGRO v5.42 (Jones et al., 1989), PNUTGRO v1.02 (Boote et al., 1989), BEANGRO v1.01 (Hoogenboom et al., 1991) were published by the University of Florida and later reprinted by IBSNAT. The SUBSTOR-Potato document (Griffin et al., 1993) was published as a Research Report and served as a reference rather user guide.

For DSSAT v3, a three-volume user's reference manual (Tsuji et al., 1994) was prepared for distribution with the software. The users' guides for all crop models were combined for presentation in volume 2 of the manual with CROPSIM-Cassava (Matthews and Hunt, 1994), one of the latest models added. CROPGRO is now used as a prefix for crop models for soybean, peanut, and dry bean.

(5) *Conference and Symposia Proceedings.* Contributed papers and reports by IBSNAT scientists were published as proceedings of workshops and symposia in 1984 and between 1990 and 1993. The initial symposium was organized by IBSNAT in 1983 at ICRISAT and served as a 'ground breaking' document that included papers on the state of crop models and systems analysis (Kumble, 1984). The concept of the minimum data set was introduced to IBSNAT participants by Nix (1984). Towards the conclusion of IBSNAT, a symposium entitled 'Systems Approaches for Agricultural Development' (Penning de Vries et al., 1993) was organized in Bangkok, Thailand, in 1991 in cooperation with scientists from IRRI and the Wageningen Agricultural University in the Netherlands.

The symposium served as a final forum for scientists associated with IBSNAT to share information on crop models and applications of DSSAT with an international audience. One output of the symposium was a mutual understanding to lay the foundation for closer working relationships between system scientists from the Wageningen Agricultural University and IBSNAT. This eventually led to the conceptual beginning of the International Consortium for Agricultural Systems Applications (ICASA) in 1994.

(6) *Newsletters.* General and semi-technical materials were published in IBSNAT's newsletter, *Agrotechnology Transfer*. From 1985 to 1992, sixteen issues of the newsletter were published and distributed to a mailing list in excess of 3000. *Agrotechnology Transfer* represented a merger of two newsletters produced by projects supported by USAID – *Benchmark Sites News* of IBSNAT and *Soil Taxonomy News* of the Soil Management Support Service (SMSS) project. The newsletter provided a venue for sharing of information on advances in systems development and on applications of crop models and DSSAT by integrating soils, weather, crop and management data.

(7) *Progress and Annual Reports.* Progress reports were semi-technical in content and served two purposes. They provided a means for reporting progress achieved by network cooperators to each participant and their associated institution, and they provided general information to a broader general audience. Annual reports were prepared in accordance with guidelines provided under contractual and cooperative agreements with USAID.

Training courses

Training courses and workshops served to promote utilization of project products during the second five years of IBSNAT. This was consistent with a recommendation of an AID-appointed external review panel (MacKenzie, 1990) to establish a regularly scheduled training course for the purpose of providing users with instructions on concepts of the minimum data set, systems analysis, and simulation.

More than 600 individuals have attended training courses and workshops organized by IBSNAT and its network collaborators since the first was held in Caracas, Venezuela in 1984. Training courses and workshops provided a forum for developers and prospective users in research, education, and planning, to establish hands-on interactive experience with crop models and DSSAT as decision making tools.

Phase 1: 1982–1987. Training programs in the early years of IBSNAT were designed primarily to introduce principles of systems analysis, concepts of the minimum data sets and data standards as inputs for crop simulation models. Four training workshops were planned and conducted at selected locations to increase the global awareness of IBSNAT (Table 3). Crops of interest in these workshops had to be limited to those for which models were operational. Local institutions and organizations hosted each workshop which numbered from 20 to 60 participants.

Table 3. IBSNAT training programs, 1984 to 1986

Year	Workshop and location	Host
1984	Systems analysis and simulation of crop (maize and soybeans) growth for agro- technology transfer, Caracas, Venezuela	International Institute for Advanced Studies (IIEA), Fondo Nacional de Investigaciones Agropecuarias – Centro Nacional de Investigaciones Agropecuarias (FONAIAP-CENIAP)
1985	Systems analysis and crop (maize and wheat) simulation models, Amman, Jordan	University of Jordan, Arab Center for Studies of Arid Zones and Dry Lands (ACSAD)
1986	Crop (soybean and maize) simulation and data base management systems, Taichung, Taiwan	Chung Hsing University, Food and Fertilizer Technology Center for the Asia and Pacific Region (FFTC/ASPAC)
1986	Crop (rice and maize) simulation and database management systems, Serdang, Malaysia	Malaysian Agricultural Research and Development Institute (MARDI)

The workshops focused on the minimum data set necessary to test and evaluate CERES-maize, CERES-wheat and SOYGRO. These workshops were invaluable in providing the necessary momentum to establish 'recommended' procedures for data collection of the minimum data set and to establish standards for computerized data input and output formats. Both the procedures developed and the standards established for the minimum data set and the crop models proved to be essential 'building blocks' in assembling DSSAT into a system that offered compatibility between databases and simulation models.

These training courses and workshops served as 'proving grounds' of sorts to model developers. In order to have a unified system with a suite of crop models for users to choose from, common formats between models and databases were essential. Biases and idiosyncrasies of model developers had to be set aside or eliminated in this sometimes painful exercise of sharing and combining of model inputs and outputs into an integrated system.

The first workshop was held at the International Institute for Advanced Studies (IIEA) in Caracas, Venezuela in December 1984 through the invitation of CENIAP-FONAIAP, Maracay. The 10-day workshop, entitled 'Systems Analysis and Simulation of Crop Growth for Agrotechnology Transfer' was attended by 27 participants. Three 'portable' microcomputers weighing nearly 20 kg were hand-carried into Venezuela for the course as none was available to rent at that time. The small monitors or screens allowed no more than two participants to share access to each unit. Maize and soybeans were the principal crops studied by participants from Venezuela, Jordan, Zambia, Bangladesh, Pakistan, India, Thailand, Malaysia, Indonesia, Philippines, Costa Rica, and Panama. Many of these participants formed the nucleus of the growing network of IBSNAT collaborators.

Jordan was the site of the second workshop held almost a year later in November 1985. Both ACSAD (Arab Center for Studies of the Arid Drylands

Table 4. Annual training courses held in the United States

Date	Location	Number of participants
15–26 May, 1989	IFDC, Muscle Shoals, Alabama	20
13–24 Aug, 1990	University of Florida, Gainesville, Florida	25
6–17 May, 1991	IFDC, Muscle Shoals, Alabama	25
16–29 June, 1992	University of Hawaii, Honolulu, Hawaii	20
10–21 May, 1993	IFDC, Muscle Shoals, Alabama	25
8–19 August, 1994	University of Florida, Gainesville, Florida	35
8–19 May, 1995	IFDC, Muscle Shoals, Alabama	36
27 May–6 June, 1996	University of Georgia, Athens, Georgia	35

in Damascus, Syria) and the University of Jordan served as hosts. Principal focus of the workshop was on the minimum data set and both the CERES-maize and CERES-wheat crop models. ACSAD provided support for participants from Egypt, Syria, Iraq, Tunisia, Algeria, Morocco, Sudan, Saudi Arabia, and Palestine. The host institution, the University of Jordan, supported eleven participants from Jordan. IBSNAT, with USAID support, invited and supported participants from Argentina, Venezuela, Pakistan, Zambia, and Burundi. Representatives from Cyprus and Germany covered their own costs. By late 1985, computers were available for rental in Jordan.

The third and fourth workshops were both held in June 1986 in Taiwan and Malaysia, respectively. Both the Food and Fertilizer Technology Center for the Asia and Pacific Region (FFTC/ASPAC) and the National Chung Hsing University hosted a two week course in Taichung, Taiwan. Participants included representatives from Japan, Korea, the Philippines, Indonesia, and Thailand, along with representatives from the National Taiwan University, the Asian Vegetable Research and Development Center (AVRDC), the sugar industry and the host institution. There were 60 participants and 25 microcomputers available for the course in which soybeans and rice were the principal crops of interest. All participants' costs were covered by FFTC/ASPAC and the National Chung Hsing University.

The Malaysian Agricultural Research and Development Institute (MARDI) hosted a similar two-week workshop at their headquarters in Serdang. A total of 25 participants from Malaysia, Thailand, the Philippines, and the International Rice Research Institute (IRRI) attended. Maize and rice were the principal crops of interest. Computers from MARDI were used during the training program.

Phase 2: 1987–1995. Concepts, principles, and application of crop modeling and systems analysis were topics covered in training courses and workshops held during the second phase of IBSNAT. Training courses have been held annually (Table 4) in the United States since 1989. Sessions included both lectures and 'hands-on' training in using the databases, crop simulation models and strategy evaluation programs in DSSAT. In many instances, participants

Table 5. Typical training course schedule (Bowen, 1995).

DAY 1:
- Registration and welcome
- Introduction of program goals
- DSSAT v3 – Distribution and registration forms
- Presentation and overview of DSSAT v3
- Basic exercises

DAY 2:
- Introduction to the systems approach
- Overview of modeling concepts
- Introduction to phenology and growth processes
- Minimum data set for potential production
- Creating file X for potential production
- Exercises on potential production

DAY 3:
- Phenology, growth, and genetic coefficients
- Calibration for genetic coefficients
- Calibration exercises

DAY 4:
- Water balance
- Creating file X for simulating water balance
- Exercises with water balance
- Nitrogen balance
- Creating file X for simulating N balance
- Exercises with N balance

DAY 5:
- Setting up data files
- Weather data
- Soil profile data
- Experimental data files

DAY 6:
- Uncertainty and risk concepts
- Seasonal analysis
- Creating file X for seasonal analysis
- Exercises with seasonal analysis

DAY 7:
- Addressing sustainability issues with models
- Sequence analysis
- Creating file X for sustainability analysis
- Exercises with sequence analysis

DAY 8:
- Case study exercise
- Present case study results

DAY 9:
- Simulating spatial variability
- Exercises on geostatistics, spatial interpolation, and spatial analysis

DAY 10:
- Applications and future plans
- Need for follow-up
- Assemble brief summaries
- Departure formalities

Table 6. Overseas workshops organized by IBSNAT, 1987 to 1996.

Date	Description anad location	Host and organizers
1987	Collection and Management of the IBSNAT MDS for Crop Modeling, Honolulu, Hawaii	IBSNAT, University of Hawaii
1987	Collection of Minimum Data Set for IBSNAT, Nausori, Fiji Islands	University of the South Pacific, Suva, Fiji
1988	Biological N-fixation technology; Hamakuapoko, Maui, Hawaii	Nitrogen Fixing Tropical Agricultural Legumes (NiFTAL), IBSNAT
1988	Sorghum and pearl millet modeling, Patancheru, India	International Crops Research Institute for the Semi-Arid Tropics (ICRISAT), IBSNAT
1988	Agroclimatology for Asian grain legumes growing areas and regional legume network, Patancheru, India	ICRISAT
1989	Agrotechnology transfer in Bangladesh, Dhaka, Bangladesh	Bangladesh Agricultural Research Council (BARC)
1989	Case study: Sub-Saharan Africa, Saly, Senegal	Predictive assessment network for ecological and agricultural responses to human activities (PAN-EARTH) semi-arid food grain research and development (SAFGRAD)
1989	Case study: Crop model calibration, Maracay, Venezuela	PAN-EARTH, Fondo Nacional de Investigaciones Agropecuarias (FONAIAP)
1990	Modeling Pest-Crop Interactions, Honolulu, Hawaii	IBSNAT, University of Hawaii
1990	Crop Models and DSSAT, Los Banos, Philippines	Philippine Council for Agriculture, Forestry and Natural Resource Research and Development (PCARRD)
1990, 1991	International Climate Change and Crop Modeling, Washington, DC	Environmental Protection Agency (EPA), United States Agency for International Development (USAID), Goddard Institute for Space Studies (GISS), Columbia University
1991	Taro and Tanier Modeling, Honolulu, Hawaii	IBSNAT, University of Hawaii
1991	First US-Hungarian IBSNAT Workshop, Budapest and Keszthely, Hungary	Research Institute of Soil Science and Agricultural Chemistry (RISSAC); IBSNAT
1991	CONCIAM Crop Modeling Workshop, Port Louis, Mauritius	Australian National University
1992	CONCIAM crop modeling workshop, Hyderabad, India	Australian National University
1992	Agrotechnology transfer using biological modeling in Malawi, Lilongwe, Malawi	Department of Agricultural Research, Malawi; Rockefeller Foundation; IFDC
1992	Second US-Hungarian IBSNAT workshop, Budapest and Godolla, Hungary	RISSAC
1993, 1994, 1995	Modeling and decision support systems, Chiang Mai, Thailand	Multiple Cropping Center, Chiang Mai University

Continued

Table 6 Continued.

1994	Crop models and DSSAT, Tirana, Albania	IFDC
1994	Crop models and DSSAT, Bucharest, Romania	IFDC
1994	Climate change country studies program, Washington, DC	Argonne National Laboratory, GISS
1995	Climate change country studies program, Honolulu, Hawaii	Argonne National Laboratory, GISS

used their own data sets in carrying out exercises on model validation and to assess risks in planning alternative management strategies. These annual training courses continue to be financially self-sustaining. Participants are responsible for a tuition fee and their own travel and lodging costs. Tuition fees are used by organizers to defray expenses of instructors and course materials, including DSSAT v3.

An example of a typical course outline is presented in Table 5 (Bowen, 1995). A report containing summaries of applications of crop models and decision support systems from each participant was assembled for the first time at the conclusion of the course held in Gainesville in 1994 (Thornton et al., 1995). A similar report was assembled after the training course in 1995 (Bowen et al., 1996), and similar ones are planned for future courses.

A number of courses and workshops, listed in Table 6, were also organized outside of the United States during this period. Through cooperative efforts between host institutions and IBSNAT scientists, several courses and workshops were held in Bangladesh, the Philippines, Thailand, Malawi, Albania, Romania, Hungary, and Commonwealth countries in Africa, Asia and the Pacific, the Caribbean and the Indian subcontinent. Costs for these courses and workshops were borne by host institutions and workshop organizers. The program for the commonwealth countries was a major undertaking of the Australian International Development Assistance Bureau (AIDAB) in their COMCIAM project for Climate Impact Assessment and Management Program for Commonwealth Countries.

Conclusions

The IBSNAT project accomplished its objectives through the collaborative effort of a global network of participants. Such an effort could not have been possible without the support of USAID and the collective vision of an international team of systems-oriented scientists. Information dissemination through publications, software development, and training provided the necessary access to and from a range of user and client groups. Measures of acceptance were determined by the increased utilization of systems concepts in teaching and research and by the application of systems tools in decision making for agricultural and environmental planning and development.

References

Alocilja E C, Ritchie J T (1988) Upland rice simulation and its use in multi-criteria optimization. Department of Crop and Soil Sciences, Michigan State University, East Lansing, Michigan. Research report series 01, Department of Agronomy and Soil Science, College of Tropical Agriculture and Human Resources, University of Hawaii, Honolulu, Hawaii.

Boote K J, Jones J W, Hoogenboom G, Wilkerson G G, Jagtap S S (1989) PNUTGRO V1.02: user's guide for peanut crop growth simulation model. Department of Agronomy and Department of Agricultural Engineering, University of Florida, Gainesville, Florida; University of Hawaii, Honolulu, Hawaii.

Bowen W T (1995) Personal communication.

Bowen W T, Thornton P K and Wilkens P W (Eds.) (1996) Applying crop models and decision support systems – Volume 2. Summaries from the Training Program on Computer Simulation for Crop Growth and Nutrient Management, held at the International Fertilizer Development Center, Muscle Shoals, Alabama, 8–19 May, 1995. IFDC Special Publication SP-23, Muscle Shoals, Alabama, USA.

Godwin D C, Ritchie J T, Singh U, Hunt L A (1989) CERES-Wheat V.2.10: user's guide for wheat crop growth simulation model, 2nd edition. Michigan State University, East Lansing, Michigan; International Fertilizer Development Center, Muscle Shoals, Alabama; University of Hawaii, Honolulu, Hawaii.

Griffin T S, Johnson B S, Ritchie J T (1993) A simulation model for potato growth and development: SUBSTOR-Potato V.20. Department of Crop and Soil Sciences, Michigan State University, East Lansing, Michigan. Research report series 02, Department of Agronomy and Soil Science, College of Tropical Agriculture and Human Resources, University of Hawaii, Honolulu, Hawaii.

Hoogenbooom G, White J W, Jones J W, Boote K J (1991) BEANGRO V.1.01: user's guide for dry bean crop growth simulation model. Department of Agronomy and Agricultural Engineering, University of Florida, Gainesville, Florida; Bean Program, Centro Internacional de Agricultura Tropical (CIAT); University of Hawaii, Honolulu, Hawaii.

International Benchmark Sites Network for Agrotechnology Transfer (IBSNAT) (1989) Decision Support System for Agrotechnology Transfer, version 2.1. IBSNAT Project, Department of Agronomy and Soil Science, University of Hawaii, Honolulu, Hawaii.

Jones J W, Boote K J, Hoogenboom G, Jagtap S S, Wilkerson G G (1989) SOYGRO V.5.42: user's guide for soybean crop growth simulation model. Department of Agronomy and Department of Agricultural Engineering, University of Florida, Gainesville, Florida; University of Hawaii, Honolulu, Hawaii.

Kitchell R E (1987) Personal communication.

Kumble V (Ed.) (1984) Proceedings of the international symposium on 'Minimum Data Sets for Agrotechnology Transfer', 21–26 March 1983. ICRISAT Center, Patancheru, India.

MacKenzie D R (1990) Report of the midterm external evaluation of the international benchmark sites network for agrotechnology transfer (IBSNAT) project, 9–14 July 1990. Department of Agronomy and Soil Science, College of Tropical Agriculture and Human Resources, University of Hawaii, Honolulu, Hawaii.

Matthews R B, Hunt L A (1994) GUMCAS: a model describing the growth of cassava (*Manihot esculenta* L. Crantz). Field Crops Research 36:69–84.

Nix H A (1984) Minimum data sets for agrotechnology transfer. Pages 181–188 in Kumble V (Ed.) Proceedings of the international symposium on minimum data set for agrotechnology transfer. Patancheru, India, 21–26 March 1983. ICRISAT Center, Patancheru, India.

Otter-Näcke S, Ritchie J T, Godwin D C, Singh U (1991) CERES-Barley V.2.10: user's guide for barley growth simulation model. International Center for Agricultural Research in the Dry Areas, Syria; Michigan State University, East Lansing, Michigan; International Fertilizer Development Center, Muscle Shoals, Alabama; University of Hawaii, Honolulu, Hawaii.

Penning de Vries F W T, Teng P S, Metselaar K (Eds.) (1993) Systems approaches for agricultural development. Proceedings of the international symposium on 'Systems Approaches for Agricultural Development', Bangkok, Thailand, 2–6 December 1991. Kluwer Academic Publishers, Dordrecht, The Netherlands.

Ritchie J T, Singh U, Godwin D C, Hunt L A (1989) CERES-Maize V.2.10: user's guide for maize crop growth simulation model, 2nd edition. Michigan State University, East Lansing, Michigan; International Fertilizer Development Center, Muscle Shoals, Alabama; University of Hawaii, Honolulu, Hawaii.

Silva J A (Ed.) (1985) Soil-based agrotechnology transfer. Benchmark Soils Project, Hawaii Institute of Tropical Agriculture and Human Resources, University of Hawaii, Honolulu, Hawaii.
Singh U, Ritchie J T, Godwin D C (1993) CERES-Rice V2.10: user's guide for rice crop growth simulation model. Michigan State University, East Lansing, Michigan; International Fertilizer Development Center, Muscle Shoals, Alabama; University of Hawaii, Honolulu, Hawaii.
Singh U, Ritchie J T, Alagarswamy G (1994) CERES-Sorghum V2.10: user's guide. International Fertilizer Development Center, Muscle Shoals, Alabama.
Singh U, Prasad H K, Goenaga R (1995) SUBSTOR-Aroids V2.10: user's guide for aroids crop growth simulation model. International Fertilizer Development Center, Muscle Shoals, Alabama; Tropical Agriculture Research Station, USDA; University of Hawaii, Honolulu, Hawaii; University of the South Pacific, Fiji.
Swindale L D (Ed.) (1978) Soil-resource data for agricultural development. Hawaii Agricultural Experiment Station, University of Hawaii, Honolulu, Hawaii.
Thornton P K, Bowen W T, Wilkens P W, Jones J W, Boote K J, Hoogenboom G (Eds.) (1995) Applying crop models and decision support systems. Summaries from a training program on Computer simulation of Crop Growth and Management Responses, University of Florida, Gainesville, Florida. IFDC Special Publication SP-22, IFDC, Muscle Shoals, Alabama, USA.
Tsuji G Y, Uehara G, Balas S (Eds.) (1994) DSSAT v3: Volumes 1, 2, and 3. IBSNAT Project, Department of Agronomy and Soil Science, University of Hawaii, Honolulu, Hawaii.
Uehara G (1989) Technology transfer in the tropics. Outlook on Agriculture 18:38–42.
Uehara G, Tsuji G Y (1993) The IBSNAT Project. Pages 505–513 in Penning de Vries F W T, Teng P S, Metselaar K (Eds.) Systems approaches to agricultural development. Kluwer Academic Publishers, Dordrecht, The Netherlands.

Crop simulation models as an educational tool

R.A. ORTIZ
School of Natural and Exact Sciences, Universidad Estatal a Distancia (UNED), Sabanilla, San José, Costa Rica

Key words: SOYGRO, distance learning, training, expert system

Abstract
Crop growth simulation models are educational tools of great potential in distance education. A project to study the utility of models as complementary instructional material for courses in Agriculture Administration at the School of Natural and Exact Sciences of the Universidad Estatal a Distancia (UNED) in Costa Rica was initiated in 1989. Using the soybean crop growth simulation model SOYGRO, the initial study was carried out over two years in three stages: (1) testing of the model and getting acquainted with its applications and functions, (2) creation of examples and class practices based on the model, and (3) evaluation of these instructional materials with distance education groups. A modified Spanish version of the User's Guide was printed in 1991 for UNED students using the model. SOYGRO has been used regularly in the Basic Grains course since 1992, and is currently viewed as an important training tool in distance education. It has provided an opening for other projects in the field of expert systems and computer-mediated teaching and learning processes at UNED.

Introduction

The soybean crop growth simulation model, SOYGRO, is a process-based computer model that simulates vegetative and reproductive growth and yield for the soybean crop (Jones et al., 1988; IBSNAT, 1993). When SOYGRO was developed, it is unlikely that anyone imagined SOYGRO's possible use as a teaching tool. In 1983, O'Shea and Self stated (quite correctly) that computers would cause great changes in education (O'Shea and Self, 1983). This chapter describes an interesting application of computer modeling as a teaching-learning tool in college distance education.

Distance learning

The concept of computer-managed instruction systems as vehicles for relieving the classroom teacher of clerical tasks associated with individualized education was described by Baker (1978). Simultaneously, following the increased development of computer-based educational systems, simulation appeared as one of the more widely used techniques of computer-assisted learning. Once a program that models a process or system becomes available, the student is able to use and study the performance of the program to better understand the process or

G.Y. Tsuji et al. *(eds): Understanding Options for Agricultural Production, 383–388*

system that is being modeled (O'Shea and Self, 1983). This is a clear example of the interactive use of computers by students for explicit learning purposes.

In distance education, even more so than in other forms of education and training, it is necessary and customary to use a systematic approach to teaching (Gastkemper, 1990). The students must follow an independent learning or self-study curriculum. Courses are designed to fulfill this need. From this perspective, computer modeling is viewed as a median to help accomplish this educational task.

Computer-managed learning usually refers to the management of learning of the individual student's work and covers some functions of recording, marking, assessing or recommending, and reporting (Fielden and Pearson, 1978). Finding a modeling system that could be adapted to cover all these areas and provide the student with more responsibility for making decisions concerning the nature and learning of his or her own course in distance education are difficult tasks.

The possibility of adapting and using the SOYGRO V5.41 model as an educational tool in distance education was considered for the first time in 1989 by the School of Natural and Exact Sciences of the Universidad Estatal a Distancia (UNED), the Open University in Costa Rica. UNED is an institute of higher education that uses the distance education methodology to develop professionals in different disciplines. It offers a complete set of different career options at the undergraduate level and some at graduate level. The University was created by the Costa Rican Congress in February 1977. Entry is available to all persons who have graduated from high school without limitation, regardless of area of work or geographical location.

Distance methodology breaks the time and space barrier, providing instructional packages to the student in any part of the country. The instructional package is essentially composed of written materials, audio-visual programs and multimedia, study guides, and tutorial sessions. Distance education was created to facilitate student learning efforts through an individualized and self-sufficient teaching-learning process. The student learns by himself or herself with the help of efficient utilization of the materials provided.

Because distance learning is a completely individualized teaching-learning process, it allows the student to study at home or anywhere else, as a complement to other activity such as work. There are 28 University Centers distributed around the country to facilitate this process. These centers provide administrative and teaching services, and constitute meeting points to meet tutors, carry out administrative procedures, and use multimedia services. They also play an important role in cultural and community activities.

In 1991, when use of the models was being considered, local information on the application of computer technology to the development of computer-based educational materials in distance education was limited (Benko, 1989; Hidalgo and Brenes, 1990). SOYGRO V5.41 seemed an obvious choice in helping to accomplish the teaching goals of distance education, because it offered the possibility of simulating potential and actual yields in the field based upon real-world experimental data. The student would have the opportunity to simulate and assess soybean growth under all sorts of local and regional conditions. The implementation process of adapting SOYGRO V5.41 for use as a teaching-learning tool in distance education at UNED is described below.

Crop models as a research and teaching tool

The IBSNAT project aimed to utilize simulation and systems analysis as a means to transfer technology for agricultural development. The soybean model SOYGRO V5.41, a component of the Decision Support System for Agrotechnology Transfer (DSSAT) version 2.1 (IBSNAT, 1989) was originally designed as a tool for research and agrotechnology transfer. As with all the crop models in DSSAT, SOYGRO could be used for the following: (1) validation of the model using data derived from historical field trials; (2) conducting sensitivity analyses by interactively selecting combinations of soils, weather, crop, and management factors; and (3) conducting risk analysis studies by simulating crop management scenarios through time and space under varying weather and soil conditions (Jones et al., 1988).

For distance education, three strategies were developed to implement a program to use SOYGRO. They were the following:

(1) testing and evaluation of the model and its functions;
(2) searching for the necessary informational elements to develop educational materials for students;
(3) adapting and creating a crop model user's guide for distance education in the Spanish language (Ortiz, 1990). This guide was based upon the original user's guide by Jones et al. (1988).

The adapted user's guide contained three minimum data sets with which to validate SOYGRO and two practicals for students. The Costa Rican minimum data sets used for model validation were obtained from the Centro Agronómico Tropical de Investigación y Enseñanza (CATIE) in Turrialba. Selected students from the Agriculture Administration Program and the Computer Science degree program of the Ciudad Neily Academic Center of UNED participated in the evaluation of the user's guide. The Center is located 315 km from UNED headquarters in San José, Costa Rica.

Evaluation and testing were carried out during several meetings. Various corrections were made to the guide following suggestions from students and responses to questionnaire surveys. The main purpose was to allow each student to work independently by following the instruction guide and the computer format.

A major constraint in testing the user's guide and the model was the lack of adequate computer hardware. First, none of the available computers had a hard disk, so all of the testing and evaluation was carried out using a version of SOYGRO on diskettes. Second, all UNED computers had graphic cards which were incompatible with the software. This problem was solved by the incorporation of an emulator.

Another constraint was computer literacy. Many students experienced difficulties in executing instructions on the computer. An introductory program for the simulation model was created using BASIC to facilitate the student's interactive communication with the computer. This procedure allowed the installation of all relevant program files to one diskette, and helped to save time and work for the student. In addition, some minor instructions were corrected and adapted in the user's guide to facilitate program operation. Apart from the costs of computer equipment (hardware and software) and the low availability of personnel trained in computer technology, resistance to technological innovations is commonly encountered in the use of computers in public education (Ministério da Educaçao, 1985). In spite of these constraints, results of evaluations of the software and user's guide were always positive. Based on these results, a training workshop on the use of both the guide and the SOYGRO model was organized for faculty members of the Natural Resources and Exact Sciences School of UNED.

The use of SOYGRO V5.41 was initiated with the cooperation of voluntary students at four educational centers as part of the Basic Grains course of UNED in 1991. This allowed UNED students to study a number of management strategies applied to the soybean crop with examples and practicals required in the course by using the multiple options included in SOYGRO. This also encouraged students to modify model conditions and inputs according to their own interest. The use of SOYGRO was finally implemented as a mandatory practical in the Basic Grains course in 1992 as a support element to the theoretical aspects of the course. It has been used continuously since then and has had wide acceptance by teachers and students as a teaching-learning tool in distance education (Ortiz, 1991).

Simulation models and expert systems

Expert systems are computer programs that utilize the knowledge and experiences of human experts to provide 'expert' prescriptions or diagnosis to symptoms of suspected problems. In view of the successful introduction of a crop model to the learning-teaching process of the Agricultural Business Administration Program, the use of expert systems was subsequently developed for application in the Fruit Crops course of the School of Natural and Exact Sciences. The first topic to be selected for development in an expert system format was disease of banana (Arrieta and Ortiz, 1992), given the quantity of information that exists and the relative importance of the crop to the economy of Costa Rica.

The creation of the banana crop diseases expert system was initiated through the use of the commercial software called VP Expert (Educational Version). This work was developed with the team work of a UNED crop production expert and a computer expert. Development of the expert system was initiated in 1991 and there have been various iterations of development and technical review since then. Once the whole development process is concluded, it is anticipated that UNED students will have easier access to information to identify banana crop diseases and to select alternative control strategies directly from the computer. Such development demonstrates how computer technology can become a powerful teaching-learning tool in distance education.

The microcomputer and these computerized science tools are complementary aids to the regular classroom and field programs. These advances in the use of computers as a teaching tool in UNED were catalyzed by the initial work with SOYGRO. Moreover, the SOYGRO model provided an unexpected outcome by establishing a baseline for future advances in utilization of systems and simulation models in distance teaching education.

Despite the success in applying SOYGRO as a teaching-learning tool, a caution sounded by Galvis (1987) is worth noting: 'The educational sector cannot leave the transfer of educational computing technology to the discretion of the suppliers, nor can it indiscriminately accept technology transfer. Either alternative would be an open door both for increasing technological dependency (in computing as well as in education) and for widening the technological and educational gap among countries.' To what extent projects such as IBSNAT could contribute to such a situation as this is unknown. On the other hand, there have been many enriching interactions, and much feedback, among IBSNAT participants and contributors around the world through an interactive network.

Conclusions

Simulation models and expert systems are promising tools in distance education for implementation and adaptation by UNED and probably by other educational institutions around the world. The general consensus of UNED students, teachers and administrators is that such education tools are resources that can increase students' knowledge and understanding of biological processes. At the same time, UNED is currently involved in computer conferencing and other computer-managed instructional projects with equal success. The use of SOYGRO inspired and accelerated the use of computers for educational purposes such as this.

The experience gained through the use of simulation models and expert systems in crop production courses could provide a basis for introducing these resource tools to other courses where computer instructional materials can be adapted or created for distance education. The objective of the IBSNAT project, the use of simulation and systems analysis as a means of transferring technology for agriculture development, was well served in the Costa Rican educational context. It also helped to create an effective and useful educational tool in college distance education.

References

Arrieta J, Ortiz R A (1992) Expert systems for tropical crops course teaching in distance education. Agronomy Abstracts 12.

Baker R (1978) Computer managed instruction. Theory and practice. Educational Technology Publications, New Jersey.

Benko A (1989) Algunas consideraciones sobre la aplicación del microcomputador en la educación. Informe de Investigaciones Educativas 3(2):51–66.

Fielden J, Pearson P K (1978) The cost of learning with computers. Council for Educational Technology, London, UK.

Galvis A H (1987) Technology transfer in educational computing: toward a non-magical approach. Pages 11–23 in Proceedings of the Meeting on the Use of Microcomputers Applications in Education and Training in Developing Countries. Board on Science and Technology for International Development, Westview Press, Boulder, Colorado.

Gastkemper F (1990) Computer aided learning and interactive video in distance education. Pages 82–93 in Bates A W (Ed.) Media and technology in European distance education. European Association of Distance Teaching Universities, Heerlen, The Netherlands.

Hidalgo G, Brenes F (1990) Dos enfoque de la producción de materiales de Instrucción apoyados por la computadora. Revista Iberoamericana de Educación a Distancia 3(1):9–22.

International Benchmark Sites Network for Agrotechnology Transfer (IBSNAT) (1989) The decision support system for agrotechnology transfer: user's guide. Department of Agronomy and Soil Science, College of Tropical Agriculture and Human Resources, University of Hawaii, Honolulu, Hawaii.

International Benchmark Sites Network for Agrotechnology Transfer (IBSNAT) (1993) The IBSNAT decade. Ten years of endeavor at the frontier of science and technology. Department of Agronomy and Soil Science, College of Tropical Agriculture and Human Resources, University of Hawaii, Honolulu, Hawaii.

Jones J W, Boote K J, Jagtap S S, Hoogenboom G, Wilkerson G G (1988) SOYGRO V.5.41: User's guide for soybean crop growth simulation model. Department of Agronomy and Department of Agricultural Engineering, University of Florida, Gainesville, Florida.

Ministério de Educaçao (1985) Educaçao e Informática. Fundaçao Centro Brasileiro de Televisao Educativa, Centro de Informática. Rio de Janeiro, Brazil.

O'Shea T, Self J (1983) Learning and teaching with computers. Artificial Intelligence in Education. Prentice-Hall, New Jersey.

Ortiz R A (1990) Gu¡a para la utilización del modelo de crecimiento del cultivo de la soya 'SOYGRO V.5.41' en educación a distancia. Sistema de Educación Computarizada, Universidad Estatal a Distancia, San José, Costa Rica.

Ortiz R A (1991) Using SOYGRO V5.41 soybean crop growth simulation model in distance education. Agronomy Abstracts, p 5.

Synthesis

G. UEHARA
Department of Agronomy and Soil Science, 1910 East West Road, University of Hawaii, Honolulu, Hawaii 96822

Key words: understanding research, prediction research, control research

Abstract
One of the greatest tragedies of science and technology is that the knowledge scientists take for granted is not accessible to people who need it most. The aim of the IBSNAT project was to make knowledge accessible to users by fulfilling the three purposes of research. They are (1) to understand processes and mechanisms, (2) to synthesize from our understanding a capability to make predictions and (3) to enable our customers to apply this predictive capability to control outcomes. The decision support system assembled by the IBSNAT project empowers users by enabling them to evaluate options and control outcomes.

Introduction

The IBSNAT project has often been called a modeling project. This is understandable given the project's heavy emphasis on assembling simulation models. But it was never its intent to build models as ends in themselves. The models were simply the means by which the knowledge scientists have and take for granted could be placed in users' hands. In this regard, IBSNAT was a project on system analysis and simulation as a way to provide users with options for change.

One criticism of science is that it does not do enough to make the knowledge it generates accessible to people who need it. This situation stems from the fact that scientists are most often recognized and rewarded for fulfilling only the first of three research purposes. If there is anything unique about the IBSNAT project, it was its effort to integrate the three purposes of research. The three purposes, as defined for the IBSNAT project by its Technical Advisory Committee, were to:

(1) *Understand* ecosystem processes and mechanisms.
(2) Synthesize from an understanding of processes and mechanisms, a capacity to *predict* outcomes.
(3) Enable IBSNAT clientele to apply the predictive capability to *control* outcomes.

While it is control that users want, it is understanding that science does best. Prediction is the bridge that connects understanding to control.

G.Y. Tsuji et al. *(eds): Understanding Options for Agricultural Production, 389–392*

The challenge for IBSNAT was to fulfill all three research purposes at the systems level. Experts in a particular discipline often use understanding to predict and control outcomes, but most problems people face are systems problems not solvable by knowledge from a single discipline. IBSNAT was not a climate, crop, pest, soil or socioeconomic project but a project to integrate them all. The aim was to develop a solid foundation for dealing with the soil-plant atmosphere continuum so that strong links between the socioeconomic and biophysical components could be forged. The predictive capability would not be confined to components within a system, but would show how perturbations in one part, affect all other parts of the system. The task was to find individuals who were at once scientifically competent and committed to team work and a system approach to solving problems.

An underlying premise of system research is that it is more convenient to use models to study systems than to conduct experiments on the systems themselves. One must also operate on the premise that every farm and every farmer is unique. This implies that the systems approach must be able to address production problems on a site specific basis and enable individual farmers to exercise choice in matters that affect their livelihood. But to exercise choice, one must have options. And in the end, it is having options that make changes for the better possible. Change for the better is possible when understanding leads to prediction and prediction ultimately allows outcomes to be controlled. While understanding research will always need to be replenished, the following shows that the IBSNAT project was able to accomplish much with existing knowledge.

Understanding research

It is safe to say that the IBSNAT project invested very little time or effort on this purpose. In fact, the project was established on the premise that a large body of knowledge already existed. The aim therefore was not to do more understanding research, but to utilize existing knowledge to build prediction models. In doing so, the researchers invariably encountered knowledge gaps, and only then, were they compelled to undertake gap-filling, understanding research. This turned out to be an efficient way to prioritize basic research.

Prediction research

Building, refining, and validating prediction models require a special mix of skills and attitude that is rare among scientists. It requires a combination of skills in knowledge engineering and an attitude that enables scientists to enjoy and be excited about work driven by user needs. The latter quality is critical for IBSNAT-type projects as researchers not constrained by user needs tend to build detailed, mechanistic models that are more useful for understanding

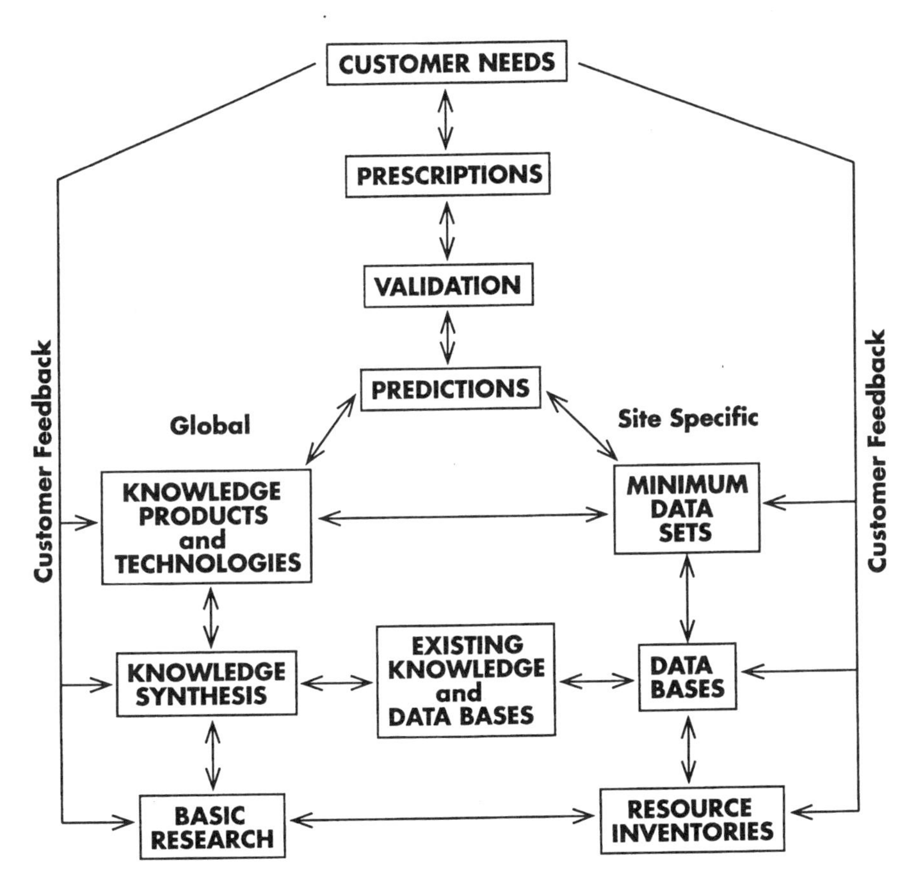

Figure 1. Diagram of a systems approach to meet client needs.

processes than for predicting outcomes of management decisions. Models built to meet user needs must be more functional than mechanistic. In the final analysis the IBSNAT project will be judged not by the mechanistic elegance of its models, but by the number of customers who will benefit from their use.

Control research

From its beginning, the research team and the Technical Advisory Committee agreed that the project would not assemble models, no matter how powerful or accurate they might be, if the customers for whom they were intended were unwilling to use them. To ensure customer acceptance, the research team spent considerable time and effort designing and implementing the concept of the minimum data set or MDS. The IBSNAT models, like others, require local input data for soil, weather and crop to predict the agronomic, economic and environmental outcomes of management decisions on a site-specific basis. A minimum rather than a maximum or even optimum data set for prediction was chosen to make it easier for users to obtain the input data required to operate the models.

The manner by which the three purposes of research are linked is shown in Figure 1. If we start with a blank sheet of paper and had to build these links to the needs of our customer, we would start with the box at the top, CUSTOMER NEEDS. Next, at the bottom left box, BASIC RESEARCH; the foundation of science rests on basic or *understanding* research. Then, the storehouse of existing knowledge, RESOURCE INVENTORIES at the bottom right, allows synthesis of *prediction* models which can be validated and used for diagnosing and prescribing cures to allow customers to *control* outcomes associated with their problems. Linkages are shown by lines and additional boxes. Because the prediction models are based on processes and mechanisms, they apply both locally (site specific) and globally to meet customer needs.

Conclusion

There are good reasons to believe that systems analysis and simulation will become an integral part of agricultural research and development. The rapidly advancing fields of information science and computer technology virtually guarantee it. Another reason for its assured growth is that systems analysis and simulation are knowledge-intensive. Knowledge unlike natural resources does not degrade with its use and is unlimited in supply. Systems analysis and simulation enable people to use knowledge as a resource for understanding options for agricultural production in the most efficient and cost-effective way.

Index

G.Y. Tsuji et al. *(eds): Understanding Options for Agricultural Production, 393–399*

Systems Approaches for Sustainable Agricultural Development

1. Th. Alberda, H. van Keulen, N.G. Seligman and C.T. de Wit (eds.): *Food from Dry Lands*. An Integrated Approach to Planning of Agricultural Development. 1992 ISBN 0-7923-1877-3
2. F.W.T. Penning de Vries, P.S. Teng and K. Metselaar (eds.): *Systems Approaches for Agricultural Development*. Proceedings of the International Symposium (Bangkok, Thailand, December 1991). 1993 ISBN 0-7923-1880-3; Pb 0-7923-1881-1
3. P. Goldsworthy and F.W.T. Penning de Vries (eds.): *Opportunities, Use, and Transfer of Systems Research Methods in Agriculture to Developing Countries*. Proceedings of a International Workshop (The Hague, November 1993). 1994 ISBN 0-7923-3205-9
4. J. Bouma, A. Kuyvenhoven, B.A.M. Bouman, J.C. Luyten and H.G. Zandstra (eds.): *Eco-regional Approaches for Sustainable Land Use and Food Production*. Proceedings of a Symposium (The Hague, December 1994). 1995 ISBN 0-7923-3608-9
5. P.S. Teng, M.J. Kroppf, H.F.M. ten Berge, J.B. Dent, F.P. Lansigan and H.H. van Laar (eds.): *Applications of Systems Approaches at the Farm and Regional Levels*. 1996 ISBN 0-7923-4285-2
6. M.J. Kroppf, P.S. Teng, P.K. Aggarwal, J. Bouma, B.A.M. Bouman, J.W. Jones and H.H. van Laar (eds.): *Applications of Systems Approaches at the Field Level*. 1996 ISBN set volume 5 & 6: 0-7923-4287-9; ISBN 0-7923-4286-0

KLUWER ACADEMIC PUBLISHERS – DORDRECHT / BOSTON / LONDON